VOLUME TWO HUNDRED AND TWELVE

ADVANCES IN IMAGING AND ELECTRON PHYSICS INCLUDING PROCEEDINGS CPO-10

VOLUME TWO HUNDRED AND TWELVE

ADVANCES IN
IMAGING AND ELECTRON PHYSICS INCLUDING PROCEEDINGS CPO-10

Edited by

PETER W. HAWKES and MARTIN HŸTCH
CEMES-CNRS
Toulouse, France

ELSEVIER

ACADEMIC PRESS
An imprint of Elsevier

Cover photo credit:
M. Berz, Organizer of CPO-10.

Academic Press is an imprint of Elsevier
125 London Wall, London EC2Y 5AS, United Kingdom
525 B Street, Suite 1650, San Diego, CA 92101, United States
50 Hampshire Street, 5th Floor, Cambridge, MA 02139, United States
The Boulevard, Langford Lane, Kidlington, Oxford OX5 1GB, United Kingdom

ISBN: 978-0-12-817475-3
ISSN: 1076-5670

For information on all Academic Press publications
visit our website at https://www.elsevier.com/books-and-journals

Publisher: Zoe Kruze
Acquisition Editor: Jason Mitchell
Editorial Project Manager: Peter Llewellyn
Production Project Manager: Divya KrishnaKumar
Designer: Miles Hitchen

Typeset by VTeX

Contents

Contributors

Seitkerim B. Bimurzaev
Almaty University of Power Engineering and Telecommunication, Almaty, Kazakhstan

Shin Fujita
Innovation Centre, Shimadzu Asia Pacific Pte Ltd., Singapore, Singapore

Volker Gerheim
CEOS GmbH, Heidelberg, Germany

Thomas Juffmann
Universitaet Wien, Vienna, Austria

Frank Kahl
CEOS GmbH, Heidelberg, Germany

Mark Kasevich
Physics Department, Stanford University, Stanford, CA, United States of America

Brannon B. Klopfer
Physics Department, Stanford University, Stanford, CA, United States of America

Vladimir Kolarik
Delong Instruments, Brno, Czech Republic

Stewart A. Koppell
Physics Department, Stanford University, Stanford, CA, United States of America

Martin Linck
CEOS GmbH, Heidelberg, Germany

Haoning Liu
Munro's Electron Beam Software Ltd, London, United Kingdom

Marian Mankos
Electron Optica, Palo Alto, CA, United States of America

Heiko Müller
CEOS GmbH, Heidelberg, Germany

Eric Munro
Munro's Electron Beam Software Ltd, London, United Kingdom

Catherine Rouse
Munro's Electron Beam Software Ltd, London, United Kingdom

John Rouse
Munro's Electron Beam Software Ltd, London, United Kingdom

Richard Schillinger
CEOS GmbH, Heidelberg, Germany

Khashayar Shadman
Electron Optica, Palo Alto, CA, United States of America

Stephan Uhlemann
CEOS GmbH, Heidelberg, Germany

Leon van Kouwen
Department of Imaging Physics, Delft University of Technology, Delft, Netherlands

Preface

In October 2018, the Tenth International Conference on Charged Particle Optics (CPO-10) was held in Key West, Florida. This was followed by the 13th International Computational Accelerator Physics Conference (ICAP-18), the two meetings being organized by Martin Berz and Kyoko Makino of Michigan State University. The object of the CPO meetings is to bring together specialists of electron optics, accelerator optics and mass spectrometer optics, who use the same equations of motion in very different ways.

The first part of this volume contains six papers from CPO-10 – others can be found in a special issue of the *International Journal of Modern Physics A*. These cover time-of-flight mass spectrometer design, electron gun optics, a post-column imaging filter for electron microscopes, the study of electron mirrors using differential algebra and an algorithm for simulation of the properties of particle-optics devices. Two other papers, by E. Valetov and M. Berz, unfortunately arrived just too late to be included as did a second paper by S. Bimurzaev; they will be printed in volume 213 and can be consulted on the Advances website (articles-in-press). I (PWH) am particularly pleased that these CPO articles appear here as Karl Brown, Hermann Wollnik and I launched the series in 1980 in Giessen shortly before I became editor of these Advances.

The second part of the volume contains a very full study of the physics of ion sources, and in particular, of nano-aperture sources by L. van Kouwen. After introducing the subject, he examines nano-fluidic flow, the physics and in particular the optics of ion emission, ion scattering and simulation of the emission process. This clear and scholarly account of a difficult subject will certainly be appreciated by readers from the field of ion physics.

We thank Martin Berz and Kyoko Makino warmly for their hard work in organizing a very successful pair of conferences and Leon van Kouwen for his very readable text.

Peter W. Hawkes
Martin Hÿtch

PART ONE

Papers from the Tenth International Conference on Charged Particle Optics edited by P.W. Hawkes and M. Berz

CHAPTER ONE

Planar multi-reflecting time-of-flight mass-spectrometer of a simple design

Seitkerim B. Bimurzaev
Almaty University of Power Engineering and Telecommunication, Almaty, Kazakhstan
e-mail address: bimurzaev@mail.ru

Contents

1. Introduction

A characteristic feature of planar multi-reflecting time-of-flight mass spectrometers is their ability to operate in the mode of analyzing ions in the full mass range. A planar multi-reflecting time-of-flight (MR-TOF) mass spectrometer of a simple design with high resolution was first proposed in Nazarenko, Sekunova, and Yakushev (1992). It consists of an ion source, an ion detector and an analyzer, which is a system of two two-dimensional electrostatic mirrors parallel to each other. The mirror electrodes are a pair of parallel plates separated by straight gaps and located at the same distance from the plane of symmetry of the field, called the middle plane of the system. Some disadvantage of this mass spectrometer is a relatively low sensitivity, caused by the absence of spatial focusing of ions in the middle plane, as the two-dimensional electrostatic mirrors, acting as an analyzer, provide spatial focusing of ions only in one direction perpendicular to the middle plane. This leads to some loss of ions in the beam, which reduces

Advances in Imaging and Electron Physics, Volume 212
ISSN 1076-5670
https://doi.org/10.1016/bs.aiep.2019.08.001

the sensitivity of the device as a whole. However, this disadvantage can be relatively easily removed by proper formation of the initial phase volume of the ion beam, for example, using an orthogonal gridless accelerator (Yavor & Verentchikov, 2004).

In Yavor and Verentchikov (2004), Yavor et al. (2008), to increase the sensitivity of the planar MR-TOF spectrometer, the authors used the method increasing the sensitivity of its analyzer. For this purpose, an additional element was included in the analyzer - a set of two-dimensional electrostatic lenses located between two parallel two-dimensional electrostatic mirrors. However, the presence of such lenses leads, on the one hand, to a significant complication of the device design, as to obtain spatial focusing of ions in two mutually perpendicular planes (a necessary condition for increasing sensitivity), the electrodes of mirrors and lenses, having the same shape, are arranged so that their middle planes lie in mutually perpendicular planes. On the other hand, the presence of such lenses reduces the sensitivity of the planar MR-TOF several times as compared with standard time-of-flight mass spectrometers with an orthogonal accelerator, which is caused by a small ion beam length ($\sim$ 2 mm) in the orthogonal accelerator needed to minimize the time-of-flight aberrations of these lenses (Hasin et al., 2006).

This work is devoted to the development of a planar MR-TOF of a simple design with high resolution and sensitivity using highly dispersive two-dimensional electrostatic mirrors and an orthogonal ion accelerator with a non-uniform two-dimensional symmetrical field.

2. Orthogonal ion accelerator with a non-uniform two-dimensional symmetrical field

2.1 Formulas for calculations

Fig. 1 shows a schematic diagram of one of the possible variants of an orthogonal accelerator with a non-uniform electrostatic field with a two-dimensional symmetry. It differs from conventional orthogonal accelerators with a uniform field only in that the ion acceleration area (from the grounded grid to electrode 4 inclusively) includes gridless electrodes (2–4) made in the form of a pair of parallel plates separated by straight gaps and symmetrically located relative to the plane of the field symmetry.

In this work, the role of ion accelerator is performed by a four-electrode system of electrodes (electrodes 1–4), the first electrode of which has a flat

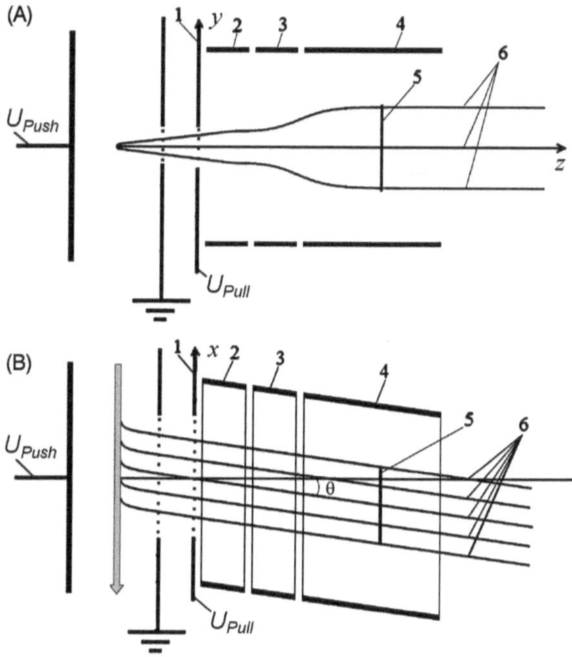

Figure 1 Schematic diagram of the orthogonal accelerator with a non-uniform electrostatic field of two-dimensional symmetry: (A) – in the *yz* plane, (B) – in the *xz* plane. 1, 2, 3, 4 are the electrodes of the ion accelerator; 5 is the plane of time-of-flight focusing of ions by energy; 6 is the course of the trajectories; θ is the angle between the direction of the parallel ion beam and the optical axis *z*.

shape, and the other electrodes are made in the form of a pair of parallel plates separated by straight gaps and located at the same distance from the plane of field symmetry, called the middle system plane. Such a system of electrodes forms a non-uniform electrostatic field with a two-dimensional symmetry. Let us introduce a rectangular coordinate system x, y, z, the z axis of which coincides with the optical axis of the two-dimensional system, and planes xz and yz coincide with its middle plane and the plane perpendicular to it, respectively. Then the equation for the inclination of the trajectory to the optical axis z for the considered system can be written as (Yakushev, Bimurzaev, & Kholodov, 2016):

$$y' = y_u p'(z) + \frac{2\sqrt{\Phi_0}}{\Phi'_u} y'_u g'(z), \tag{1}$$

where $p = p(z)$ and $g = g(z)$ are partial solutions of the equation

$$2\Phi y'' + \Phi' y' + \Phi'' y = 0. \tag{2}$$

Hereinafter $\Phi = \Phi(z)$ is the axial distribution of the electrostatic potential, the subscripts "u" and "0" denote the values at the point of emission of ions $z = z_u$ and the space free from the field, respectively, and the primes denote differentiation with respect to z variable, p is an analytical function, and function g can be represented as

$$g = \sqrt{\Phi} q, \tag{3}$$

where $q = q(z)$ is an analytical function satisfying the equation

$$2\Phi q'' + 3\Phi' q' + 2\Phi'' q = 0, \tag{4}$$

derived from (2) and (3).

Functions p and q satisfy the following initial conditions:

$$p_u = q_u = 1, \quad p'_u = -\frac{\Phi''_u}{\Phi'_u}, \quad q'_u = \frac{2}{3} p'_u. \tag{5}$$

The collimator mode of the ion accelerator operation, the condition of parallelism in the space of images of the particle beam ($y' = 0$), coming from a point source ($y_u = 0$), is determined, as it is seen from (2), by the equality

$$g' = 0. \tag{6}$$

The position of the primary temporary focus (the plane of time-of-flight focusing of ions by energy) $z = z_T$, created by the orthogonal accelerator with a non-uniform field is determined by the formula (Bimurzaev, 2015):

$$z_T = z_u + \frac{2\Phi_0}{\Phi'_u} \left\{ 1 + \frac{1}{2} \int_{z_u}^{z} \left[\sqrt{\frac{\Phi_0}{\Phi}} \left(\frac{\Phi' - \Phi'_u}{\Phi} \right) + \frac{\Phi'_u}{\Phi_0} \right] dz \right\}. \tag{7}$$

The axial potential distribution of a two-dimensional system consisting of N electrodes arranged in the form of a pair of parallel plates separated by straight gaps and located at the same distance from the middle plane, is determined by the formula (Glikman & Yakushev, 1967):

$$\Phi(z) = V_1 + \frac{2}{\pi} \sum_{i=1}^{N-1} (V_{i+1} - V_i) U(z - z_i), \tag{8}$$

where V_i is the potential on the i-th electrode, and

$$U(z - z_i) = arctg \exp\left[\pi (z - z_i)/d\right], \tag{9}$$

Here d is the distance between the plates of the electrodes, and z_i is the coordinate of the center of the i-th gap.

2.2 Results of calculations

The characteristics of the four-electrode ion accelerator were calculated as follows. For the given potentials on the electrodes 1 and 4, the lengths and potentials of the middle electrodes 2 and 3 were found, providing the conditions for the collimator mode (6) in the plane yz simultaneously with time-of-flight focusing of ions at the boundary of the accelerator field. At the same time, the telescopic mode is provided in the middle plane xz due to the two-dimensionality of the field – the incident parallel beam remains parallel after passing through the field of the system. Thus, the orthogonal ion accelerator with a non-uniform field of two-dimensional symmetry forms a parallel beam in two mutually perpendicular planes.

The results of calculation are presented in Table 1 and in Fig. 2. The geometric dimensions are given in units of the distance d_{acc} between the plates of the accelerator electrodes, and the potentials on the electrodes in units of V_4. The origin of coordinates coincides with the electrode plane with the potential U_{pull} (Fig. 1).

Here, in order to be able to use these data for any values of the pulling potential U_{pull} and the accelerating potential V_4, the ratios of potentials ratios are given in dimensionless values V_i/V_4 $(i = 1, 2, 3, 4)$ at a zero value of the potential V_1 at the first electrode. When the electrode 1 of the considered ion accelerator is combined with the grid with the pulling potential U_{pull} of the orthogonal ion accelerator (Fig. 1), the potentials on the electrodes 1-4 will be determined by the equality:

$$V_i^* = U_{pull} + V_i \quad (i = 1, 2, 3, 4). \tag{10}$$

Table 1 Conditions of time-of-flight characteristics of the four-electrode ion accelerator of two-dimensional symmetry.

V_1/V_4	V_2/V_4	V_3/V_4	l/d_{acc}	z_T/d_{acc}
0	0.251	4.33	0.73	2.69

As can be seen from Fig. 2, the axial potential distribution has an intermediate acceleration. It should be noted that only with such an axial

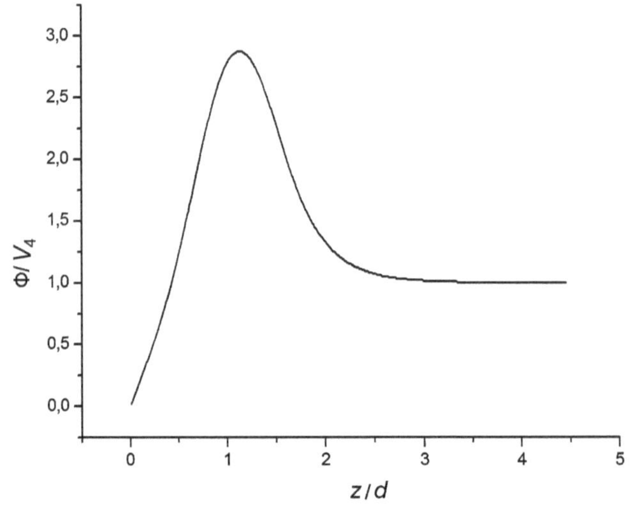

Figure 2 Axial distribution of the electrostatic potential of the ion accelerator.

distribution of the potential, the time-of-flight focusing of ions by energy can be achieved at the boundary of the field of the ion accelerator (Fig. 1).

3. Three-electrode two-dimensional electrostatic mirror

The axial potential distribution of the three-electrode two-dimensional mirror, the electrodes of which are paired parallel plates separated by straight gaps and located symmetrically with respect to the middle plane of the mirror, is described by formula (6), in which instead of the real potentials V_1, V_2, V_3 on the electrodes their effective values determined by the relations

$$V_i^* = V_i - V_1 \sin^2\theta \quad (i = 1, 2, 3) \tag{11}$$

are used, where θ is the angle between the direction of the ion beam and the optical axis z.

The use of effective potentials enables us to describe the focusing properties of a mirror in the same way for any angle of incidence θ, i.e., the focusing properties of such a mirror do not depend on the angle at which the ion enters the field of the mirror.

In this work, the conditions of time-of-flight focusing of ions by energy up to the third order inclusively in a three-electrode two-dimensional electrostatic mirror are determined by numerical calculations for two modes of its operation: (1) a flat mirror mode; (2) a collimator mode.

An electrostatic mirror is called flat when the incident parallel beam of charged particles remains parallel after reflection from the mirror. The deflecting properties of two-dimensional flat electrostatic mirrors were first studied in Kelman, Fedulina, and Yakushev (1971). It was shown that two-dimensional flat mirrors can be of two types. In the first type of flat mirror, the direct and inverse branches of the trajectory coincide, while in the second type, the direct and reverse branches of the trajectory are symmetrical with respect to the median plane of the mirror. The first type of the flat mirror is determined by the condition $p' = 0$, and the second by $g' = 0$, where $p = p(z)$ and $g = g(z)$ are the same partial solutions of Eq. (2), described by equalities (3)–(5). The results of calculations for the two types of flat mirrors are shown in Table 2, and the course of the trajectories in such mirrors is shown in Fig. 3.

Table 3 and Fig. 4 show the results of calculations and the trajectories in the mirror for two types of the collimator mode. The first type of the collimator mode corresponds to p-solution, and the second type – to g-solution.

Table 2 Conditions of time-of-flight focusing for a three-electrode two-dimensional flat mirror.

Mirror type	V_2^*/V_1^*	$-V_3^*/V_1^*$	l/d_M	$-z_T/d_M$	L/d_M
1	0.0458	0.472	0.793	5.38	6.42
2	0.00781	0.148	1.09	27.5	30.0

Table 3 Conditions of time-of-flight focusing for a three-electrode two-dimensional mirror in the collimator mode.

Mirror type	V_2^*/V_1^*	$-V_3^*/V_1^*$	l/d_M	$-z_T/d_M$	L/d_M
1	0.0428	0.451	0.804	5.72	6.80
2	0.00767	0.146	1.09	28.0	30.5

Here, the geometrical dimensions are given in units d_M – the distances between the plates of the mirror electrodes, and the potentials on the electrodes are given in units V_1 – the potential of the first electrode. The numbering of the electrodes starts from 1 to 3 in the direction of the ion beam incident on the mirror. The origin of coordinates coincides with the center of the gap between the electrodes 2 and 3.

As can be seen from these data, the effective drift distance of the mirror L, hence, the specific time-of-flight dispersion (in d_M units) of the

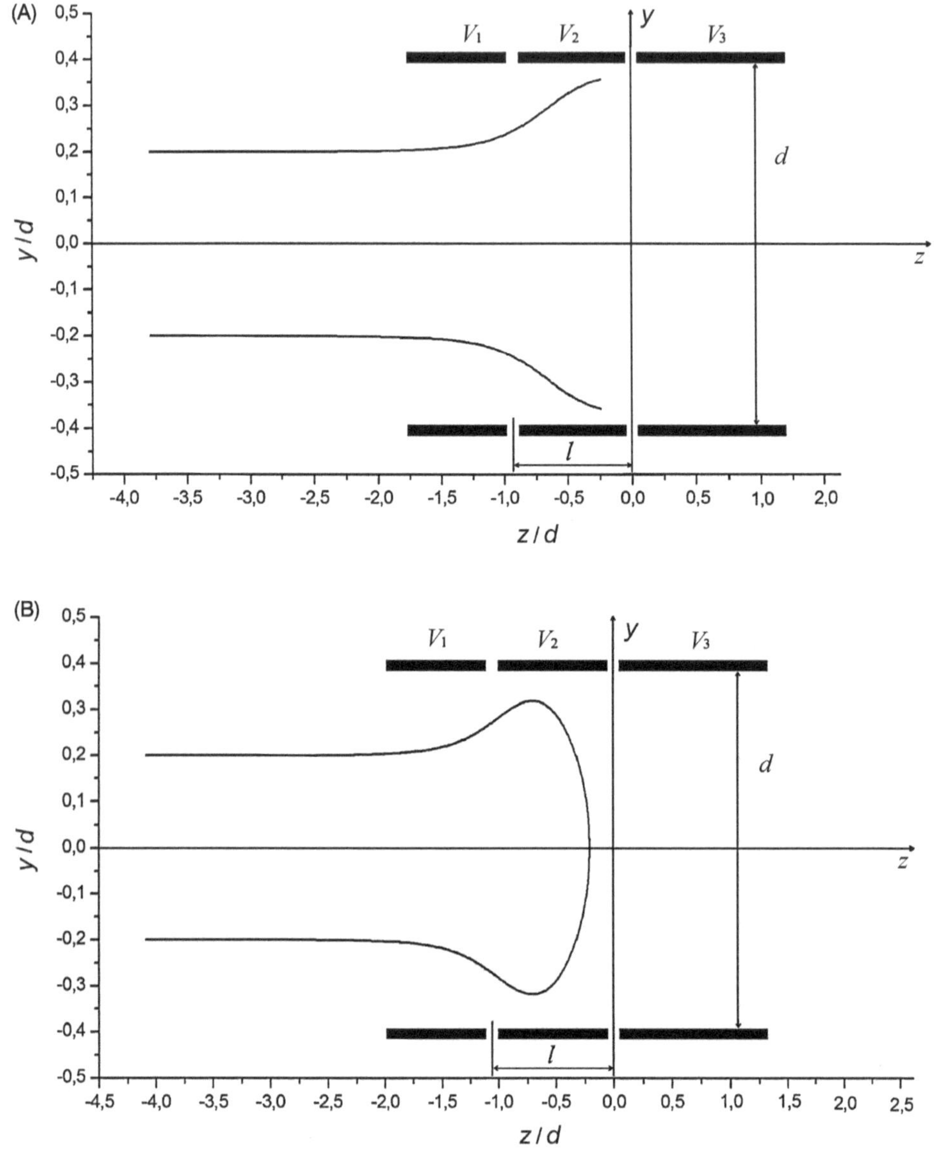

Figure 3 The course of the trajectories in the mode of a flat mirror: (A) – the first type; (B) – the second type.

mirror by mass (Bimurzaev, Bimurzaeva, & Sarkeev, 1991) is several times (more than four times) higher in the second type of mirrors than in the first type for both modes.

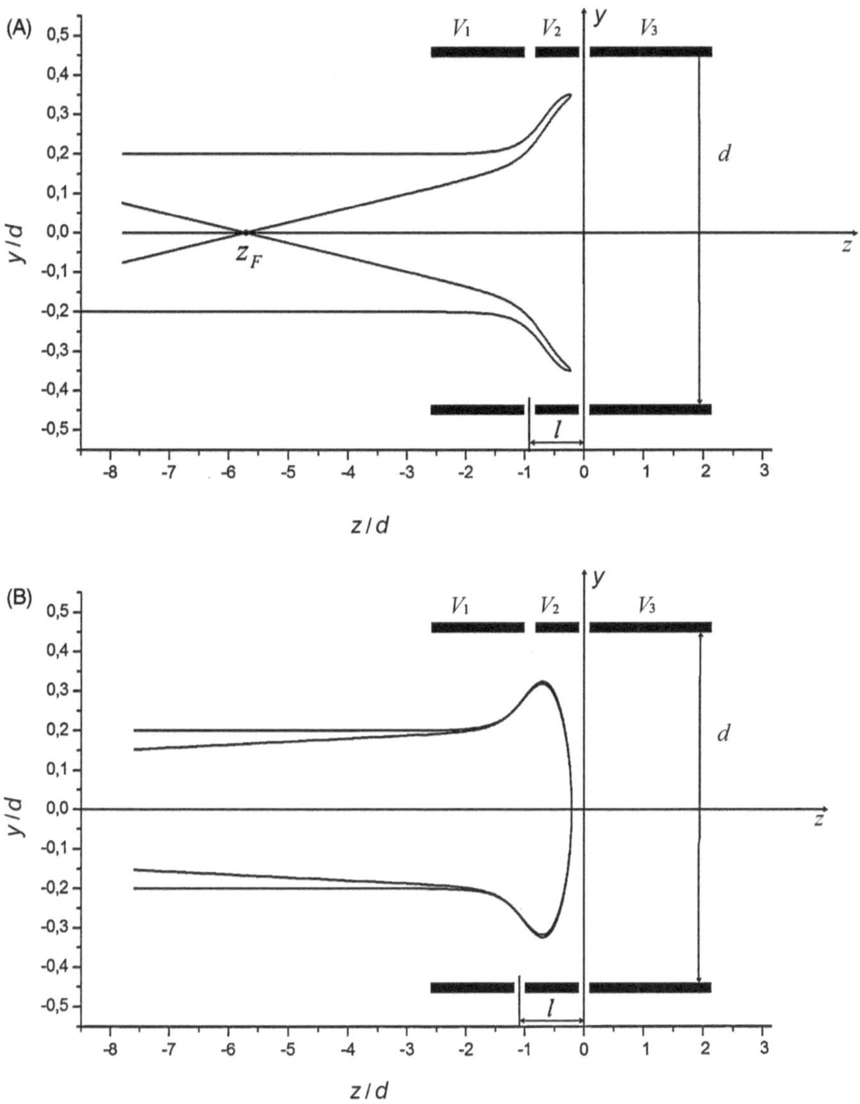

Figure 4 The course of the trajectories in the collimator mirror mode: (A) – the first type; (B) – the second type.

4. Planar multi-reflecting time-of-flight mass-spectrometer of a simple design

Fig. 5 shows a schematic diagram of one of the possible variants of the planar multi-reflecting time-of-flight mass spectrometer with a simple

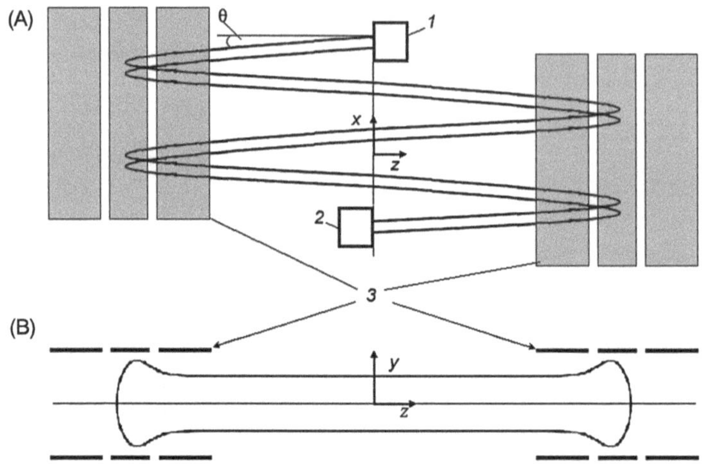

Figure 5 A schematic diagram of a planar multi-reflecting time-of-flight mass spectrometer of a simple design: (A) – in *XZ* plane, (B) – in *YZ* plane. 1 – orthogonal ion accelerator; 2 – detector; 3 – mirrors.

design based on the use of an orthogonal ion accelerator with a two-dimensional non-uniform field in the collimator mode and three-electrode two-dimensional mirrors in the flat mirror mode with a high time-of-flight mass dispersion.

5. Conclusion

The use of an orthogonal ion accelerator with a non-uniform field forming a parallel ion beam in combination with highly dispersive two-dimensional mirrors enables us to achieve a simultaneous increase in the resolution and sensitivity of a planar multi-reflecting time-of-flight mass spectrometer, maintaining the simplicity of its design. It should be noted that the use of effective potentials allows us to change the number of reflections to increase the resolution of the device without disturbing its focusing properties. In addition, the use of two-dimensional electrostatic mirrors with a high time-of-flight dispersion by mass makes it possible to significantly reduce the size of the device without degradation of its resolution.

Acknowledgments

I am sincerely grateful to Dr. E.M. Yakushev for attention to this work and useful advice.

This work was supported in part by the Ministry of Education and Science of the Republic of Kazakhstan, grant No. AP05132483.

References

Bimurzaev, S. B. (2015). Time-of-flight mass-spectrometer with high resolution and sensitivity via elimination of chromatic TOF aberrations of higher orders. *International Journal of Mass Spectrometry*, *376*, 23–26.

Bimurzaev, S. B., Bimurzaeva, R. S., & Sarkeev, B. T. (1991). Spatial and time-of-flight focusing in an electromagnetic system with two planes of symmetry. *Radiotehnika i Elektronika*, *36*, 2186–2195.

Glikman, L. G., & Yakushev, E. M. (1967). Field of an electrostatic cylindrical three-electrode lens. *Zhurnal Tekhnicheskoi Fiziki*, *37*, 2097–2104.

Hasin, Ju. I., Gavrik, M. A., Kozlov, B. N., Muradymov, M. Z., Yavor, M. I., & Verentchikov, A. N. (2006). Increasing sensitivity and mass accuracy of a multi-reflecting time-of-flight mass spectrometer with a continuous ion source. *Nauchnoje Priborostroenie*, *16*(4), 11–18.

Kelman, V. M., Fedulina, L. V., & Yakushev, E. M. (1971). Deviation of parallel beams of charged parts by a flat electrostatic mirror. *Zhurnal Tekhnicheskoi Fiziki*, *41*, 1825–1831.

Nazarenko, L. M., Sekunova, L. M., & Yakushev, E. M. (1992). *SU Patent 1725289 A1*.

Yakushev, E. M., Bimurzaev, S. B., & Kholodov, M. A. (2016). On the definition of cardinal elements and angular characteristics of cathode lenses with rotational symmetry. *Vestnik of K. Zhubanov ARGU*, *2*, 104–111.

Yavor, M., Verentchikov, A., Hasin, Ju., Kozlov, B., Gavrik, M., & Trufanov, A. (2008). Planar multi-reflecting time-of-flight mass analyzer with a jig-saw ion path. *Physics Procedia*, *1*, 391–400.

Yavor, M. I., & Verentchikov, A. N. (2004). Planar multi-reflecting time-of-flight mass analyzer operating without a mass range limit. *Nauchnoje Priborostroenie*, *14*(2), 38–45.

CHAPTER TWO

Generalization of paraxial trajectory method for the analysis of non-paraxial rays

Shin Fujita
Innovation Centre, Shimadzu Asia Pacific Pte Ltd., Singapore, Singapore
e-mail address: fujita@shimadzu.com.sg

Contents

1. Introduction

In the designing process of electron optical columns (EOCs) such as transmission electron microscopes, scanning electron microscopes, and microfocus X-ray tubes, simulation software is now extensively used. One major reason the EOC analysis programs are so useful is that they can offer the results in terms of the *optical parameters* that are amenable to interpretation by the designers on the basis of a well established theory, the paraxial trajectory method (PTM). A clear physical picture of column characteristics is obtained exploiting cardinal optical parameters like the focal lengths, the principal plane positions, and the magnification factors, as well as several aberration coefficients that describes deviations of the system's behavior from linear approximation.

Fig. 1 presents an example of optical concepts used in the PTM. In the first linear approximation, the effects of lens field on electron trajectories is summarized as a linear transformation relation between the trajectory condition on the upstream side $(\xi, \tan\alpha)$ at $z = z_c$ and that of the downstream

Advances in Imaging and Electron Physics, Volume 212
ISSN 1076-5670
https://doi.org/10.1016/bs.aiep.2019.08.002

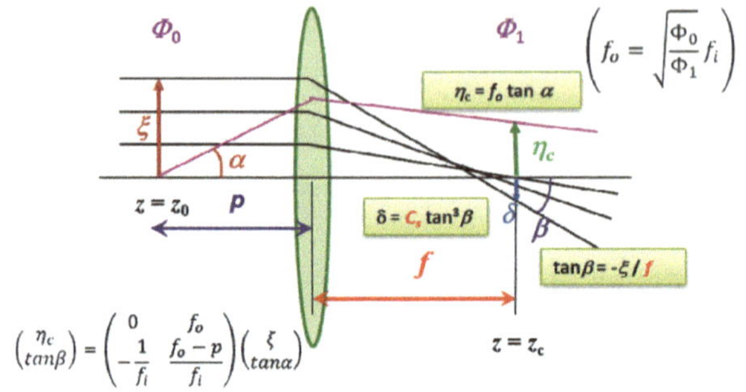

Figure 1 Optical concepts and parameters obtained with Paraxial Trajectory Method.

side $(\eta_c, \tan\beta)$ at $z = z_1$:

$$\begin{pmatrix} \eta_c \\ \tan\beta \end{pmatrix} = \begin{pmatrix} 0 & f_o \\ -\frac{1}{f_i} & \frac{f_o - p}{f_i} \end{pmatrix} \begin{pmatrix} \xi \\ \tan\alpha \end{pmatrix}. \tag{1}$$

Here, ξ and η designate the trajectory coordinates and α and β are the inclination angles at the respective positions. The lens focal lengths f_0 and f_i represent deflection strength of the lens field, and the trajectory condition relation can be specified with the help of distance parameter p that designate the reference position where the upstream condition is given.

An important point to note is the fact that one does not need to study individual electron trajectories to understand their ensemble behavior. An optical parameter, the lens focal length f, suffices to know the trajectory conditions at $z = z_c$, $(\eta_c, \tan\beta)$, if one has the conditions at $z = z_0$ $(\xi, \tan\alpha)$. It can easily be imagined that the perspective of lens design improves significantly by understanding its properties in terms of the optical parameters.

Higher order effects that cannot be dealt with by the linear approximation are taken into account in terms of the *aberrations* coefficients, which are obtainable from perturbation corrections to the linear approximation. We assert that the PTM has made the lens design processes, including aberration considerations, substantially efficient by making it possible to evaluate its characteristics with a few optical parameters.

On the other hand, electron gun designs have depended mainly on ray-tracing calculation. No systematic interpretation of ray-tracing results has so far been established. Fig. 2 gives an example of what one normally conducts in the characterization of an electron gun. Here, the crossover characteristics of a triode thermionic gun is investigated. The electron rays are calculated from the cathode surface to the field free space. The initial

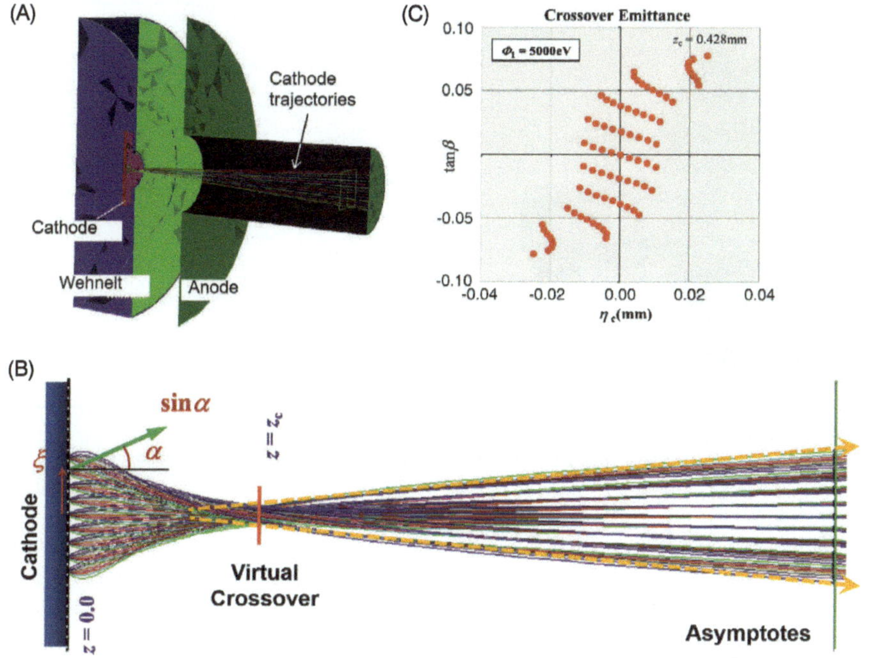

Figure 2 (A) A triode electron gun model. (B) Cathode ray behavior inside triode electrode gun. (C) Emittance diagram of the source.

ray conditions on the cathode are distributed appropriately so that the ray emanation angle covers $-\pi/2 < \alpha < \pi/2$, and their emission points spread over the cathode surface area where the electric field is in the extracting direction. In the example, the initial kinetic energy was assumed to be fixed at $E_0 = 0.2$ eV. By drawing asymptotes to each ray calculation result, one can deduce the virtual crossover position $z = z_c$ where the spatial spread of the asymptotes becomes most constricted. The emittance diagram at the virtual crossover, i.e. the ensemble of the coordinate and slope points $(\eta_c, \tan\beta)$, provides both the source size (the horizontal spread of the emittance diagram) and the angular spread (the vertical extension) of the electron source.

Conventionally, it has been necessary to calculate all the relevant electron rays from the cathode to the field free space for obtaining an emittance diagram. It is not only a time-consuming process, but it is sometimes non-straightforward to interpret the diagram for gun design purposes. For example, the electron source angular intensity is not obvious from the diagram alone. These facts make simulation evaluations of the electron gun inefficient, compared with the lens design cases.

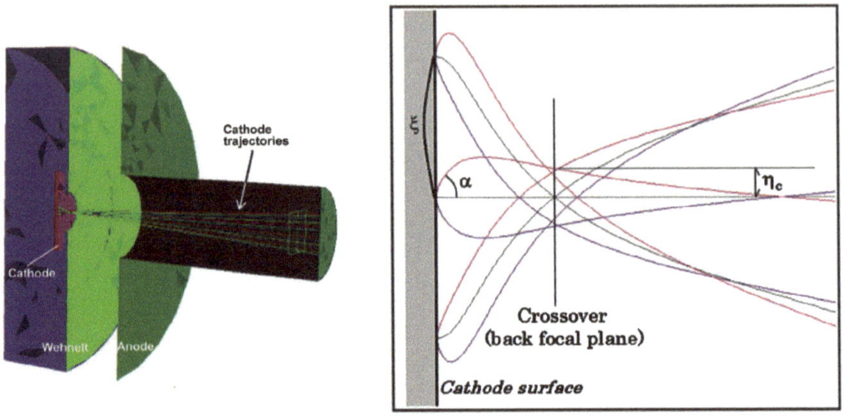

Figure 3 Crossover formation at the pseudo back focal plane in an electron gun.

Major difficulty of the PTM, when applied to electron gun characterizations, is that it cannot deal with large starting angle α of cathode rays. The conventional theory assumes that both the trajectory coordinate and slope are small enough for the PTM condition to be valid. The theory breaks down from the beginning if it is applied to cathode rays.

Ray tracing results, however, suggest some of the optical concepts derived in paraxial condition still hold in the electron gun as well, such as crossover formation at the back focal plane (see Fig. 3).

In this article, the optical parameters valid with large slope cathode rays will be derived from the investigation of the ray condition mapping between the cathode surface and the (virtual) crossover plane:

$$\begin{pmatrix} \xi \\ u = \sin\alpha \end{pmatrix} \Rightarrow \begin{pmatrix} \eta_c \\ v = \sin\beta \end{pmatrix}. \tag{2}$$

It will be shown that a simple change in the ray slope specification from $\tan\alpha$ to $\sin\alpha$ makes Paraxial Trajectory Method functionable for rays with large slope angles. Some optical parameters, which one has beneficial use of in conventional lens designs, are extended to cases of electron gun characterization.

2. Canonical mapping and optical parameters

Fig. 4 gives a schematic diagram of the Canonical Mapping (CM) scheme that interconnects the ray conditions on the cathode surface

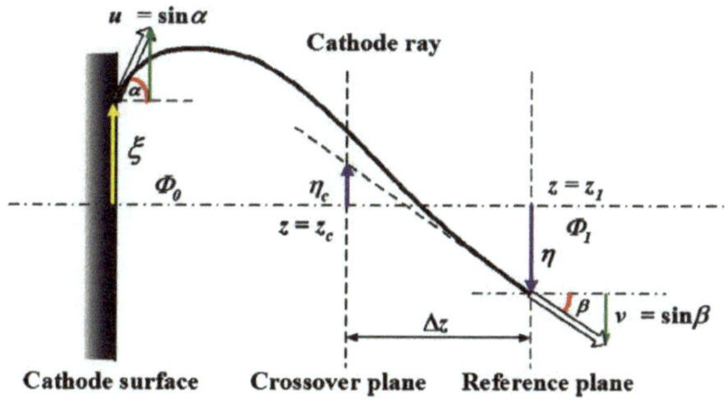

Figure 4 Schematic representation of the Canonical Mapping of cathode ray conditions.

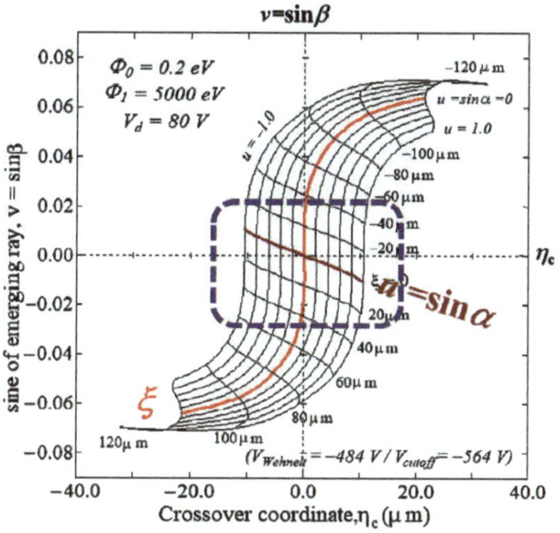

Figure 5 The Canonical Mapping Transformation (CMT) diagram of the triode gun model.

$(\xi, u = \sin\alpha)$ to those on the (virtual) crossover plane $(\eta_c, v = \sin\beta)$. It is assumed that the electron energy is specified by setting the electron kinetic energy on the cathode at $E_0 = e\Phi_0$.

The transformation can be expressed in a 2D to 2D mapping as shown in Fig. 5, which will be designated as the CM diagram. The overall figure of the CM diagram provides both the spatial and angular spread of an electron source (crossover characteristics) and is similar in shape to emittance diagrams conventionally used in electron gun characterizations, except for the adoption of $\sin\beta$, instead of $\tan\beta$ in the vertical axis. However, the uniqueness of the CM diagram is that we will attempt to approximate the transformation relation given by Eq. (2) by a linear and higher order polynomial mapping. The diagram states, for each ray condition at the crossover emittance, where and at what direction the electron was started on the cathode. Some of the polynomial coefficients found in the mapping approximation can be used as *electron gun optical parameters*.

Use of the trigonometric function $\sin\alpha$, instead of $\tan\alpha$, for the ray slope specification is the key to success of the CM diagram. It will then enable generalization of the PTM to rays with large angles with respect to the axis. With this slope angle specification, the *Lagrange differential invariant* relation holds (Fujita & Shimoyama, 2005):

$$\frac{\partial \eta_c}{\partial \xi}\frac{\partial \nu}{\partial u} - \frac{\partial \eta_c}{\partial u}\frac{\partial \nu}{\partial \xi} = \sqrt{\frac{\Phi_0}{\Phi_1}}, \tag{3}$$

where Φ_1 represents the electron kinetic energy (in eV) in the field free space from which the asymptotes to the crossover are drawn. We can make use of the relation in approximating the Canonical Mapping.

It can easily be shown that a simple linear relation satisfies the Lagrange differential invariant:

$$\eta_c(\xi, u) \cong f\sqrt{\frac{\Phi_0}{\Phi_1}}u, \tag{4.1}$$

$$\nu(\xi, u) = \sin\beta \cong -\frac{\xi}{f} + mu. \tag{4.2}$$

As one can see in Fig. 5, the parallelogrammatic grids in the mapping around the origin enclosed by a dotted square indicate that the transformation can well be approximated by a linear relation. The linear approximation given by Eqs. (4) is a good representation of the Canonical Mapping as long as the rays emanate from near axis points on the cathode. The linear approximation works satisfactorily for rays with the initial slope as large as $\alpha = \pi/2$. The reason for the validity of linear approximation will be explored later in the dissuasion section.

Eqs. (4.1) and (4.2) show that one parameter, *electron gun focal length* f, prescribes both the spatial and angular spread of the crossover. When the spatial spread is made small to have a fine source size (short f), the angular spread will inevitably increase, leading to a lower angular current intensity. The mutually inverse dependence of the two source properties comes from the Lagrange differential invariant and is accordingly a very fundamental property.

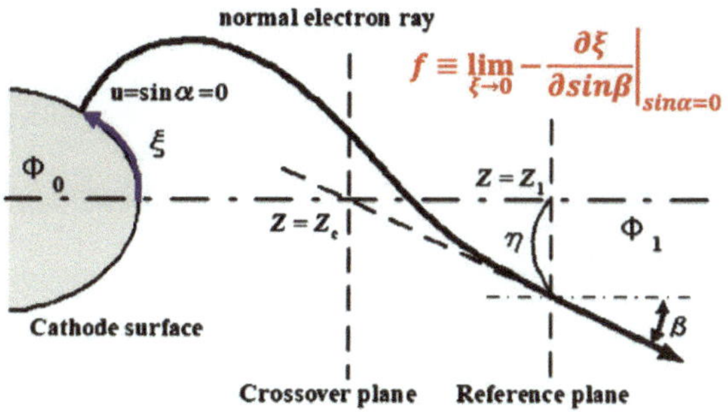

Figure 6 A general definition of the electron gun focal length.

The *electron gun focal length* defined below (see Fig. 6) is the most important parameter in the Canonical Mapping:

$$f \equiv \lim_{\xi\to 0} -\frac{\partial \xi}{\partial \sin\beta}\Bigg|_{\sin\alpha=0} \tag{5}$$

It is defined as the proportional factor between the ray starting coordinate ξ and the asymptotic ray slope $\sin\beta$ and can extensively be used in electron gun characterization.

3. Electron source characterization

One can derive from Eqs. (4.1) and (4.2) two fundamental electron source characteristics using the electron gun focal length f, source size d_{co} and angular intensity J_Ω:

$$d_{co} = 2f\sqrt{\frac{kT}{e\Phi_1}}, \tag{6.1}$$

$$J_\Omega = j_s f^2. \tag{6.2}$$

In the crossover size formula above, the thermionic energy kT has replaced the initial energy $e\Phi_0$. The cathode current density j_s is used in the angular current intensity formula.

We can see that the same optical parameter, electron gun focal length, is relevant in specifying both the source characteristics. A short foal length is required in order to achieve a very fine source size, but it automatically leads to a low angular current intensity.

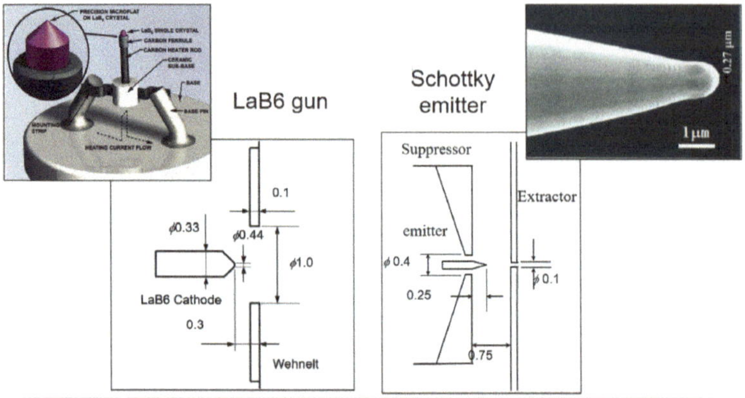

	LaB6 gun	Schottky emitter
Cathode current load, j_s	10.0 A/cm^2	5,000 A/cm^2
Reduced brightness, R_0	2.05x10^5 A/m^{2}srV	1.03x10^8 A/m^{2}srV
Crossover size, d	12,900 nm	17.9 nm
Angular intensity, J_Ω	3,600 mA/sr	0.14 mA/sr
Electron gun focal length, f	5,980 μm	-1.69 μm

Figure 7 Comparison of the source characteristics.

Very different nature in source characteristics between thermionic (LaB6 cathode) and point cathode (Schottky emitter) guns stems from difference in electron gun focal length. Fig. 7 compares their source characteristics. Typical cathode current densities for LaB6 and Schottky emitter are assumed in the estimates. The cathode temperature is set at $T = 1800$ K ($e\Phi_0 = 0.155$ eV) in both cases. The respective electron gun focal lengths were calculated using models shown in the inserts according to the definition given in Fig. 6. The reduced brightness is calculated using the following formula:

$$R_0 = \frac{J_\Omega}{\pi\left(d_{co}/2\right)^2}\frac{1}{\Phi_1} = \frac{1}{\pi}\frac{ej_s}{kT}. \tag{7}$$

The difference in source size comes from the three-orders-of-magnitude disparity in the electron gun focal length between the two types of electron guns. A very short focal length of the Schottky emitter comes from its finely pointed tip geometry. A diverging electric field produced at the tip makes cathode rays spread significantly in direction right after the emanation from the cathode. A very short electron gun focal length is obtained in accordance with its definition in Fig. 6. Note that the negative sign in the focal length is due to the diverging nature of the cathode rays. Absence of the crossover formation gives a negative electron gun focal length as defined in Eq. (5).

Relatively low angular current intensity of the Schottky emitter originates from its short electron focal length as well. Since the intensity is proportional to square of the focal length, even a high cathode current density does not compensate. Low angular current intensity is general characteristics of all the point cathodes. One needs to be careful about EOC designs when using a point cathode, especially for high beam current applications. An intrinsically high brightness of the point cathode can very easily be lost due to aberrations inside the electron gun.

There is one important point to note regarding how the source characteristics are determined. While the source intrinsic brightness R_0 is solely given by cathode material properties such as current density and operation temperature, the source size d_{co} and angular current intensity J_{Ω} depend on the electron gun focal length, which in turn is given by emitter geometries. Optics design, i.e. electrode geometry deign, counts in deciding the source size and angular current intensity but not the intrinsic beam brightness.

4. Generalized paraxial trajectory method

Deviation from the linear approximation of the Canonical Mapping given by Eqs. (4.1) and (4.1) becomes important when one attempts to get a high beam current from the electron gun. An enlarged beam acceptance angle required entails the *apparent* increase in the source size due to the aberrations. The CM diagram deformation in the upper and lower portions, as is presented in Fig. 5, has an effect of making the source apparently larger than the original size defined by the diagram width at $v = \sin\beta = 0$. The mechanism of the size increase is the same as the blurring of probe spots by the spherical aberration encountered in probe formation by conventional lens systems.

The apparent increase in the source size results in the *lowering of the apparent brightness* as well. The axial brightness (intrinsic brightness) formula given by Eq. (7) assumes the source size is solely defined by the CM diagram width. When the beam acceptance angle is small under a low beam current condition, the diagram width corresponds well with the apparent, i.e.

effective, source size. The axial brightness gives a good indication of experimentally obtained brightness obtained from the electron gun. However, as the beam acceptance angle is increased for higher currents, the experimental brightness starts to be lowered due to blurring of the apparent source size.

The brightness lowering is an especially serious issue with point cathodes such as Schottky emitter and cold field emitter. As we studied in the previous section, their intrinsic source size is extremely small and the angular current intensity relatively low due to very short electron gun focal lengths. The point cathode guns are extremely vulnerable to the aberration influences in the gun and their intrinsically high brightness can easily be lost in high beam current applications unless proper cares are taken.

The Generalized PTM is able to take the aberration effects inside electron guns into account by very similar procedures as exploited in the conventional PTM. Its mathematical approach is based on the perturbation theory (Fujita, Takebe, Ushio, & Shimoyama, 2010), starting from a paraxial trajectory. In the following, we present the step-by-step procedures of the Generalized PTM to approximate the Canonical Mapping of an electron gun with the aberration effects included.

Fig. 8 schematically presents the processes using the same triode electron gun as an example.

The unit axial potentials are first calculated from an electron gun model. Here, the axial potential distributions are obtained with each electrode (the Wehnelt and the anode) potential being in turns set at $V = 1$, respectively. The cathode potential is taken as the reference and is always assumed to be $V = 0$. Any combination of the electrode voltages can arbitrarily be set later in the generalized PTM calculations.

Principal trajectories, $g(z)$ and $h(z)$, are then calculated using the same equation as the conventional PTM based on the total axial potential distribution. Some of the important optical parameters relevant to electron source characteristic are found from the principal trajectory $g(z)$:

$$\text{Crossover position } z_c: \quad g'\left(z_{\text{asymp}}\right)\left(z_c - z_{\text{asymp}}\right) + g\left(z_{\text{asymp}}\right) = 0, \tag{8.1}$$

$$\text{Electron gun focal length } f: \quad f = -\frac{1}{g'\left(z_{\text{asymp}}\right)}, \tag{8.2}$$

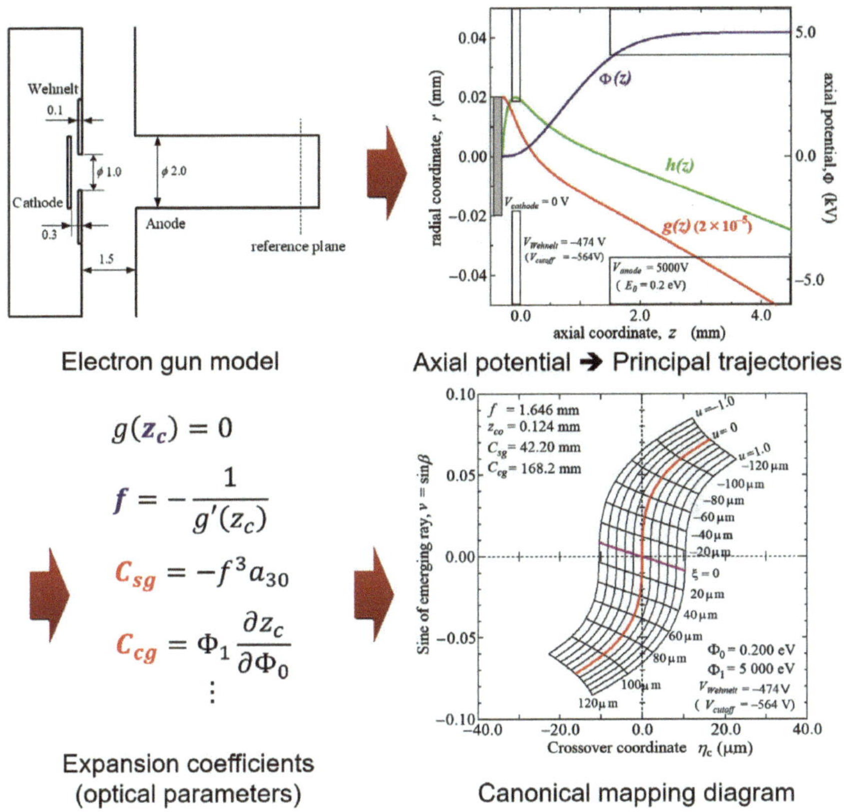

Figure 8 Step-by-step procedures of the Generalized PTM for electron gun analysis with aberrations.

where the asymptotic coordinate and slope, $g\left(z_{\text{asymp}}\right)$ and $g'\left(z_{\text{asymp}}\right)$, should be taken in field free space where the principal trajectory is a straight line.

As was explained in the previous section (Eqs. (6.1) and (6.2)), the intrinsic source size d_{co} and the angular current intensity J_{Ω} can be derived from the electron gun focal length f, together with the cathode temperature T and the cathode current density j_s.

The third order polynomial coefficients to approximate the Canonical Mapping beyond the linear transformation can then be estimated using the aberration integrals. Please refer to the original papers (Fujita et al., 2010) for the calculation details. They are in principle calculated using the same mathematical formulas as the conventional PTM.

The only differences in the aberration calculations is that the polynomial expansions are attempted both in the coordinate $\eta_c(\xi, u)$ and in the slope $v(\xi, u)$ because one wants here to have an approximation to the CM diagram, which defines the transformation between $(\xi, u = \sin\alpha)$ and $(\eta_c, v = \sin\beta)$. In conventional PTM aberration calculations, in contrast, one is usually concerned only with the perturbation in the trajectory coordinate.

In the Generalized PTM, a higher order approximation to Canonical Mapping is obtained by adding the third order polynomial terms. The polynomial coefficients, a_{kl} and b_{kl}, represent aberrations:

$$\eta_c(\xi, u) = f\sqrt{\frac{\Phi_0}{\Phi_1}}u + \sum_{k+l=3} a_{kl}\xi^k u^l, \tag{9.1}$$

$$v(\xi, u) = -\frac{1}{f}\xi + mu + \sum_{k+l=3} b_{kl}\xi^k u^l. \tag{9.2}$$

Note that the expansion formula uses $u = \sin\alpha$ instead of x' in accordance with the principles of the Generalized PTM as expounded in Section 2.

Two of the useful optical parameters that represent major aberrations and that one can derive from the expansion formula, Eq. (9.1), are:

$$\text{Spherical aberration } C_{sg}: \quad C_{sg} = -f^3 a_{30}, \tag{10.1}$$

$$\text{Chromatic aberration } C_{cg}: \quad C_{cg} = \Phi_1\frac{\partial z_c}{\partial \Phi_0}. \tag{10.2}$$

The CM diagram that is approximated by the Generalized PTM is drawn by use of the polynomial expansion Eqs. (9.1) and (9.2) in the last step of the calculation. An example of the polynomial-expanded CM diagram (green grid lines) is given in Fig. 9 together with the ray-tracing results (red circle marks). It is seen that deformation of the CM diagram due to aberrations are reproduced in the polynomial approximations fairly well. The S-shape curvature of the diagram center line, which is the diagram representation of the gun spherical aberration, is accurately reproduced by the approximation.

Please note that there are no fitting processes involved in the comparison. The polynomial expansion of the transformation and drawing of the approximate CM diagram by the Generalized PTM are carried out independently of calculation of the ray-tracing diagram. It exemplifies a degree of confidence with which one can evaluate the cathode ray characteristics by use of the Generalized PTM calculations.

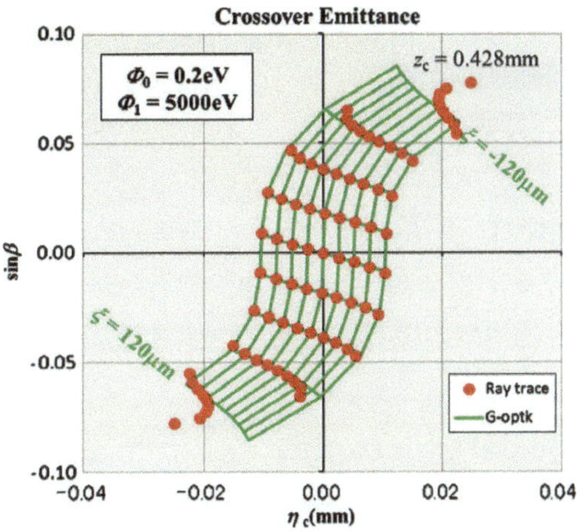

Figure 9 Polynomial approximation of the CMT diagram with aberrations.

5. *G-optk* program

G-optk program was developed to provide calculation capability for the Generalized PTM. Calculation flow of the program follows the methodologies described in the previous section.

The GUI of the program is presented in Fig. 10. The calculation condition is defined in the upper window. The cathode position and the electron initial energy are given there. The axial potential files, which have been evaluated in advance by a field computation software, are specified and their positions in the global coordinate system given. One can also specify the excitation strengths (the electrode voltage and coil ampere-turns) of all the optical elements in the window. CM diagram drawing condition is controlled by specifying the cathode ray emanation condition in terms of the range in ξ and $u = \sin\alpha$ for which the CM diagram at the crossover will be drawn.

The upper window also shows some of the calculated optical parameters such as *crossover position*, *electron gun focal length*, and *principal aberrations*. The parameters can be saved in a spread sheet as well.

The lower left window displays the primary trajectory calculation results together with the total axial potential and the total axial magnetic flux density. The lower right window gives the CM diagram at the crossover. Together, the *G-optk* program provides an intuitive understanding of the electron gun properties both in graphics and in the optical parameters.

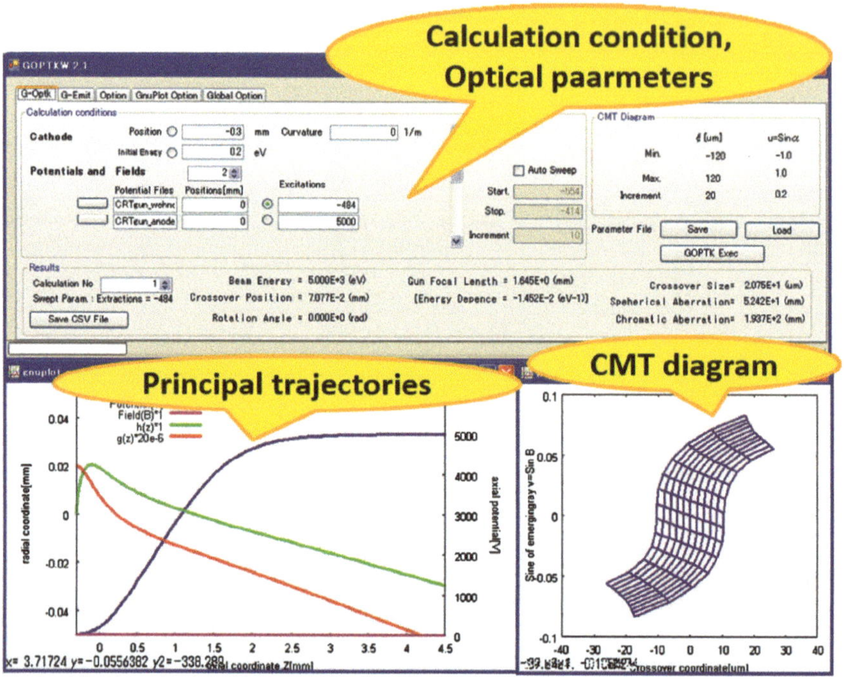

Figure 10 GUI of *G-optk* program for the calculation of Generalized PTM.

One advantage of the program is that *G-optk* can incorporate the lens system analysis into that of electron guns seamlessly thanks to the Generalized PMT's capability to deal with the both systems in a consistent way. Fig. 11 shows calculation results conducted on a microfocus X-ray tube with a LaB6 cathode gun and double magnetic lenses.

The optical elements of the X-ray tube are given in Fig. 11A. It has a LaB6 electron gun of the triode type operable up to $V = 160$ keV. The cathode is a single crystal LaB6 tip whose top is a tapered cone with a truncation of 15 μm diameter plane that exposes (100) crystal face. The electrons are assumed to emit from the truncated face with an initial energy of $E_0 = 0.155$ eV ($T = 1800$ K). The first magnetic lens is used for beam current control. The lens strength determines the amount of beam that goes through the beam defining aperture located in the polepiece gap of the second lens. The beam current setting is the lowest when Lens 1 excitation is null and augments as the lens ampere-turns is increased until it reaches the maximum when the beam is focused at the aperture position. The second

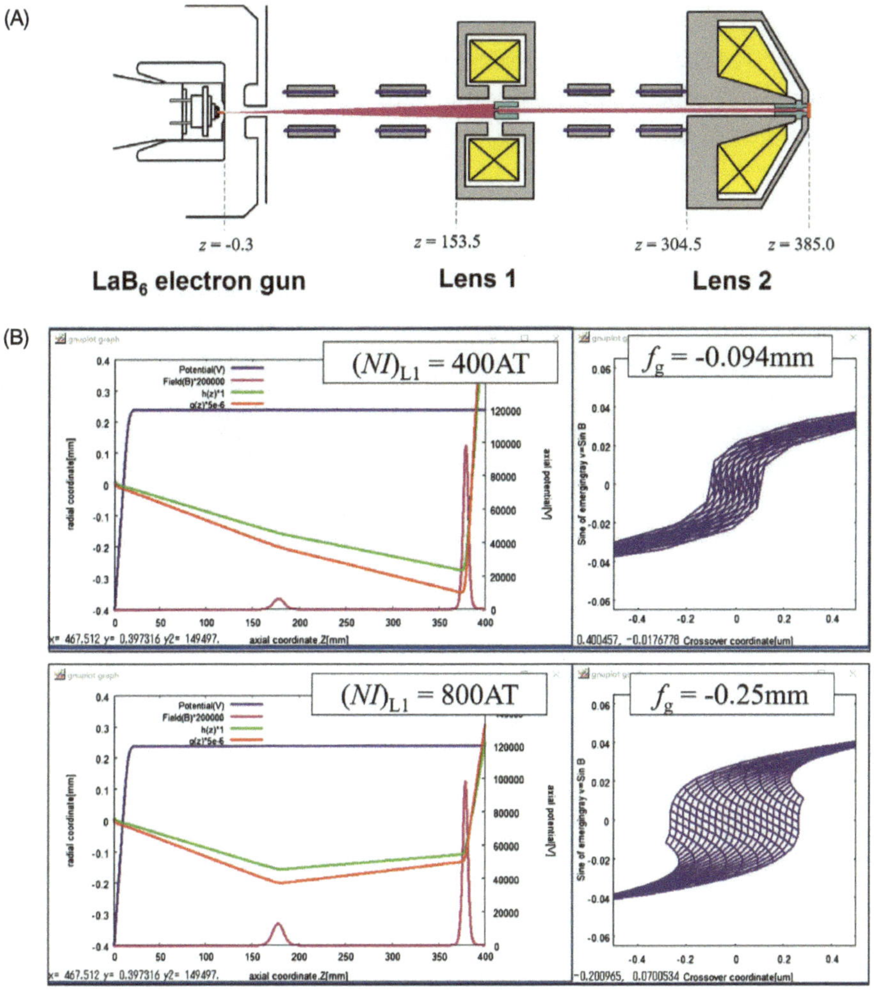

Figure 11 *G-optk* calculation results for a microfocus X-ray tube. (A) Model of microfocus X-ray tube with a LaB6 cathode gun and the double magnetic lenses. (B) *G-optk* graphical output for two lens excitation conditions.

lens focuses the electron beam on a tungsten target positioned on the outer face of the Lens 2 to produce a fine X-ray source.

In *G-optk* calculations, the probe formed at the target by Lens 2 is regarded as the gun crossover. When the optical analysis is carried out starting from the cathode, as in the case of the Generalized PMT, probes focused by lens systems are positioned in a plane conjugate to the gun (virtual) crossover plane.

Fig. 11B presents the results of the Generalized PMT at two different Lens 1 excitations. The first result corresponds to a low beam current condition. The deflection of the trajectory at Lens 1 is nominal and the electron gun length is shorter, $|f_g| = 0.094$ mm. The intrinsic crossover size at the target is correspondingly small, $d_{target} = 0.2$ μm. Note, however, that the spherical aberration effect ($C_{sg} = 12.8$ mm) can be seen in the CM diagram as the S-shape curvature. One must take the aberration effects into account when the beam defining aperture size exceeds $\sin\beta > 0.01$. With an increased acceptance angle, the virtual crossover starts to show 'tails' due to aberration that sprawl out of the intrinsic width.

When Lens 1 excitation is increased to achieve a higher beam current, the electron gun focal length also increases to $|f_g| = 0.25$ mm, and the corresponding intrinsic crossover diameter becomes 0.53 μm. Even though the spherical aberration coefficient is about the same, $C_{sg} = 11.9$ mm as in the previous case (it is mainly dictated by Lens 2 aberration whose operation condition remains practically the same in both low/high beam current conditions), its effect is less conspicuous at the high beam current condition because (1) the intrinsic crossover size is now relatively larger and (2) the beam convergence angle for a given beam current is more confined at a high Lens 1 excitation.

G-optk program can provide not only the electron source characteristics but also the crossover properties of the entire electron optical column that comprises a gun, condensers, and objective lens. As far as the author knows it is the first program that can analyze the whole electron optical system in a consistent way based on the paraxial trajectory calculation principles.

6. Discussion

A brief discussion is given to explore reasons why simple change in the ray slope specification from $\tan\alpha$ to $\sin\alpha$ makes the Paraxial Trajectory Method valid even for cathode rays with large slope angles. We here attempt to present only an intuitive physical foundation of the Generalized PTM. The readers are advised to refer to the original papers (Fujita et al., 2010) for rigorous mathematical derivations.

For this discussion we will focus on electron ray behavior in space very close to cathode surface. Fig. 12 gives a simplified calculation model and the representative cathode rays. It is assumed that the electric field F is uniform, and the trajectory accordingly takes a parabolic shape.

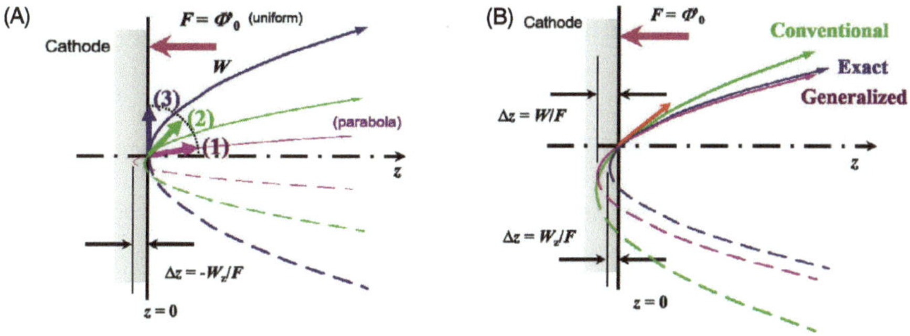

Figure 12 A simplified calculation model and the representative cathode rays. (A) Movement of the apex position of cathode ray with emission angle. (B) A new assignment strategy of the paraxial trajectory in the Generalized PTM calculation.

Three trajectories having the same total initial kinetic energy W that are emitted from cathode with increasing initial angles α with respect to the axis are presented in Fig. 12A. When the electron emission angle changes, the apex position of the parabola should move in accordance with the change in the kinetic energy in the axial direction, W_z:

$$\Delta z = -W_z/F \qquad \left(eW_z = \frac{1}{2}mv_{0z}^2\right), \tag{11}$$

where m is the electron mass and v_{0z} is the axial component of the electron initial velocity.

The apex of the parabola trajectory moves forward as the initial angle α increases. The conventional PTM calculation, however, cannot reproduce this apex shift since the paraxial trajectory equation implicitly assumes that the axial initial kinetic energy W_z remains constant irrespective of the ray slope angle.

Because of this difficulty, when a trajectory that has the same slope angle as the exact ray is taken as an approximating paraxial trajectory, the overall discrepancy of the two trajectories becomes increasingly large as the ray starting angle α increases (see Fig. 12B). Since the parabola apex position of the exact ray and that of the paraxial trajectory increasingly separate from each other, the paraxial trajectory parabola tends to have a significantly wider spread in transverse direction than the exact ray parabola if the ray slopes are matched. We can also say that the transverse momentum of the assigned paraxial trajectory, $p_t^{(P)}$ becomes larger than the exact transverse momentum $p_t^{(exact)}$, $p_t^{(P)} > p_t^{(exact)}$.

The definition of the transverse momentum in the paraxial approximations is given by:

$$p_x^{(P)} = \sqrt{2me\Phi_0}x' = \sqrt{2me\Phi_0}\tan\alpha, \tag{12.1}$$

because the axial momentum is assumed to be $p_z^{(P)} = \sqrt{2me\Phi_0}$ regardless of the ray slope.

However, the exact ray has the transverse momentum of:

$$p_x^{(exact)} = \sqrt{2me\Phi_0}\frac{x'}{\sqrt{1 + x'^2 + y'^2}} = \sqrt{2me\Phi_0}\sin\alpha. \tag{12.2}$$

Mathematically, the difference between the two is due to change in the Lagrangian when one moves from the paraxial approximation to the exact dynamic system (Fujita et al., 2010).

In the Generalized PTM calculation a new assignment strategy of the paraxial trajectory is adopted. Here, the paraxial trajectory that has the same transverse momentum as the exact trajectory momentum, rather than the same initial ray slope, is assigned as an initial approximation.

By using $\sin\alpha$ instead of $\tan\alpha$ in ray slope specification, we become able to assign a proper paraxial trajectory to a large angle cathode ray that has the matching transverse momentum. The assignment of a paraxial trajectory that has a different initial slope than the exact ray is possible because the slope difference (more exactly speaking, it is a difference in the momentum) can be taken into account in the aberration calculation step. Any discrepancies in initial condition between exact and paraxial trajectories will be accounted for as perturbation correction.

It is worth noting here that in the framework of the paraxial trajectory method complemented with aberration perturbation corrections, it is not necessary to match the initial position and ray direction between the exact ray and a paraxial trajectory assigned to it. The difference in the condition can later be accounted for in the aberration calculation step. While there is rarely a positive motivation to assign a paraxial trajectory that starts from a different spatial position than the exact ray, there are many cases, like in the treatment of large slope cathode rays, where a paraxial trajectory with an appropriately modified slope is a more suitable choice. In the cathode ray cases we have been dealing with here, it turns out that the perturbation correction becomes *simpler* if the transverse momentum is matched instead of the ray slope.

7. Conclusions

An extension of the Paraxial Trajectory Method (PTM) to deal with cathode rays that have a very large initial slope with respect to the axis was discussed, where the conventional PTM breaks down from the beginning.

The concept of the Canonical Mapping was introduced. The mapping of the ray conditions (coordinate, slope as expressed in sine), between cathode surface to crossover plane was studied. A polynomial approximation of the mapping gives, in the form of its coefficients, some optical parameters that are useful in electron gun designs.

Among them, the Electron gun focal length is the most important optical parameter. It defines both the (intrinsic) source size and the angular intensity of the electron gun.

It was shown that the optical parameters as the third order polynomial coefficients of Canonical Mapping can be calculated in the same fashion as the conventional Paraxial Trajectory Method.

G-optk program was developed to conduct the Generalized PTM calculations. The program capability was demonstrated using a LaB6 gun model for use with a microfocus X-ray tubes. Thanks to the calculation methodology of the Generalized PTM, it is possible to analyze the entire electron optical columns (EOCs) starting from cathode in the electron gun through the transfer lenses to the final objective lens in a single step. Electron guns can now be handled as a part of ordinary optical elements in the EOC.

A brief discussion was given on physics underlining the success of replacing $\tan\alpha$ by $\sin\alpha$ in dealing with large angle cathode rays. It is suggested that matching the transverse momentum between the exact ray and the paraxial trajectory instead of their initial slopes (direction) will result in a better trajectory assignment. Allocating a paraxial trajectory with different initial slope than the exact ray would not raise any issues in the paraxial trajectory method (with aberration corrections), and is sometimes more advantageous than conventional ray condition matching.

Acknowledgments

This work was conducted under the supervision of Prof. Emeritus Shimoyama at Meijo Univ. Japan. We would like to express our deep gratitude.

References

Fujita, S., & Shimoyama, H. (2005). Theory of cathode trajectory characterization by canonical mapping transformation. *Journal of Electron Microscopy*, *54*, 331–343.

Fujita, S., Takebe, M., Ushio, W., & Shimoyama, H. (2010). Generalization of the paraxial trajectory method for the analysis of non-paraxial rays: simulation program G-optk for electron gun characterization. *Journal of Electron Microscopy*, *59*, 3–15.

CHAPTER THREE

Test and characterization of a new post-column imaging energy filter

Frank Kahl*, Volker Gerheim, Martin Linck, Heiko Müller, Richard Schillinger, Stephan Uhlemann
CEOS GmbH, Heidelberg, Germany
*Corresponding author. e-mail address: kahl@ceos-gmbh.de

Contents

1. Introduction

Since the early experiments of Rudberg (1930) and Ruthemann (1941) it is known that the energy spectrum of electrons reflected from surfaces or transmitted through thin foils carries valuable information about the chemical composition of the specimen. Hillier and Baker (1944) and Möllenstedt (1949) already designed and used electron-optical spectrometers in the first half of the 20th century. Initially, the aim was to record the energy spectrum only. Later also electron filters for post-specimen energy-selection have been applied mostly for low-energy beams.

Advances in Imaging and Electron Physics, Volume 212
ISSN 1076-5670
https://doi.org/10.1016/bs.aiep.2019.08.005

Over the years Transmission Electron Microscopes (TEM) with higher beam energies became more widely available. For these instruments spectrometers have been developed combining spatial resolution with analytical information from the same sample.

Castaing and Henry (1962) integrated an energy filter into the ray path of a TEM capable of both energy-filtered imaging and spectroscopy. Their design was based on a combination of a magnetic prism and an electrostatic electron mirror. This marked the birth of energy-filtered transmission electron microscopy. Later on Rose and Plies (1974) reinvestigated this concept, which is nowadays known as in-column energy filter. They used a purely magnetic design with an Ω-shaped double-symmetric course of the optical axis and additional means for aberration correction. This concept finally have been commercialized by Zeiss company (Bihr, 1991; Rose & Krahl, 1995). The currently most advanced in-column energy filter is known under the acronym MANDOLINE (Uhlemann, 1994). An experimental characterization was published by Essers et al. (2010).

For the Scanning Transmission Electron Microscope (STEM) a single-prism spectrometer is sufficient in most cases. Marton (1944) already used such a device. Based on this Wittry (1969) developed the basic concept of what is known today as post-column spectrometer. Later Isaacson and Johnson (1975) and Egerton (1978) designed more advanced magnetic spectrometers. Based on the post-column filter principle Krivanek and Swann (1981) developed the first commercially available spectrometer for serial recording, known as Gatan EELS Model 607. With the advent of pixelated detectors the first parallel recording spectrometers (Egerton & Crozier, 1987; Krivanek, Ahn, & Keeney, 1987; Haider, 1989) and thereafter a post-column imaging energy filter (Krivanek, Gubbens, Dellby, & Meyer, 1992) have been developed. Their commercial versions are known as Gatan PEELS Model 666 and the GIF, respectively. A post-column imaging energy filter has a quadrupole optics behind the 90°-sector magnet to either magnify the energy-selection plane or the achromatic image plane of the spectrometer onto the pixelated detector. The up to now most advanced commercially available device of this type for which we found published information about its performance is the GIF Quantum (Gubbens et al., 2010). In 2018 Gatan introduced the Continuum series as successor to the Quantum design with improved detector performance and acquisition speed (Hart et al., 2017).

Usually all techniques involving energy filtering in a TEM are subsumed under the term Energy-Filtered Transmission Electron Microscopy

(EFTEM). Modern in-column and post-column energy filters used for EFTEM support both energy-filtered imaging and spectroscopy. Those two operation modes are commonly referred to as Electron Spectroscopic Imaging (ESI) and Electron Energy-Loss Spectroscopy (EELS), respectively.

The inherent disadvantage of the rather small intrinsic dispersion of a single 90°-sector magnet at the energy-selection plane can at least partially be reduced by using a magnification optics for the dispersive section between the sector and the spectrum plane as proposed by Egerton and Crozier (1987). Unfortunately, this technique only magnifies the spectrum but does not add additional dispersion. This means that the sampling of the spectrum can be improved but not the energy resolution. In-column filters provide larger dispersion due to multiple sector magnets connected in a row. This is one of the reasons why a post-column filter is always inferior to an in-column filter with respect to spectral acceptance (Uhlemann & Rose, 1996).

Another reason is that the post-filter magnification for an in-column filter is much larger than for a post-column filter. The beam size inside an in-column filter is about one order of magnitude smaller than for a post-column filter. Therefore, an in-column filter with the energy selection slit retracted does not restrict the field of view of a TEM while a post-column filter always does. For spectroscopy with large spectral range or energy-filtered imaging at low beam energies the situation is different, since the post-filter projector lenses after the in-column filter introduce spectrum rotation, dispersion of higher degree and other artifacts due to the Larmor rotation and their unavoidable chromatic and spherical aberration. A post-column filter avoids these issues by using only quadrupoles for post-filter magnification. Their two-fold symmetry provides the flexibility to optimize the optical aberrations for the dispersive section appropriately. To overcome these restrictions also for an in-column filter Gerheim (2001) proposed a quadrupole projective between the in-column filter and the detector in a TEM.

Nowadays, post-column filters are standard equipment for most high-end TEMs. In-column filters are used mostly in life science for zero-loss or most-probable energy-loss filtering due to their superior optical properties in ESI mode, alignment stability, and ease of use. In Cryo-TEM and tomography energy-filtering is used to reduce the image degradation caused by inelastic scattering in fully automized series-acquisition workflows running for hours or even days. This requires that the imaging performance of the filter must be stable over the recording period without any need for human intervention.

In material science spectroscopy is used to analyze a wide range of electron energy losses such as plasmons and core-loss excitations. Those spectra carry valuable information about electronic and optical properties and the local composition of a sample. STEM-EELS data cubes as well as ESI stacks provide detailed information about the chemical composition at high spatial resolution. Thickness determination and elemental mapping is possible by means of ESI using the two- or three-window technique as well as by STEM Spectral Imaging (SI). The availability of fast CMOS detectors dramatically reduced the mapping times, making STEM SI more attractive. Nonetheless, ESI still is an invaluable tool for in-situ work, time-resolved microscopy, high-throughput analysis, and beam sensitive materials. Atom-by-atom spectroscopy of very thin samples has become possible by combining EELS with aberration-corrected STEM (Muller et al., 2008; Suenaga & Koshino, 2010).

More extensive information about history and background of EFTEM can be found in the text book by Egerton (2011). Reviews of this subject from a personal perspective have been published by Egerton (2012) and Colliex (2017).

2. Optics design

2.1 Requirements

From the point of view of electron optics the design of an energy filter and a multipole aberration corrector require very similar mathematical methods and tools for optical modeling and field calculation (Müller, Uhlemann, Hartel, & Haider, 2008). In contrast to an aberration corrector the optical axis inside an energy filter is curved so that its theory is even more complex compared to that of an aberration corrector (Rose, 1987). These facts motivated us to apply our modeling and design tools and the expertise acquired from designing aberration correctors also to the development of an imaging energy filter. Technologically, the design of the first commercially available post-column imaging energy filters dates back to the times before aberration correctors became widely available. Therefore, we were quite confident that modern technologies and production methods successfully employed for our aberration correctors could improve the performance and usability of energy filters compared to existing designs.

One important difference between correctors and filters is that the latter have many fundamentally different alignments and users often want to switch between pre-aligned operation modes like imaging and spectroscopy at different dispersions with little or no need for re-tuning. In contrast, the alignment of a corrector basically never changes during normal operation except for fine-tuning. Therefore, our key design goals were stability, reproducibility and ease of use.

The optical performance of an energy filter is determined by the four characteristic properties

a. non-isochromaticity (NI),
b. geometrical distortion (DA),
c. chromatic distortions (CA), and
d. spectrum shape.

The non-isochromaticity is caused by the so-called off-axial aberrations in the slit plane. These aberrations describe how the arrival position of an electron in the slit plane depends on its starting position in the aperture plane. A large non-isochromaticity is equivalent to large off-axial aberrations causing a significant broadening and distortion of the zero-loss caustic in the slit plane. Since in spectroscopy mode the zero-loss peak is essentially an image of this caustic, the non-isochromaticity also determines the focusing quality of the zero-loss peak in EELS mode.

The chromatic distortion describes the change of the geometrical distortion in the image of the entrance aperture for small energy changes. All these characteristics are the same for in-column and post-column filters. For in-column filters additional second-order aberrations beyond pure distortions must be considered and corrected (Rose & Krahl, 1995). For post-column filters these aberrations are negligible owing to the much higher intermediate magnification. In recent years chromatic aberration correction has enabled EFTEM with large energy windows (Linck et al., 2016; Kahl et al., 2016) and spatially resolved EELS across interfaces (Wang et al., 2018). Aberration correctors reduce the loss of information due to the chromatic focus spread in EFTEM imaging dramatically and then the chromatic distortion of the imaging energy filter becomes a critical parameter.

The energy resolution in spectroscopy is determined by the shape of the zero-loss peak (ZLP). The narrower its profile in dispersive direction is, the better the obtainable energy resolution. Two key measures to achieve this are to focus the ZLP at the center of the detector in dispersive direction and to correct for the energy dependent change of ZLP focus commonly known as spectrum inclination. Equally important is to make sure that the ZLP is not tilted with respect to the dispersive direction since this would broaden the integrated ZLP line scan. This is commonly referred to as y-focus. The linear dispersion determines the energy range over the detector while dispersion coefficients of higher order are responsible for any kind of spectrum non-linearity. Additional unwanted spectrum properties are an energy dependent change of the spectrum height, and any rotation or bending of the spectrum with respect to the pixel grid of the detector.

A post-column filter is much easier to integrate into any existing (S)TEM and its more flexible post-slit optics provides advantages for spectroscopy. Therefore, we decided to accept its fundamental disadvantages with respect to spectral acceptance and field of view compared to an in-column filter. Nevertheless, our goal was to push the limits with respect to maximum field of view and best attainable isochromaticity. In order to facilitate a trade-off between non-isochromaticity and field of view from the very beginning we designed the post-slit optics with a variable magnification. Effectively, this simulates an entrance aperture of variable size. This allows to switch between a setting with a very small non-isochromaticity at the price of a smaller field of view which is only comparable to that of existing filters, and a setting with a by more than a factor of two larger field of view at the expense of a larger non-isochromaticity. Even in the latter case the non-isochromaticity of our filter is still significantly smaller than the results published by Gubbens et al. (2010) for a much smaller field of view.

The substantial improvements in aberration-corrected STEM, monochromated electron sources and low-voltage TEM achieved in recent years called for a post-column energy filter with enhanced optical performance. The introduction of improved gun and ground-potential monochromators stimulates the development of imaging energy filters with sub-10 meV energy resolution in spectroscopy mode for plasmon and vibrational spectroscopy. For those applications both very large dispersion and improved electronics stability especially of the sector current are required. For the latter a very simple relationship between the electronics noise of the sector current and the minimum achievable energy width of the zero-loss peak (ZLP) can be derived under the assumption that all other effects on the ZLP width are negligible

$$\frac{\Delta E(\mathrm{fwhm})}{\sqrt{8\ln(2)}} = \frac{\sigma(I)}{I} E \frac{2 + E/E_0}{1 + E/E_0} . \tag{1}$$

Here $E_0 = m_0c^2$ is the rest energy of the electron and $\sigma(I)$ denotes the standard deviation of the current noise for the relevant frequency range. From an electronics point of view the most demanding task is to achieve the required stability also in the frequency range below 10 Hz. For a combination of an electro-static gun monochromator (Kahl & Rose, 1998) and the MANDOLINE in-column filter Essers et al. (2010) demonstrated an energy resolution of 43 meV and 87 meV at 200 kV for an acquisition time of 100 ms and 100 s, respectively. Mukai, Morishita, Sawada, and Suenaga (2016) at 30 kV achieved 15 meV and 22 meV for 2 ms and 500 ms, respectively, with a Wien-filter gun monochromator and a Gatan energy filter. Recently, Lovejoy, Corbin, Dellby, Hoffman, and Krivanek (2018) have demonstrated 5 meV ZLP-fwhm at 30 kV for 100 ms single-shot exposure using a ground-potential monochromator (Krivanek et al., 2009) combined with a custom designed spectrometer. Ultra-stable current drivers with relative stability in the 10 ppb range as employed in imaging correctors for the chromatic aberration (Haider et al., 2008) may help in the future to enable similar or even better performance for general purpose imaging energy filters. This would allow us to use an almost standard (S)TEM platform with all its extra functionality for applications which previously required highly specialized and custom-designed spectrometers as developed by Geiger, Nolting, and Schröder (1970), Ibach and Mills (1982), and Fink (1989).

In order to enable a variety of EELS applications, the spectrometer has to provide a wide range of spectrum magnifications. For high-resolution EELS with monochromated electron beams a very high dispersion equivalent to a range of 16 eV or less over the entire detector is essential. At the other extreme, EELS modes with a range of several keV allow for a large spectrum overview. In between, several intermediate EELS magnifications have to be provided. Moreover, a flexible choice of the spectrum height in EELS mode together with a detector featuring a sufficiently large number of independent pixels is crucial for angularly-resolved low-loss spectroscopy with high momentum resolution (Midgley, 1999).

2.2 Hardware

The design of the CEOS Energy-Filtering and Imaging device CEFID is shown schematically in Fig. 1. The filter body is mounted by means of a rotatable flange to the bottom of the pre-filter camera housing. The entrance aperture of the filter has a pneumatic actuator. The user can select between the imaging aperture with a diameter of 12.5 mm, the alignment

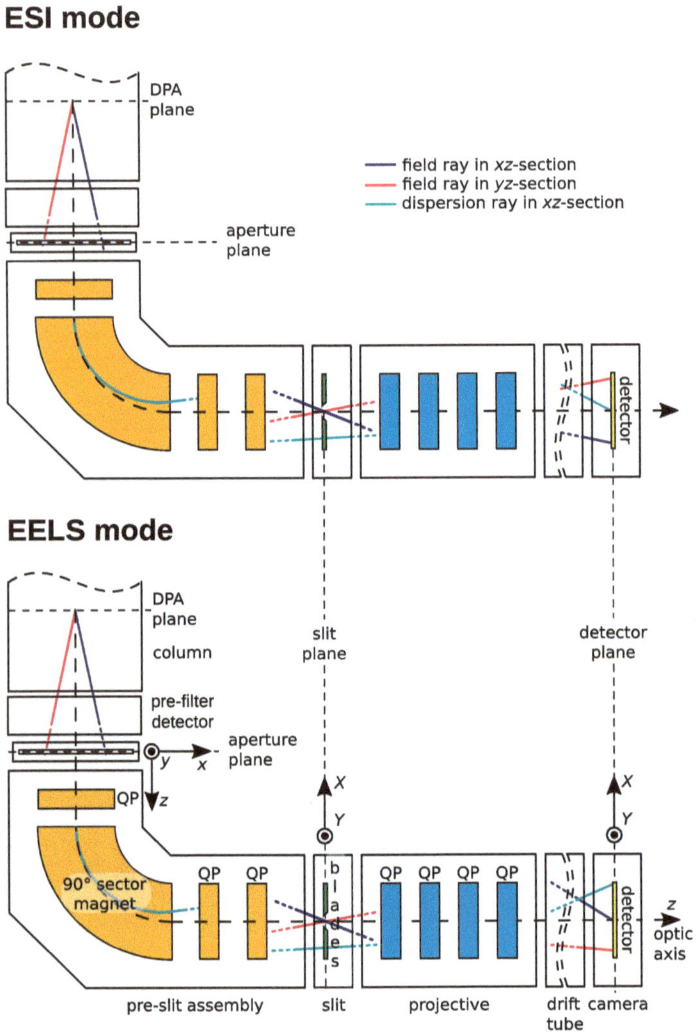

Figure 1 Schematic design of the CEOS Energy-Filtering and Imaging device CEFID. In the bottom and top cross-sections the course of the field rays together with the dispersion ray are shown at the DPA plane, the aperture plane, the slit plane and the detector plane for ESI and EELS mode, respectively. Please note that only the position of those quadrupole fields are shown which determine the course of the field rays. The two quadrupole fields right in front of and behind the sector are not generated by real physical quadrupole elements but by dodecapoles.

mask consisting of a quadratic array of 11 × 11 holes, the standard EELS aperture with a diameter of 5 mm and an alternative EELS aperture with only 2 mm. The sector magnet with a bending radius of 135 mm has been designed very carefully to minimize its deflection aberrations. In combination with one dodecapole at its entrance and one dodecapole at its exit face it acts like a round lens (with long focal-length) imaging the cross-over close to the last TEM projective lens into the energy-selection plane. Since the selection slit is located in this plane we will refer to it also as slit plane. The imaged cross-over is often called DPA cross-over due to its position close to the Differential Pumping Aperture of the TEM which separates the column vacuum from the vacuum in the camera chamber. Further multipoles located between the sector and the slit plane are used for adjusting the dispersion and correcting the spectrum inclination introduced by the 90°-deflection.

The slit mechanics consists of two independent, piezo-driven slit blades with sub-μm positioning accuracy. For energy filtering both blades are used to define a central energy window while for spectroscopy only the lower blade is used as a beam stop for blocking the intense ZLP, if necessary. The dispersion in the slit plane can be adjusted to 3.6 μm/eV at 300 kV. This allows for a minimum energy selection window well below 1.0 eV. This dispersion corresponds to 30.2 μm/eV at the lowest supported acceleration voltage of 30 kV.

The part of the filter from the entrance aperture to the slit plane is referred to as pre-slit optics while the part between the slit plane and the final detector plane is denoted as post-slit optics. The post-slit optics consists of four strong quadrupoles and a set of additional correction elements. In its standard configuration the filter is equipped with a 4k × 4k CMOS camera with a pixel size of 15.5 μm. Other detector geometries are possible by changing the length of the drift tube between the post-slit optics and the detector housing as shown in Fig. 1.

2.3 Optical alignment

The excitation of the optical elements in the pre-slit optics is almost identical between ESI and EELS settings. In ESI mode the post-slit optics is aligned such that the entrance aperture is transferred with minimum geometrical distortion to the detector plane. Three different magnification settings are provided corresponding to virtual entrance apertures of 8 mm, 10 mm, and 12 mm. The quadrupoles in the post-slit optics are adjusted such that the achromatic plane of the sector is exactly conjugated to the detector plane in order to avoid residual dispersion.

For EELS operation the post-slit optics is set up to image the slit plane into the detector plane with variable magnification in the *xz*-section covering a very large dispersion range. For 200 kV with a 4k × 4k camera the dispersion can be changed at least between 250 meV/px and 2 meV/px. By giving up the cross-over at the slit plane even lower dispersions down to 1.0 eV/px are available. Hence, the filter currently supports EELS ranges between 8 eV and 4096 eV. For beam energies below 80 kV the spectral range of the EELS overview mode has to be reduced. Due to the large dispersion already in the slit plane the post-slit optics provides sufficient flexibility to adjust also the spectrum height over a certain range. For the 5 mm EELS aperture usually a spectrum height of about 20% of the detector chip is used. However, the height can be reduced well below 5% to allow for a tiled readout of the CMOS detector to maximize read-out speed or it can be increased to almost the full height of the detector for instance for momentum-resolved low-loss EELS. In this special mode of operation the entrance aperture plane is conjugated to the detector plane with respect to the non-dispersive *yz*-section. In Section 4 we characterize the optical performance of a subset of EELS modes. This subset covers the entire supported dispersion range, different spectrum heights and a special mode for momentum-resolved EELS.

The Continuum series as the most recent GIF generation offers scintillator-based detector options up to a sensor size of 2k × 2k with 18 μm pixel size. Since we have the opportunity to use a 4k × 4k camera with 15.5 μm pixel size we have the choice to either increase the post-magnification or increase the entrance aperture of the filter. Both strategies have their pros and cons. The pre-slit optics fully-corrects the intrinsic and parasitic NI aberrations up to third order and can systematically minimize the aberrations of fourth and fifth order. Therefore, the residual non-isochromaticity – which limits the energy resolution – is small but increases rather steeply with increasing size d_a of the virtual entrance aperture. The non-isochromaticity for ESI 12 mm is about one order of magnitude larger than that for ESI 8 mm.

On the other hand Appendix A, Eq. (A.9) shows that even for an ideal aberration-free systems the energy resolution for a given number of equally resolved image pixels is limited by diffraction. In contrast to the non-isochromaticity this limit is reduced by increasing the virtual aperture diameter. There is an optimum size of the virtual aperture for which the total energy resolution arising from both effects combined is best. At this size the non-isochromaticity roughly equals the limit given by Eq. (A.9).

3. Characterization

In order to describe the optical properties of an energy filter we consider an electron with an energy deviation of ΔE starting from the DPA-crossover plane such that it passes through the aperture plane at the position (x, y). At any subsequent plane perpendicular to the curved optical axis at position z we can derive the perturbation series

$$\begin{pmatrix} X \\ Y \end{pmatrix} = \begin{pmatrix} X_{00} \\ Y_{00} \end{pmatrix} + \begin{pmatrix} X_{10} & X_{01} \\ Y_{10} & Y_{01} \end{pmatrix} \begin{pmatrix} x \\ y \end{pmatrix} + \begin{pmatrix} X_{001} \\ Y_{001} \end{pmatrix} \Delta E + \begin{pmatrix} \Delta X(x, y, \Delta E) \\ \Delta Y(x, y, \Delta E) \end{pmatrix} \tag{2}$$

with the non-linear term

$$\begin{pmatrix} \Delta X \\ \Delta Y \end{pmatrix} = \sum_{R>1} \sum_{k=0}^{R} \sum_{\substack{m=0 \\ n=R-k-m}}^{R-k} \begin{pmatrix} X_{nmk} \\ Y_{nmk} \end{pmatrix} x^n y^m \Delta E^k . \tag{3}$$

In the following X_{nm}, Y_{nm} are referred to as geometric distortion aberrations DA of order $N = n + m$ with k implicitly set to zero and X_{nmk}, Y_{nmk} with $k > 0$ are the chromatic aberrations CA of rank $R = n + m + k$ and degree k.

For a perfect system the beam shape is symmetric with respect to the xz-section $(y = 0)$:

$$\begin{pmatrix} X(x, y, \Delta E) \\ Y(x, y, \Delta E) \end{pmatrix} = \begin{pmatrix} X(x, -y, \Delta E) \\ -Y(x, -y, \Delta E) \end{pmatrix} . \tag{4}$$

This condition is only compatible with the *intrinsic* aberration coefficients

$$X_{n,2m}, \; Y_{n,2m+1}, \; n, m \in \mathbb{N} . \tag{5}$$

All *parasitic* aberration coefficients

$$X_{n,2m+1}, \; Y_{n,2m}, \; n, m \in \mathbb{N} \tag{6}$$

vanish for the ideal system since they violate Eq. (4). In practice, of course, we have to measure and correct for *intrinsic* as well as *parasitic* aberrations, since mid-section symmetry can only be expected within the mechanical tolerances of the system.

3.1 Aberrations at slit plane

In the energy-selection plane the dispersion $D_s = X_{001}$ is large, e.g. for 300 kV the design value is 3.6 μm/eV. The dispersion for any other beam energy E can be calculated from the identity relation

$$\frac{D_s(E)}{E} \times \frac{1 + E/E_0}{2 + E/E_0} = \text{const.}, \tag{7}$$

where $E_0 = m_0 c^2$ is the rest energy of an electron.

Electrons with an energy offset of ΔE passing through the entrance aperture at (x, y) are displaced on the energy scale by

$$E(x, y, \Delta E) = \Delta E + \sum_{r \geq 0} \sum_{\substack{m=0 \\ n=r-m}}^{r} E_{nm} x^n y^m, \tag{8}$$

where the aberration-induced spatial displacement in the X-direction has been translated into an energy shift. The coefficients $E_{nm} = X_{nm}/D_s$ are referred to as the non-isochromaticity aberrations NI. The intensity profile in the dispersive direction of the caustic at the slit plane is proportional to the energy distribution of the beam electrons $p = p(E)$ with an approximately Gaussian shape and shifted due to the NI aberrations.

In ESI mode we observe an image of the aperture plane. This means that electrons coming from different spots in the aperture plane are again spatially separated in the detector plane so that they are registered independently from each other. If we move the lower slit blade upwards the intensity at the center of the detector eventually begins to darken. Increasing the beam energy by ΔE then shifts the caustic at the slit plane by $D_s \Delta E$ towards the blade. The larger the energy offset is, the more electrons are blocked by the blade so that the intensity on the detector gradually fades, yielding different attenuation curves for different segments of the detector as illustrated in Fig. 2. These curves are of the same shape but are shifted with respect to each other according to Eq. (8). If we assume a gaussian shaped energy distribution of the source the shape of each attenuation curve is

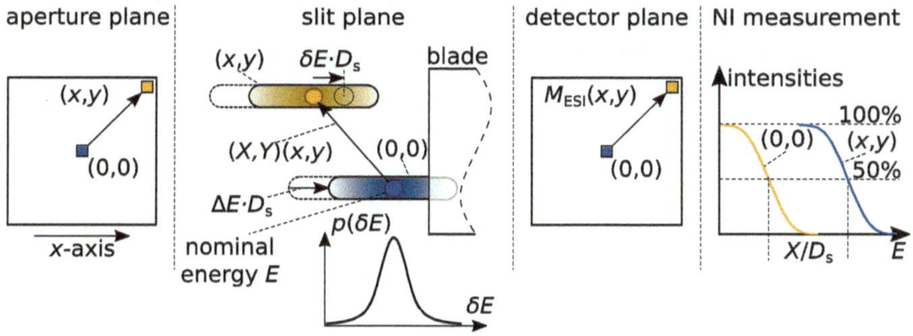

Figure 2 Measurement principle for non-isochromaticity aberrations. $D_S = X_{001}$ denotes the dispersion at the slit plane.

given by

$$I(\Delta E) = \frac{I_0}{2}\operatorname{erfc}\left(2\sqrt{\ln(2)}\,\frac{\Delta E - \Delta E_{1/2}}{\delta E}\right). \tag{9}$$

Here δE is the full-width-half-maximum (FWHM) of the energy distribution of the beam electrons. For $\Delta E = \Delta E_{1/2}$ the intensity drops to 50%.

For measuring the non-isochromaticity – in the following called NI measurement – we retract all apertures and record a series of images where the energy of the beam is increased by some energy increment ΔE from image to image. We start with an energy offset for which the image is not obstructed by the blade at all and stop the series when the blade fully blocks the beam. During normal operation the projective is already very well aligned prior to an NI measurement. Therefore, in most cases we can assume an ideal isotropic imaging with the target magnification M_t of the currently active ESI mode so that we can easily map detector coordinates back to aperture coordinates. In certain situations the projective might be so poorly aligned that the assumption of an ideal isotropic imaging no longer holds. In this case the NI measurement series can optionally be recorded with inserted alignment mask. As long as the image of each hole of the mask can be unambiguously mapped to a hole of the physical mask $(X_{ij}, Y_{ij}) \mapsto (x_{ij}, y_{ij})$, the aperture coordinates for the attenuation curve of each hole are well defined even if the mask image is severely distorted.

All attenuation patterns are fitted with Eq. (9) using $\Delta E_{1/2}$, δE and I_0 as fit parameters. As a result we obtain a list of tuples $(x_{ij}, y_{ij}, \Delta E_{ij})$, $-N/2 < i, j < N/2$. Here $N = 3, 5, 7, 9, 11$ is the size of the visible sub-grid of the alignment mask and the positions $x_{ij} = id_m$ and $y_{ij} = jd_m$ of the mask holes are given by the mask pitch d_m. Default sizes of the mask grid are 7×7 for ESI 8 mm, 9×9 for ESI 10 mm, and 11×11 for ESI 12 mm. The mean value of all δE_{ij} values is a good approximation for the FWHM of the energy width of the beam electrons. The peak-to-peak width and the standard deviation of all ΔE_{ij} values are measures for the maximum NI(max) and the root-mean-square NI(rms) non-isochromaticity, respectively. It should be noted that the non-isochromaticity is entirely determined by the filter optics and is not related to the energy width of the source at all.

Inserting the set of values $(x_{ij}, y_{ij}, \Delta E_{ij})$ into Eq. (8) yields the overdetermined linear system

$$\Delta E_{ij} = \sum_{r=0}^{R} d_\mathrm{m}^{r} \sum_{\substack{m=0 \\ n=r-m}}^{r} E_{nm} i^n j^m \tag{10}$$

for the slit aberration coefficients E_{nm} with $n + m \leq R$. The coefficients are obtained by a least squares fit with maximum rank R. The additional representation

$$\mathcal{E}_{nm} = E_{nm} r^{n+m}, \qquad r = l\, d_\mathrm{m} \tag{11}$$

specifies the NI-contribution in electron Volts of each coefficient if solely present. The index l takes the value 3, 4 or 5 if the visible sub-grid for the particular ESI mode is of size 7×7, 9×9 or 11×11, respectively.

3.2 Geometrical distortions at the detector plane

According to Eq. (2) ESI imaging in linear approximation is described by

$$\begin{pmatrix} X \\ Y \end{pmatrix} = \begin{pmatrix} X_{00} \\ Y_{00} \end{pmatrix} + \begin{pmatrix} X_{10} & X_{01} \\ Y_{10} & Y_{01} \end{pmatrix} \begin{pmatrix} x \\ y \end{pmatrix} . \tag{12}$$

The four linear coefficients of this affine transformation can be expressed by the more commonly used properties isotropic magnification M, shear angle γ, aspect ratio violation A and image rotation ϕ as

$$\begin{pmatrix} X_{10} & X_{01} \\ Y_{10} & Y_{01} \end{pmatrix} = M \begin{pmatrix} \cos(\phi) & -\sin(\phi) \\ \sin(\phi) & \cos(\phi) \end{pmatrix} \begin{pmatrix} 1 + \frac{A}{2} & \tan(\gamma) \\ \tan(\gamma) & 1 - \frac{A}{2} \end{pmatrix} , \tag{13}$$

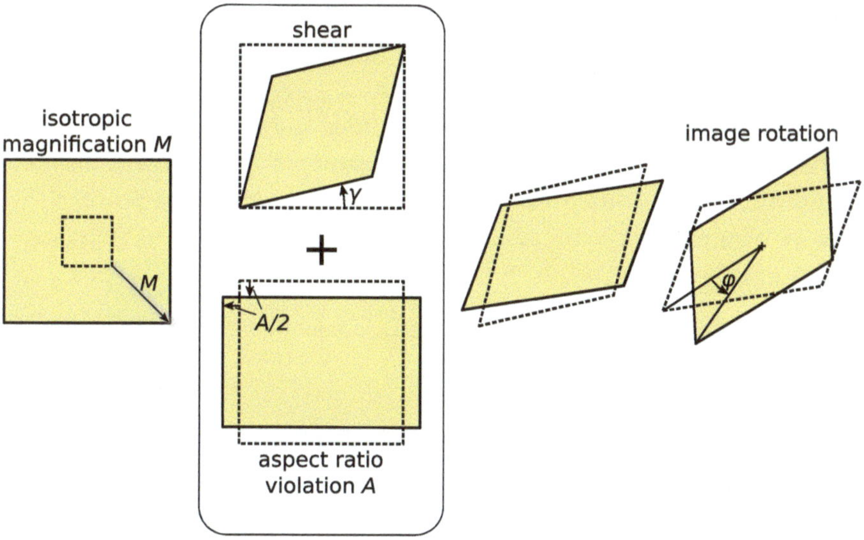

Figure 3 Definition of isotropic magnification *M*, shear angle γ, aspect ratio violation *A* and image rotation ϕ and their sequence of application illustrated by imaging a square. The superposition of shear and aspect ratio violation in the rounded box illustrates the effect of the right matrix of Eq. (13).

where the characteristic properties are given by

$$\begin{aligned}
M &= \frac{1}{2}\sqrt{(X_{10}+Y_{01})^2+(Y_{10}-X_{01})^2} && \approx \frac{1}{2}(X_{10}+Y_{01}) \, , \\
\tan(\gamma) &= \frac{1}{2M^2}(X_{10}X_{01}+Y_{01}Y_{10}) && \approx \frac{1}{2M}(Y_{10}+X_{01}) \, , \\
A &= \frac{\left(X_{10}^2-X_{01}^2\right)-\left(Y_{01}^2-Y_{10}^2\right)}{2M^2} && \approx \frac{1}{M}(X_{10}-Y_{01}) \, , \\
\tan(\phi) &= \frac{Y_{10}-X_{01}}{X_{10}+Y_{01}} && \approx \frac{1}{2M}(Y_{10}-X_{01}) \, . \qquad (14)
\end{aligned}$$

The effect of this transformation and the sequence of application is illustrated in Fig. 3. Shear and aspect ratio violation are applied before the image rotation. During normal operation the distortions are always small enough compared to the isotropic magnification to allow for using the approximation formulas on the right hand sides of Eq. (14).

The isotropic magnification is adjusted such that the selected region of the entrance aperture is matched to the detector. Residual image rotation ϕ

and image shift $\vec{R}_0$ can be ignored since they do not distort the image, but both shear γ and aspect ratio violation A must be corrected.

For measuring the distortion coefficients X_{nm}, Y_{nm} the alignment mask is inserted. Then an image of the mask is recorded and the center positions $\vec{R}_{ij}$ of the mask holes in the image are reconstructed. If we insert the hole positions into the geometrical part of Eq. (2) and restrict the power series to R we obtain an approximation for the measured positions

$$\vec{R}_{ij} = \sum_{r=0}^{R} d_{\mathrm{m}}^{r} \sum_{\substack{m=0 \\ n=r-m}}^{r} \begin{pmatrix} X_{nm} \\ Y_{nm} \end{pmatrix} i^n j^m \,. \tag{15}$$

The distortion coefficients X_{nm} and Y_{nm} are determined by means of a least squares fit over all measured $\vec{R}_{ij}$ with order R.

The DA coefficients determined by a higher-order least squares fit can be represented in SI units or can be normalized using a reasonable length r as

$$\mathcal{X}_{nm} = \frac{X_{nm} r^{n+m}}{\sqrt{2} M_{\mathrm{t}} r}, \quad \mathcal{Y}_{nm} = \frac{Y_{nm} r^{n+m}}{\sqrt{2} M_{\mathrm{t}} r}, \quad r = l d_{\mathrm{m}} \tag{16}$$

where M_{t} is the target magnification and d_{m} is the mask pitch. The index l takes the value 3, 4 or 5 if the visible sub-grid for the particular ESI mode is of size 7×7, 9×9 or 11×11, respectively. This normalization allows to compare the size of the effects of coefficients across different orders. Additionally, a first-order fit is performed using only M, ϕ and the shift vector $\vec{R}_0$ as fit parameters while both A and γ are set to zero. From the set of difference vectors

$$\Delta \vec{r}_{ij} = \vec{R}_{ij} - \vec{r}_{ij}(\vec{R}_0, M, \phi) \tag{17}$$

between the hole positions $\vec{R}_{ij}$ predicted by the Rth-order fit and the positions $\vec{r}_{ij}(\vec{R}_0, M, \phi)$ determined by the first-order fit the maximum displacement DA(max) and the root-mean-square displacement DA(rms) are calculated.

3.3 Chromatic distortions at the detector plane

For measuring the chromatic distortion CA we determine the mutual displacements between three images recorded at the energy offsets $-\Delta E/2$, 0 eV and $+\Delta E/2$, respectively. The maximum and root-mean-square residual chromatic distortion is defined by the maximum of the left-sided, right-sided and central differences. For the CA coefficients of first and second degree we get

$$\begin{pmatrix} X_{nm1} \\ Y_{nm1} \end{pmatrix} = \frac{1}{\Delta E}\left(\begin{pmatrix} X_{nm} \\ Y_{nm} \end{pmatrix}\Bigg|_{\frac{\Delta E}{2}} - \begin{pmatrix} X_{nm} \\ Y_{nm} \end{pmatrix}\Bigg|_{-\frac{\Delta E}{2}} \right), \tag{18}$$

$$\begin{pmatrix} X_{nm2} \\ Y_{nm2} \end{pmatrix} = \frac{2}{\Delta E^2}\left(2\begin{pmatrix} X_{nm} \\ Y_{nm} \end{pmatrix}\Bigg|_{0} - \begin{pmatrix} X_{nm} \\ Y_{nm} \end{pmatrix}\Bigg|_{\frac{\Delta E}{2}} - \begin{pmatrix} X_{nm} \\ Y_{nm} \end{pmatrix}\Bigg|_{-\frac{\Delta E}{2}} \right).$$

Analogously to the geometric distortion coefficients the chromatic distortion coefficients can be presented in SI units or in a normalized fashion which allows to compare the size of the effects of aberration coefficients across various orders

$$\begin{pmatrix} \mathcal{X}_{nm1} \\ \mathcal{Y}_{nm1} \end{pmatrix} = \begin{pmatrix} X_{nm1} \\ Y_{nm1} \end{pmatrix} \frac{r^{n+m}\,\Delta E}{\sqrt{2}M_t r},$$

$$\begin{pmatrix} \mathcal{X}_{nm2} \\ \mathcal{Y}_{nm1} \end{pmatrix} = \begin{pmatrix} X_{nm2} \\ Y_{nm2} \end{pmatrix} \frac{r^{n+m}\,(\Delta E/2)^2}{\sqrt{2}M_t r}, \tag{19}$$

with $r = l d_{\mathrm{m}}$, $l = 3, 4, 5$ and the target magnification M_{t}.

3.4 Spectrum aberrations at the detector plane

In EELS mode the post-slit optics images the spectrum in the slit plane onto the detector. Also the spectrum shape and quality can be described by aberration coefficients derived from Eq. (2) evaluated at the detector plane. In order to distinguish the aberrations in EELS mode from the aberrations in ESI mode we prefix them with *S*. An overview of the relevant EELS aberrations is given in Table 1. The effect of these aberrations on the visual appearance of the spectrum is illustrated in Fig. 4.

The energy resolution ΔE_{ZLP} is defined as the FWHM of a line scan in *x*-direction across the ZLP using a scan width matching the spectrum height *H* as shown in Fig. 4.

The presence of higher-order dispersions causes a distortion of the spectrum linearity. We quantify this non-linearity of the energy scale by the half-value of the peak-to-peak deviation of the third-order dispersion polynomial $SX^{(3)}_{001}\,\Delta E + SX^{(3)}_{002}\,(\Delta E)^2 + SX^{(3)}_{003}\,(\Delta E)^3$ from the linear fit $SX^{(1)}_{001}\,\Delta E$ divided by the spectral energy range $R_E = L_x/X^{(1)}_{001}$. For the dispersion coefficients up to third order the maximum deviations over the entire detector in percent of the detector size which each coefficient would cause if solely

Table 1 Definition of the relevant EELS aberrations. Here *single ZLP* means recording the ZLP for a HT offset of 0 V. For a *ZLP series* at least five ZLPs are recorded for different HT offsets ΔE.

Aberration/ Property	Measurement method	Definition
SX_{10}	Single ZLP	x-focus, width of ZLP
SX_{01}	Single ZLP	y-focus, tilt of ZLP
$SY_{01} = \frac{H}{d_a}$	Single ZLP	Spectrum height H normalized by aperture diameter d_a
$SX_{001}^{(1)}$	ZLP series	Linear dispersion coefficient obtained from a first-rank least squares fit
$SX_{00k}^{(n)}$	ZLP series	Dispersion coefficient of kth degree obtained from a least squares fit of rank n
δSX_{001}	ZLP series	Maximum and rms deviation of dispersion over the energy range (non-linearity)
$R_E = \frac{L_x}{SX_{001}}$	ZLP series	Measured energy range over detector width
SY_{001}	ZLP series	Spectrum rotation / first-order fit of parasitic linear dispersion coefficient in y-direction
SX_{101}	ZLP series	Spectrum inclination

present are printed in addition to their values in SI units. They are computed as

$$\delta\mathcal{X}_{001} = \frac{1}{2}\left|\frac{SX_{001}^{(3)}}{D_d} - 1\right| , \tag{20}$$

$$\mathcal{X}_{00n} = \frac{1}{L_x}\left|SX_{00n}^{(3)}\right|\left(\frac{R_E}{2}\right)^n , \quad n = 2, 3 , \tag{21}$$

$$\mathcal{Y}_{00n} = \frac{1}{L_y}\left|SY_{00n}^{(3)}\right|\left(\frac{R_E}{2}\right)^n , \quad n = 1, 2, 3 . \tag{22}$$

All higher-order aberrations which might reduce the energy resolution by broadening the ZLP are generated predominantly in the pre-slit optics. This means that the NI aberrations in the slit plane are virtually identical to the higher-order EELS aberrations in the detector plane. Therefore, we can correct for those higher-order EELS aberrations first by simply measuring and correcting the NI aberrations in the slit plane using a special post-slit ESI setting which works with the EELS pre-slit configuration, before switching the post-slit optics to the desired EELS mode. The residual non-isochromaticity is equivalent to the minimum ZLP width at the detector attainable by the filter optics if we assume a perfectly monochro-

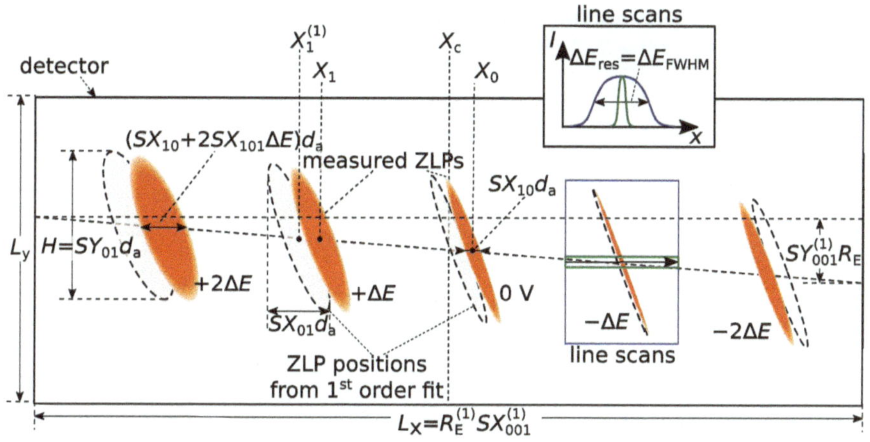

Figure 4 Effects of the EELS aberrations on the spectrum and illustration of the properties listed in Table 1.

matic source. This means the non-isochromaticity with an EELS aperture inserted is a good upper estimate for the maximum possible EELS energy resolution. The pre-slit optics also allows for correcting the spectrum inclination SX_{101}. Since the dispersion SX_{001}, the x-focus SX_{10} and the spectrum height SY_{01} as well as the parasitic aberrations y-defocus SX_{01} and spectrum rotation SY_{001} predominantly origin from behind the slit, they must be measured from the spectrum data itself and corrected using elements in the post-slit.

4. Performance

In order to demonstrate the performance of the CEFID we have installed and tested prototypes on various microscopes. Currently, our primary test platform is a JEOL 2100F with a Schottky-FEG and a maximum beam energy of 200 kV. On this system we tested the imaging performance for all three ESI modes at 30 kV, 80 kV and 200 kV. As post-filter camera we use a latest-generation TVIPS XF416 4k × 4k (Oster, Ghadimi, & Tietz, 2018) equipped with a fiber-coupled CMOS detector with a pixel size of 15.5 μm. To demonstrate the EELS performance we have chosen a 300 kV Hitachi HF3300S microscope. Since its cold-FEG has a much smaller primary energy width than the Schottky-FEG of the JEOL the Hitachi is, therefore, better suited to demonstrate the EELS performance of the CEFID. The filter on the Hitachi TEM is equipped with an

Table 2 Measured ESI properties for the three available ESI modes at 200 kV for the filter prototype installed at the JEOL 2100F. The chromatic distortion refers to $\Delta E = 50$ V.

Mode		**8 mm**	**10 mm**	**12 mm**
Mask size		7×7	9×9	11×11
Magnification		11.10	8.80	7.40
Distortion / %	max.	0.36	0.51	1.01
	sdev.	0.17	0.21	0.33
Chrom. distortion / %	max.	0.17	0.25	0.33
	sdev.	0.08	0.11	0.14
Non-isochromaticity / meV	max.	92	393	592
	sdev.	17	65	122
Shear / degree		−0.01	0.01	0.00
Image rotation / degree		−0.23	−0.23	−0.22
Aspect ratio violation / %		0.02	0.07	0.02

older-generation Gatan US1000 as post-filter camera with a fiber-coupled 2k × 2k CCD detector having a pixel size of 14 μm. In order to adapt the post-slit optics to the smaller detector size a shorter drift tube has been mounted between the post-slit optics module and the camera block. This is purely because of geometric reasons related to the detector size and has no influence on the optical performance of the filter.

The two different microscopes also require different installation options. In case of the JEM-2100F the filter is mounted in 45°-direction to the rear right and for the HF-3300S in 90°-direction to the left. Both is easily achievable. In any case the filter projective is supported by an extra rod mounted to the suspended part of the frame construction. This is useful to reduce the force on the bottom-mount flange. Water cooling is only necessary for the camera but not the filter itself. The CEFID electronics cabinet is available in two different versions, one air-cooled and one water-cooled. The latter version is expected to provide even better long-term stability if cooling water with high temperature stability is available like for aberration-corrected high-end microscopes.

The imaging performance for the three supported post-slit magnifications is summarized in Table 2. In each case the magnification is chosen such that the quadratic area of the detector projected back into the aperture plane fits a virtual circular aperture with a diameter of 8 mm, 10 mm or 12 mm, respectively. The physical size of the entrance aperture is always 12.5 mm. For high-resolution energy-filtered imaging with a Schottky FEG the ESI 10 mm mode is the optimum choice: it provides a large field of view while maintaining the geometric and chromatic distortions sufficiently small with a non-isochromaticity significantly smaller than the primary energy width of a Schottky-FEG. In contrast the ESI 12 mm mode images nearly the full aperture onto the detector but at the expense of increased non-isochromaticity with a peak-to-peak value of 0.6 eV. The maximum geometrical distortion over the field of view is about 1% of half the diagonal length of the 4k × 4k detector chip which we use as reference length. This is equivalent to a maximum displacement of 29 pixel in the corner area of the camera. The maximum chromatic distortion is 0.33% for $\Delta E = 50$ V. This adds a chromatic blur of 4 pixel for an energy window of 20 eV at 200 kV. This mode seems to be well suited for zero-loss-filtering in life science like cryo-tomography and single-particle analysis which benefit from a field of view as large as possible. Fig. 5 shows non-isochromaticy maps over the inner 90% of the detector for the ESI 8 mm and ESI 12 modes calculated from the coefficients obtained from a 5th order least squares fit as described in Subsection 3.1. For both modes the sets of energy shifts were measured at detector positions which correspond to the positions of the mask holes. In case of the 8 mm mode this corresponds to a set of 7 × 7 measurements while for the 12 mm mode the full set of 11 × 11 measurements were used. The high degree of symmetry of both maps with respect to the horizontal axis means that even for the largest aperture the parasitic aberrations are still negligible with respect to the intrinsic ones. Within the inner area of the NI map for ESI 12 mm the violation of isochromaticity is stronger compared to ESI 8 mm, because partially 2nd and 3rd order NI aberrations have to be applied in order to counter-balance higher order NI aberrations at the outer area of the NI map. This is not necessary for the smaller area of only 8 mm diameter.

Figs. 6 and 7 show images of the visible inner 7 × 7 and 11 × 11 subgrids of the alignment mask for the 8 mm and the 12 mm mode, respectively. The intersections and corners of the overlayed red grids mark the positions $\vec{r}_{ij}$ of the fitted ideal imaging according to Eq. (17). The differences of two images taken at high-tension offsets of $\Delta E = \pm 25$ V shown in Fig. 8 demonstrate the residual chromatic distortion introduced mainly by the post-slit optics of the filter for the ESI 8 mm and 12 mm mode, respectively. Also in this case the pattern is fairly symmetric. This is an indication for a well aligned system.

For optimizing the EELS energy resolution it is sufficient to minimize the width of the ZLP by correcting all NI aberrations up to third order as described in Section 3.4. Then the dispersion, the spectrum height, the

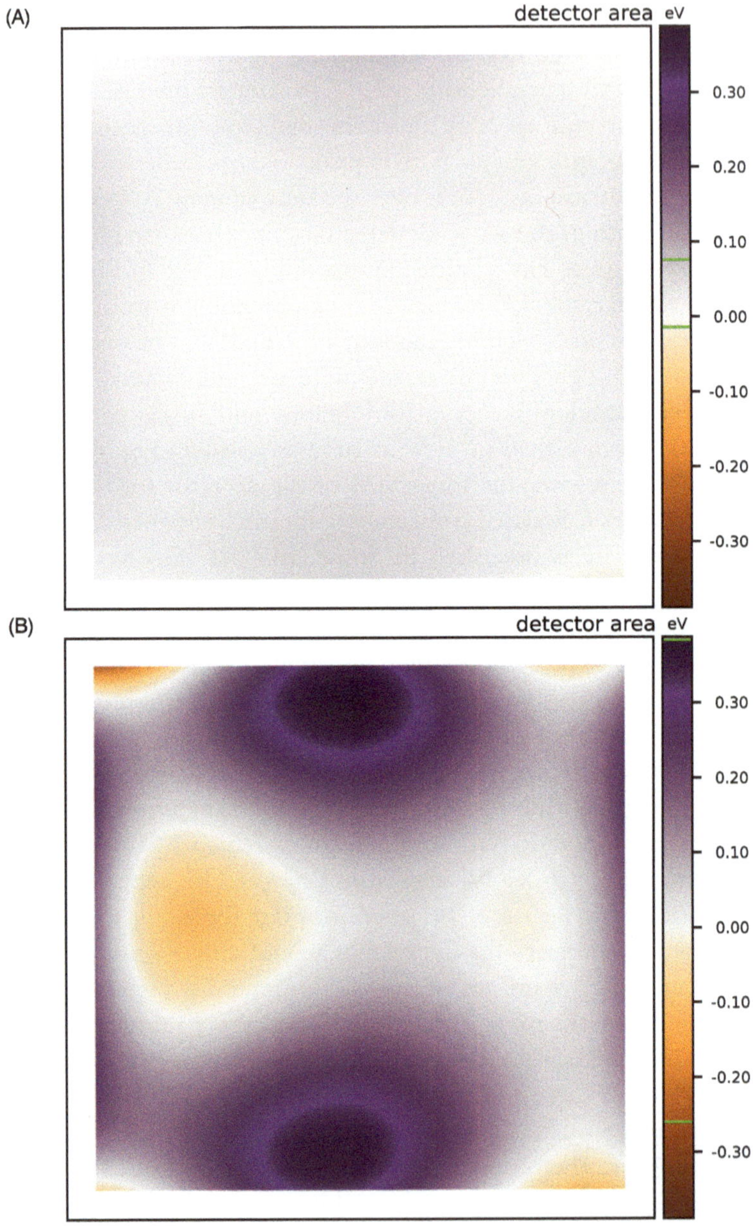

Figure 5 Non-isochromaticity maps for the 8 mm mode (A) and 12 mm mode (B) calculated from a 5th order least squares fit. The two lime green bars in the color scales mark the minimum and maximum values of the maps. Both maps use the same value range and color encoding to allow for easy visual comparison. The range is chosen symmetric with respect to zero which is represented by white.

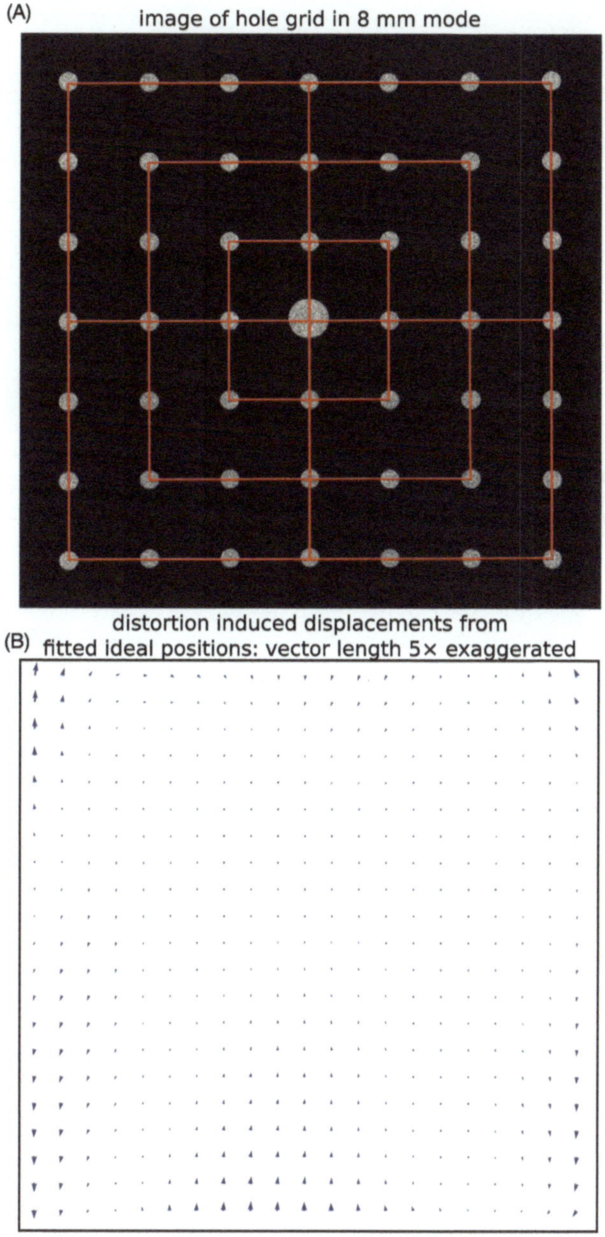

Figure 6 Residual distortions of the 8 mm ESI mode illustrated by the image of the inner 7 × 7 holes of the mask in (A) and the difference vectors between the positions calculated from a 5th order fit of the distortion coefficients and the positions calculated from the fitted ideal imaging parameters $\vec{R}_0$, M, ϕ in (B) (see Section 3.2). Those ideal positions are indicated by the red grid in (A). For improved clarity the vector length has been exaggerated by a factor of five.

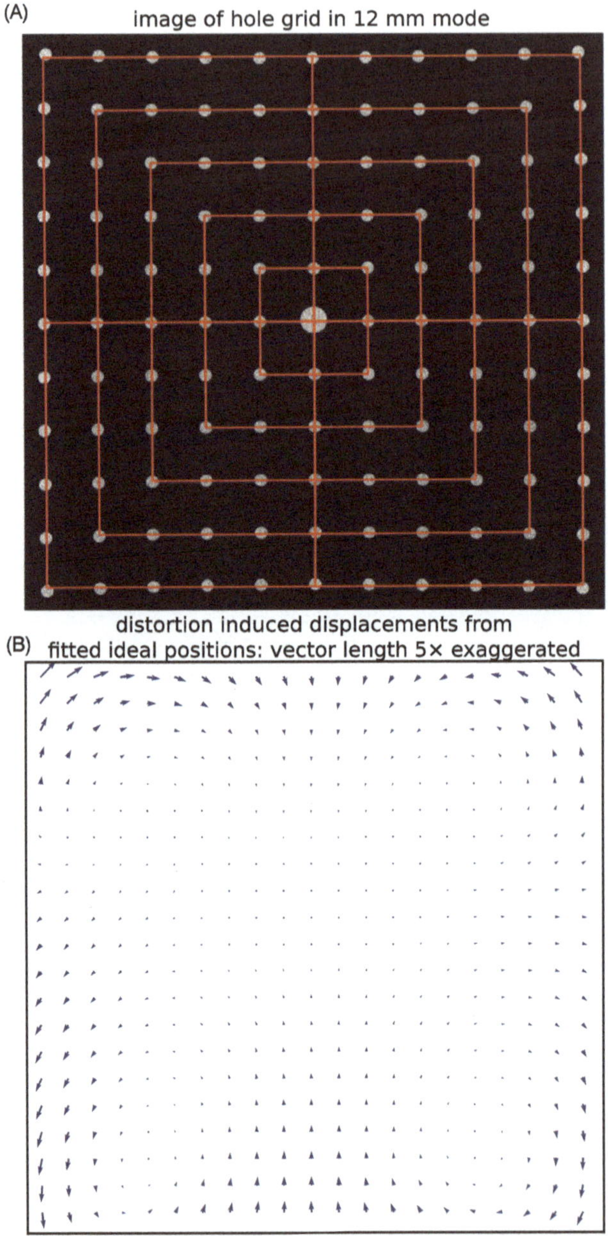

Figure 7 Residual distortions of the 12 mm ESI mode. For more details see caption of Fig. 6.

(A)

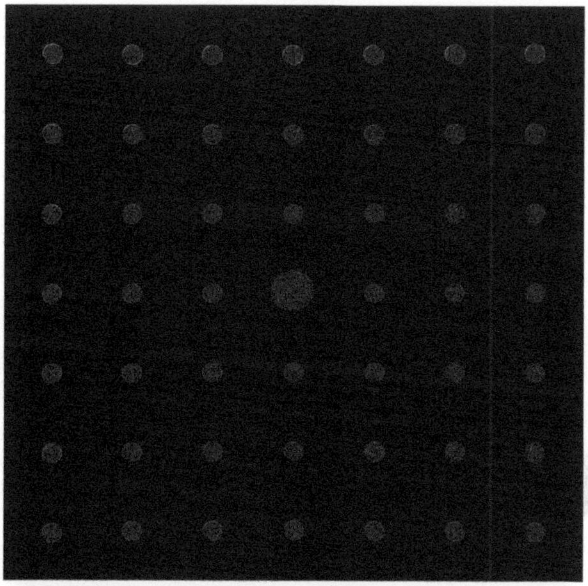

(B)

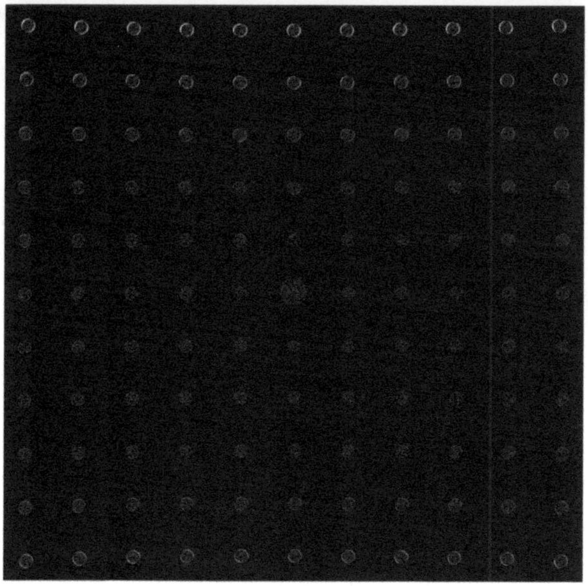

Figure 8 Residual chromatic distortions of (A) the 8 mm ESI mode and (B) the 12 mm ESI mode illustrated by the difference images between the masks recorded at +25 eV and at −25 eV. The yellow crescents are covered by the holes at +25 eV but not at −25 eV while for the light blue crescents its vice versa. Each hole was shifted by the maximum crescent diameter along the line connecting their centers of gravity.

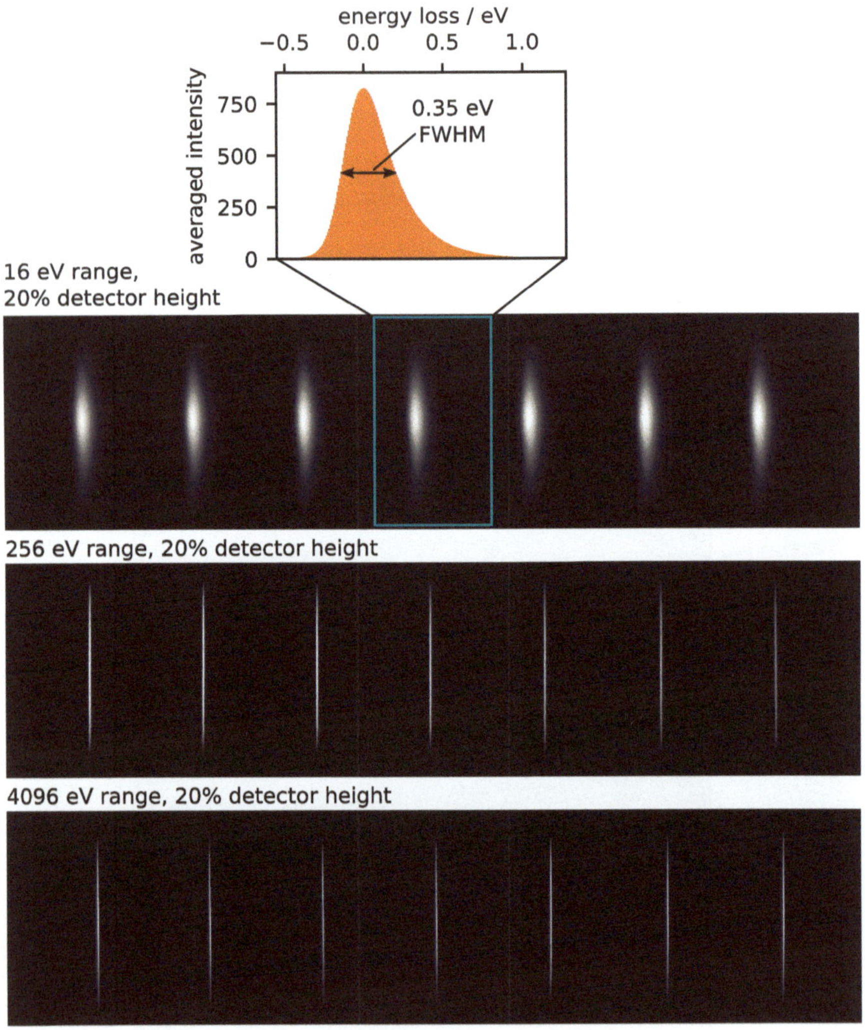

Figure 9 Images combining all ZLPs of the measurement series for the 16 eV, 256 eV and 4 keV modes with 20% spectrum height, respectively.

spectrum rotation and the ZLP *x*- and *y*-foci are measured from a ZLP series as shown in Fig. 4 and finally being compensated using elements in the post-slit.

Fig. 9 shows data sets of three EELS modes ranging from 16 eV to 4096 eV. For the EELS aberration measurements the zero-loss peaks at seven different energies are combined into one image each. The highest

Table 3 Spectrum distortions and energy resolutions at the detector center for all modes presented in Figs. 10, 11 and 12. Please note that the energy resolution of the 4096 eV mode is limited not by the optics but by the Modulation Transfer Function (MTF) of the 2k × 2k detector of around 4-5 pixels which corresponds to 8-10 eV.

Range / eV	Aperture / mm	Spectrum height / %	FWHM of ZLP for HT offset of 0 eV / eV	Distortion / %
16	5	20.0	0.35	0.38
256	5	20.0	0.68	0.14
4096	5	20.0	9.79	0.25
256	5	2.5	0.68	0.14
(ω-q) 64	2	78.0	0.56	0.13

dispersion of 8 meV/pixel is more than sufficient to exploit the full energy resolution of the cold-FEG given by the FWHM of the ZLP of 350 meV. Table 3 summarizes the measured parameters for the investigated spectroscopy modes. The EELS overview mode with a range of 4096 eV would benefit from using a 4k CMOS detector since the Modulation Transfer Function (MTF) of the 2k CCD detector severely limits its energy resolution.

The spectra of hexagonal boron-nitride (h-BN) in Fig. 10 nicely demonstrate the performance of the three modes of Fig. 9 for a real world application. All those spectra have a height of about 410 pixels which is equivalent to 20% of the 2k × 2k detector chip. Fig. 11 shows the 256 eV range with a significantly reduced spectrum height of only 2.5% of the detector size. Reducing the height helps boosting the frame rate of continuous EELS acquisition for cameras with modern CMOS detectors, since in contrast to CCDs they can read out vertical strip-shaped sub-areas much faster than full frames. Regrettably, this comes at the expense of a reduced dynamic range of the detector. For instance, the TVIPS XF416 camera can read out an area of 4096 × 256 pixels in 2.5 ms, while a full 4096 × 4096 frame takes 40 ms. This operation mode is very attractive for speeding-up STEM EELS in the low-loss region if the probe current is high enough to generate sufficient signal in the spectra.

Additionally an EELS mode with a range of 64 eV is provided which is designed for momentum resolved EELS (q-EELS). Following a commonly used terminology we refer to it as ω-q mode.

For q-EELS the microscope is switched to diffraction mode so that the y-coordinate in the aperture plane corresponds to a certain scattering angle. Each scattering angle in the y-direction is equivalent to a certain

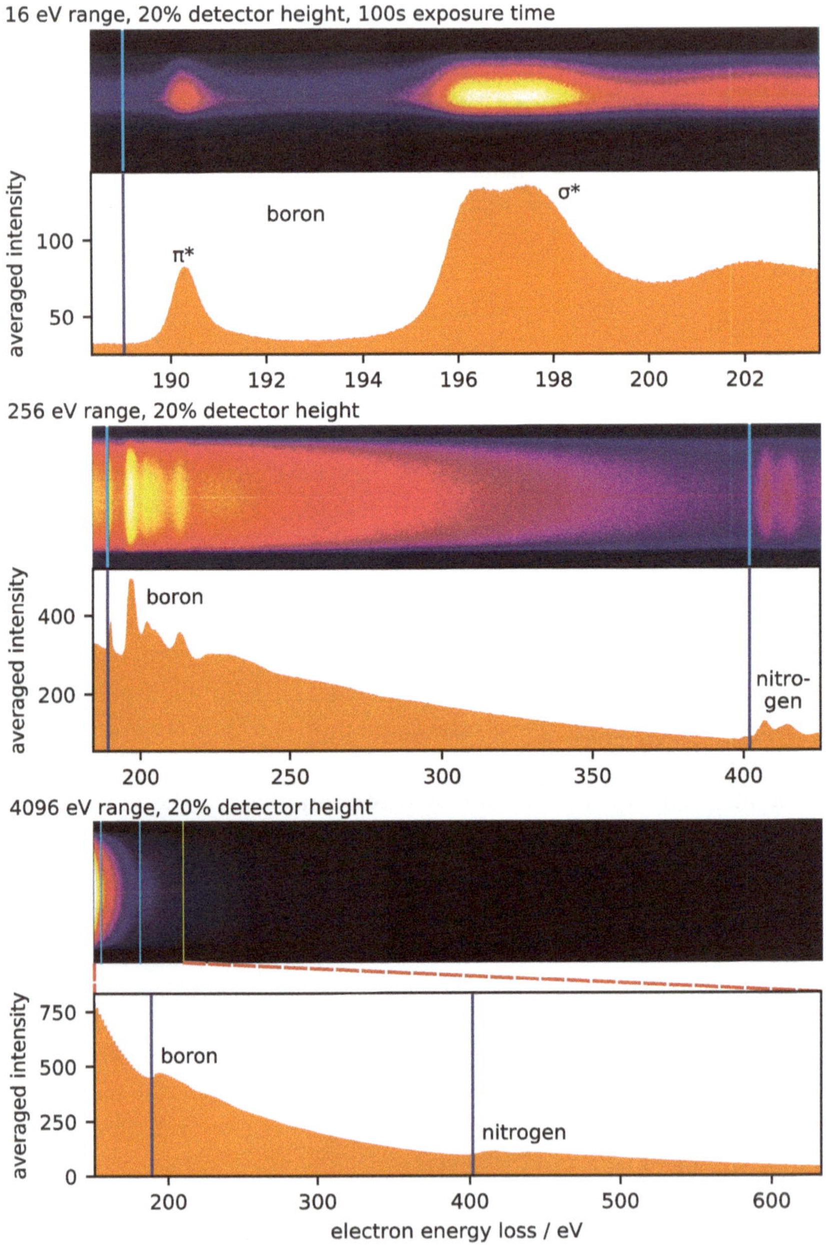

Figure 10 Spectra of h-BN for the modes with the highest, a medium and the lowest energy resolutions of 16 eV, 256 eV and 4 keV range, respectively. For the 4 keV spectrum a magnified line scan of the sub-section from the very left to the yellow line is shown. The images of the spectra have been scaled down in *y*-direction by a factor of 0.8 to fit the page.

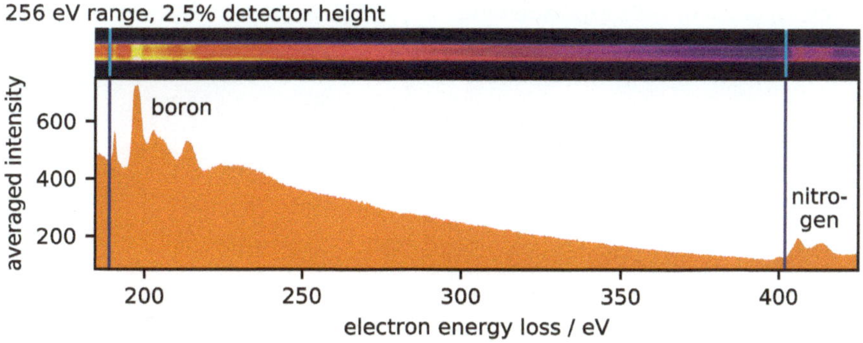

Figure 11 Spectrum of h-BN with 256 eV range and a spectrum height of 2.5%.

y-component q_y of the total momentum of the electrons. The objective of q-EELS is to measure the dependency of the EELS spectrum from the lateral momentum (q_x, q_y). For defining the q_x component a special slit shaped aperture with a rectangular profile is required (Midgley, 1999). The slit is oriented such that its narrow side extends along the x-coordinate of the aperture plane so that only a very small window of q_x can pass. The smaller the width of the slit in x-direction is chosen, the better the q_x-resolution will be. In order to maximize the q_y-resolution it is essential that in the y-direction the detector plane is conjugated to the aperture plane. The provided ω-q mode meets this criterion.

For many q-EELS applications the relevant scattering angles at 300 kV are in a quite low range of less than 0.2 mrad and the investigated energy losses are usually in the range of only a few dozen electron volts (Midgley, 1999). Since the maximum camera length of modern TEMs rarely exceeds 5 m, this range of scattering angles corresponds to y-positions in the aperture plane of less than $5\text{ m} \times 0.2\text{ mrad} = 1\text{ mm}$ in magnitude, barely filling the 2 mm aperture. For taking full advantage of the detector chip for optimum momentum resolution the ideal magnification in y-direction is $2048 \times 14\ \mu\text{m}/2\text{ mm} \approx 14.34$. The 64 eV ω-q mode has a maximum attainable magnification of 11.37 quite close to ideal one. Fig. 12 shows a spectrum of the boron k-edge recorded using this mode in combination with the 2 mm EELS aperture. In order to highlight the spectrum quality over the whole spectrum height, this example was acquired in TEM imaging mode, thus resembling a spatial resolution in y-direction instead of angular resolution.

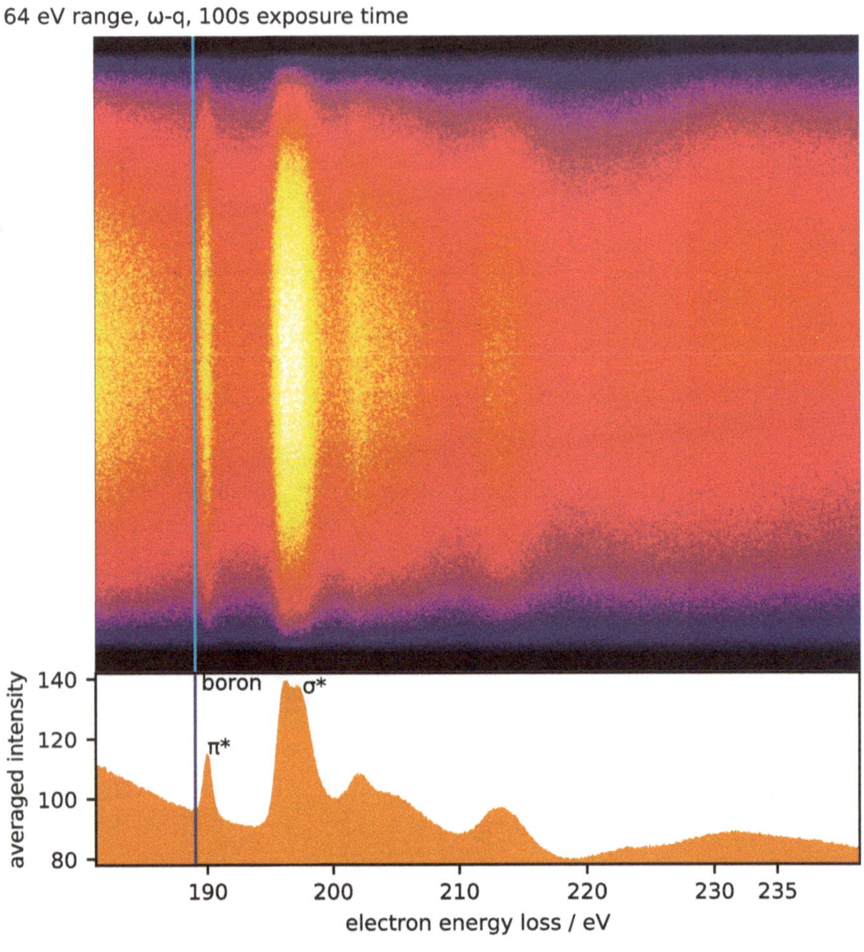

Figure 12 Spectrum of the boron *k*-edge of h-BN using the ω-q mode with 64 eV range in combination with the 2 mm aperture acquired in TEM imaging mode.

Using our preferred setup comprising a TVIPS XF416 4k × 4k CMOS camera combined with the longer drift tube allows for identical alignments but with a two times finer sampling in both directions and a much faster read-out speed.

The good stability of both the current supply and the filter hardware significantly reduces the need for retuning. After switching between ESI modes, between ESI and EELS or between EELS modes with different dispersions, normalizing those elements whose excitations have changed strongly is sufficient to restore a proper alignment. This allows to switch between modes very quickly within an acquisition workflow. All spectra presented in Section 4 were recorded without intermediate retuning.

5. Conclusion and outlook

The optical performance of our new post-column imaging energy filter CEFID shows a significant improvement compared to previous systems as published by Gubbens et al. (2010). The latest generation GIF Continuum released by Gatan in 2018 is not yet included in this survey. This performance is remarkable since the quadrupole projective (post-slit optics) implements a minimalistic approach without higher-order correction. However, we already developed improved concepts for the next CEFID generation which will allow to correct the dominant geometrical as well as chromatic distortions by means of additional higher-order multipoles in the post-slit optics. Currently, the dominant third-order distortions can be measured very reproducibly and with good accuracy so that the recorded images can be digitally dewarped either on-the-fly or a-posteriori.

However, the most remarkable improvement of the CEFID over currently existing designs is the huge reduction of the residual non-isochromaticity. For the 8 mm mode the non-isochromaticity was reduced by about one order of magnitude down to below 100 meV. Even for the 12 mm mode the non-isochromaticity is still about three times smaller than the values guaranteed for the most advanced commercially available filter today with an entrance aperture of only 9 mm.

The combination of low non-isochromaticity with a huge-field of view makes the CEFID very attractive for zero-loss filtering applications in life science, especially in combination with the large 4k × 4k CMOS detector.

In spectroscopy mode a small non-isochromaticity is equivalent to an excellent energy resolution even for a quite large entrance aperture. The filter is currently equipped with a 5 mm EELS aperture but the extremely small measured NI aberrations suggest to use even larger apertures with better collection efficiency for core-loss spectroscopy. As demonstrated in this publication, the spectrum shape is virtually ideal for all spectrum magnifications and both dispersion and spectrum height can be tuned over large ranges even for low beam energies.

The CEFID platform provides open and flexible interfaces for detector systems, control and scripting software, advanced user alignments, and data formats, making this post-column filter more flexible and extendable even for customized experimental setups.

Appendix A. Energy resolution and number of resolved image points

All conclusions of this appendix are based on the assumption of an ideal aberration free system whose spatial resolution is limited only by diffraction. Therefore, we can use the gaussian approximation for describing the geometrical optics. This means that the electron trajectory $x(z)$ along the optical axis z in the dispersive section is approximated by the linear superposition $x(z) = x_o x_\gamma(z) + \theta x_\alpha(z)$ of the axial ray x_α and the field ray x_γ. For their starting conditions in the object plane z_o we choose $x_\alpha(z_o) = 0$, $x'_\alpha(z_o) = 1$, $x_\gamma(z_o) = 1$ and $x'_\gamma(z_o) = 0$. Then θ means the angle at which the electron leaves the sample, and x_o is its off-axial starting position.

If the TEM is in imaging mode the slit plane z_s is a Fourier plane with respect to the object plane so that $x_\gamma(z_s) = 0$ holds. This means that all electrons leaving the sample at an angle of ϑ are focused into a sharp spot in the slit plane at the distance of $\vartheta x_\alpha(z_s)$ from the optical axis. Therefore, the diffraction pattern formed by all electrons with a scattering angle of $|\vartheta| < \theta$ has a diameter of $2\theta x_\alpha(z_s)$ in the slit plane. In order to allow all those electrons to pass the energy window in ESI mode adjusted by the slit blades must not be smaller than

$$\Delta E = 2\theta x_\alpha(z_s)/D_s \tag{A.1}$$

where D_s is the magnitude of the dispersion in the slit plane. Since the strength of the magnetic field of the sector is adjusted such that the curvature radius of the optical axis just matches its bending radius, Eq. (2.78) of Rose and Krahl (1995) simplifies to

$$D_s = x_\alpha(z_s) \frac{1 + E/E_0}{E(2 + E/E_0)} \int_0^{\frac{\pi}{2}} x_\gamma(\varphi)\, \mathrm{d}\varphi \tag{A.2}$$

in Sharp Cut-Off Fringe Field (SCOFF) approximation. Here $x_\gamma(\varphi)$ is the value of the field ray at the position inside the sector which corresponds to the deflection angle φ, E is the beam energy and E_0 is the rest energy of the electron. The focusing effect of a 90°-sector magnet on the field ray is gentle enough that we can replace $x_\gamma(\varphi)$ in the integral by the value $\hat{x}_\gamma$ which x_γ would assume at the center of the sector if no focusing effect were present:

$$\hat{x}_\gamma \approx \frac{L_s}{L} M\,. \tag{A.3}$$

Table A.4 Values of scaling function $F(E)$ and energy resolution limits for $N = 2000$ and $d_a = 8$ mm for various acceleration voltages.

E / kV	30	80	200	300
$F(E)$ / mm·meV	0.632	0.968	1.339	1.496
ΔE / meV	126	194	267	299

Here M is the magnification between sample and aperture plane, and L and L_s are the distances of the aperture plane and the sector center from the DPA plane, respectively. This results in the approximation

$$D_s \approx x_\alpha \frac{1 + E/E_0}{E(2 + E/E_0)} \frac{\pi}{2} \frac{L_s}{L} M \tag{A.4}$$

for the upper limit of the dispersion. This leads to

$$\Delta E \approx \frac{4\theta}{\pi} \frac{2 + E/E_0}{1 + E/E_0} \frac{L}{L_s M} E \tag{A.5}$$

for the smallest energy window possible if scattering angles up to θ should still be transferred. Due to the Abbe diffraction limit no object details smaller than

$$d_{\text{px}} = 1.22 \frac{\lambda}{\theta} \tag{A.6}$$

are resolved in the image. If we want to image a circular field of view in the object plane with a diameter of $N \times d_{\text{px}}$ the magnification must be smaller or equal to $M = d_a/(N d_{\text{px}})$ to guarantee that the field of view fits through an entrance aperture of diameter d_a. Inserting this relation and the wavelength

$$\lambda = \frac{hc}{e} \frac{1}{E_0 \sqrt{(2 + E/E_0) E/E_0}} \tag{A.7}$$

into Eq. (A.5) yields

$$\Delta E \approx F(E) \frac{L}{L_s} \frac{N}{d_a}, \quad F(E) = 1.22 \frac{4}{\pi} \frac{hc}{e} \frac{\sqrt{(2 + E/E_0) E/E_0}}{1 + E/E_0} \tag{A.8}$$

The distance L typically ranges from 500 mm to 900 mm. For the CEFID the position z_P at the center of the sector field is located at a distance of $\Delta z = L_s - L = 155$ mm behind the aperture plane. Since the available mounting space for a post-column filter is very limited, the order of magnitude of Δz cannot easily be increased. Thus, the factor L/L_s lies in the quite narrow range of 0.76…0.85. By replacing this factor by the center value 0.8 we can further simplify Eq. (A.8):

$$\Delta E \approx 0.8F(E)\frac{N}{d_a}. \quad \text{(A.9)}$$

So the only way to increase the energy resolution of a post-column filter for a given number of image pixels is to increase the diameter of the aperture. Modern cameras provide detector sizes up to 4k × 4k. Direct electron detectors of this size allow to resolve up to the Nyquist limit of $N = 2000$ pixels. For all other detector types this number is smaller due to their MTF. Table A.4 lists approximations of the maximum attainable energy resolution according to Eq. (A.9), assuming $N = 2000$ in ESI 8 mm mode for various acceleration voltages.

Acknowledgments

The authors are very grateful to their colleagues Christian Berger, Dirk Barty, Dr. Christian Elting, Anja Firsching, Klaus Hessenauer, Dr. Peter Hartel, Angelika Leibscher, Ulrich Löbau, Ingo Maßmann, Dennis Sczerba, Dr. Joachim Zach and Prof. Dr. Max. Haider at CEOS GmbH for their support and their contributions to development, construction, alignment and characterization of the first CEFID prototypes. We are also very grateful to the scientific community and our cooperation partners who stimulated and supported our work on the first CEOS-designed energy filter over many years.

References

Bihr, J. (1991). Design of an analytical TEM with integrated imaging ω-spectrometer. In *Proceedings of the 49th annual meeting of EMSA, 1991* (pp. 354–355). San Francisco Press.

Castaing, R., & Henry, L. (1962). Filtrage magnétique des vitesses en microscopie électronique. *Comptes Rendus de l'Académie des Sciences, Paris, Série B, 255*, 76–78.

Colliex, C. (2017). From a physicist's toy to an indispensable analytical tool in many fields of science: A personal view of the leading contribution of Ondrej Krivanek to the spectacular successes of EELS spectroscopy in the electron microscope. *Ultramicroscopy, 180*, 14–21.

Egerton, R. (1978). A simple electron spectrometer for energy analysis in the transmission microscope. *Ultramicroscopy, 3*, 39–47.

Egerton, R., & Crozier, P. (1987). A compact parallel-recording detector for EELS. *Journal of Microscopy, 148*(2), 157–166.

Egerton, R. F. (2011). *Electron energy-loss spectroscopy in the electron microscope*. Springer Science & Business Media.

Egerton, R. F. (2012). TEM-EELS: A personal perspective. *Ultramicroscopy, 119*, 24–32.

Essers, E., Benner, G., Mandler, T., Meyer, S., Mittmann, D., Schnell, M., & Höschen, R. (2010). Energy resolution of an omega-type monochromator and imaging properties of the MANDOLINE filter. *Ultramicroscopy, 110*(8), 971–980.

Fink, J. (1989). Recent developments in energy-loss spectroscopy. In *Advances in electronics and electron physics: Vol. 75* (pp. 121–232). Elsevier.

Geiger, J., Nolting, M., & Schröder, B. (1970). How to obtain high resolution with a Wien filter spectrometer. In *Electron microscopy–1970* (pp. 111–112).

Gerheim, V. (2001). *Ein Projektivsystem mit variabler Vergrößerung für energiefilternde Elektronenmikroskope* (PhD thesis). Technical University of Darmstadt.

Gubbens, A., Barfels, M., Trevor, C., Twesten, R., Mooney, P., Thomas, P., ... McGinn, B. (2010). The GIF quantum, a next generation post-column imaging energy filter. *Ultramicroscopy, 110*(8), 962–970.

Haider, M. (1989). A corrected double-deflection electron spectrometer equipped with a parallel recording system. *Ultramicroscopy, 28*(1–4), 190–200.

Haider, M., Müller, H., Uhlemann, S., Zach, J., Loebau, U., & Hoeschen, R. (2008). Prerequisites for a cc/cs-corrected ultrahigh-resolution TEM. *Ultramicroscopy, 108*(3), 167–178.

Hart, J., Lang, A., Leff, A., Longo, P., Trevor, C., Twesten, R., ... Taheri, M. (2017). Direct detection electron energy-loss spectroscopy: A method to push the limits of resolution and sensitivity. *Scientific Reports*, 7(1), 1–14.

Hillier, J., & Baker, R. (1944). Microanalysis by means of electrons. *Journal of Applied Physics, 15*(9), 663–675.

Ibach, H., & Mills, D. (Eds.). (1982). *Electron energy loss spectroscopies and surfaces vibrations.* NY: New Academic.

Isaacson, M., & Johnson, D. (1975). The microanalysis of light elements using transmitted energy loss electrons. *Ultramicroscopy, 1*(1), 33–52.

Kahl, F., Linck, M., Hartel, P., Mueller, H., Uhlemann, S., Haider, M., & Zach, J. (2016). Performance of the SALVE III corrector for EFTEM applications. In *European microscopy congress 2016: Proceedings* (pp. 787–788). Wiley Online Library.

Kahl, F., & Rose, H. (1998). Outline of an electron monochromator with small Boersch effect. In *Proceedings of the 14th international conference on electron microscopy, Vol. 1* (pp. 71–72).

Krivanek, O., Ahn, C., & Keeney, R. (1987). Parallel detection electron spectrometer using quadrupole lenses. *Ultramicroscopy, 22*(1–4), 103–115.

Krivanek, O., & Swann, P. (1981). An advanced electron energy loss spectrometer. In *Quantitative microanalysis with high spatial resolution* (pp. 136–140).

Krivanek, O. L., Gubbens, A. J., Dellby, N., & Meyer, C. E. (1992). Design and first applications of a post-column imaging filter. *Microscopy, Microanalysis, Microstructures, 3*(2–3), 187–199.

Krivanek, O. L., Ursin, J. P., Bacon, N. J., Corbin, G. J., Dellby, N., Hrncirik, P., ... Szilagyi, Z. S. (2009). High-energy-resolution monochromator for aberration-corrected scanning transmission electron microscopy/electron energy-loss spectroscopy. *Philosophical Transactions - Royal Society. Mathematical, Physical and Engineering Sciences, 367*(1903), 3683–3697.

Linck, M., Hartel, P., Uhlemann, S., Kahl, F., Müller, H., Zach, J., ... Kaiser, U. (2016). Chromatic aberration correction for atomic resolution TEM imaging from 20 to 80 kV. *Physical Review Letters, 117*(7), 076101.

Lovejoy, T., Corbin, G., Dellby, N., Hoffman, M., & Krivanek, O. (2018). Advances in ultra-high energy resolution stem-eels. *Microscopy and Microanalysis, 24*(S1), 446–447.

Marton, L. (1944). Electron microscopy. *Reports on Progress in Physics, 10*(1), 204.

Midgley, P. A. (1999). A simple new method to obtain high angular resolution ω–q patterns. *Ultramicroscopy, 76*(3), 91–96.

Möllenstedt, G. (1949). Die elektrostatische Linse als hochauflösender Geschwindigkeits-analysator. *Optik, 5*, 499–517.

Mukai, M., Morishita, S., Sawada, H., & Suenaga, K. (2016). Ultra high energy resolution EELS mapping using aberration-corrected low-voltage STEM equipped with monochromator. *Microscopy and Microanalysis, 22*(S3), 962–963.

Muller, D., Kourkoutis, L. F., Murfitt, M., Song, J., Hwang, H., Silcox, J., ... Krivanek, O. (2008). Atomic-scale chemical imaging of composition and bonding by aberration-corrected microscopy. *Science, 319*(5866), 1073–1076.

Müller, H., Uhlemann, S., Hartel, P., & Haider, M. (2008). Aberration-corrected optics: From an idea to a device. *Physics Procedia, 1*(1), 167–178.

Oster, M., Ghadimi, R., & Tietz, H. (2018). Performance of a novel detector for TEM employing a 4t pixel design for on-chip correlated double sampling and a selectable full-well capacity. *Microscopy and Microanalysis, 24*(S1), 896–897.

Rose, H. (1987). Hamiltonian magnetic optics. *Nuclear Instruments & Methods in Physics Research. Section A, Accelerators, Spectrometers, Detectors and Associated Equipment, 258*(3), 374–401.

Rose, H., & Krahl, D. (1995). Electron optics of imaging energy filters. In *Energy-filtering transmission electron microscopy* (pp. 43–149). Springer.

Rose, H., & Plies, E. (1974). Design of a magnetic energy analyzer with small aberrations. *Optik, 40*(3), 336–341.

Rudberg, E. (1930). Characteristic energy losses of electrons scattered from incandescent solids. *Proceedings of the Royal Society of London. Series A, Containing Papers of a Mathematical and Physical Character, 127*(804), 111–140.

Ruthemann, G. (1941). Diskrete Energieverluste schneller Elektronen in Festkörpern. *Naturwissenschaften, 29*(42), 648.

Suenaga, K., & Koshino, M. (2010). Atom-by-atom spectroscopy at graphene edge. *Nature, 468*(7327), 1088.

Uhlemann, S. (1994). The MANDOLINE filter-a new high-performance imaging filter for sub-eV EFTEM. *Optik, 96*, 163–178.

Uhlemann, S., & Rose, H. (1996). Acceptance of imaging energy filters. *Ultramicroscopy, 63*(3–4), 161–167.

Wang, Z., Tavabi, A. H., Jin, L., Rusz, J., Tyutyunnikov, D., Jiang, H., ... Zhong, X. (2018). Atomic scale imaging of magnetic circular dichroism by achromatic electron microscopy. *Nature Materials, 17*(3), 221.

Wittry, D. (1969). An electron spectrometer for use with the transmission electron microscope. *Journal of Physics. D, Applied Physics, 2*(12), 1757.

CHAPTER FOUR

Electron optics for a multi-pass transmission electron microscope

Marian Mankos[a,*], Stewart A. Koppell[b], Brannon B. Klopfer[b], Thomas Juffmann[c], Vladimir Kolarik[d], Khashayar Shadman[a], Mark Kasevich[b]
[a]Electron Optica, Palo Alto, CA, United States of America
[b]Physics Department, Stanford University, Stanford, CA, United States of America
[c]Universitaet Wien, Vienna, Austria
[d]Delong Instruments, Brno, Czech Republic
*Corresponding author. e-mail address: marian@electronoptica.com

Contents

1. Introduction

Recent advances in cryo-electron microscopy (cryo-EM) (Agard, Cheng, Glaeser, & Subramaniam, 2014; Glaeser, 2015) and direct electron detection have spurred renewed interest in the development of novel electron imaging techniques for applications in structural biology at atomic resolution (Khoshouei, Radjainia, Baumeister, & Danev, 2017). The challenge with imaging unstained biological specimens is that they provide a low scattering cross-section to the probing electrons because they are composed primarily of low atomic number elements. Hence, high electron doses are needed to obtain sufficient signal-to-noise ratios (SNR) in order to overcome the shot-noise limit. Such doses, however, severely damage the specimens. The shot-noise limit can in principle be overcome by using correlated particles (Giovannetti, Lloyd, & Maccone, 2011), as demonstrated in optical microscopy (Ono, Okamoto, & Takeuchi, 2013;

Advances in Imaging and Electron Physics, Volume 212
ISSN 1076-5670
https://doi.org/10.1016/bs.aiep.2019.08.003

Juffmann, Klopfer, Frankort, Haslinger, & Kasevich, 2016). The first proposals for quantum electron microscopy (Putnam & Yanik, 2009; Kruit et al., 2016) involved interaction-free quantum measurements of entangled electrons. Alternatively, the noise reduction can be achieved by letting a single electron interact with the sample multiple times (Giovannetti, Lloyd, & Maccone, 2006) in what is dubbed multi-pass transmission electron microscopy (MPTEM) (Juffmann et al., 2017). In transmission electron microscopy (TEM), where thin biological samples are used, the electrons scatter weakly as they pass through the sample, adopting a small phase change (Reimer, 1984). In MPTEM, the electron beam interacts elastically with the specimen multiple times so that the change in the phase accumulates before reaching the detector. The multiple interactions are accomplished by injecting the electron into a resonator formed by two electron mirrors sandwiching the specimen and the objective lens. Once the electron accumulates a strong phase change by passing through the sample many times, it is ejected and the resulting image is magnified onto the detector. Past simulations have predicted an improvement in resolution and sensitivity for a range of electron microscopy imaging techniques, and an order-of-magnitude reduction in damage at equivalent resolution (Juffmann et al., 2017). Here the electron-optical design of a practical implementation of a MPTEM that is currently under construction is examined.

2. Electron-optical layout of a multi-pass transmission electron microscope

Fig. 1 shows the key components of the MPTEM: the pulsed electron gun, the illumination optics, the resonator optics, which includes the objective lenses, mirrors and specimen, and the projection optics. The electron pulse, triggered by an ultrafast laser beam, is focused by the illumination optics and transmitted by the entrance electron mirror, rendered transparent by a voltage pulse synchronized with the laser beam. The transmitted electron pulse enters an electron resonator, bounded by the entrance and exit mirrors. The resonator includes two objective and field lenses that sandwich the specimen. The electron pulse is collimated by the upper field and objective lens onto the specimen, and refocused by the lower objective and field lens onto the exit mirror, which reflects it back symmetrically so that the electron pulse is collimated again at the specimen. This reflection is carried out multiple times until a second voltage pulse renders the exit mirror transparent to allow the electron pulse with the

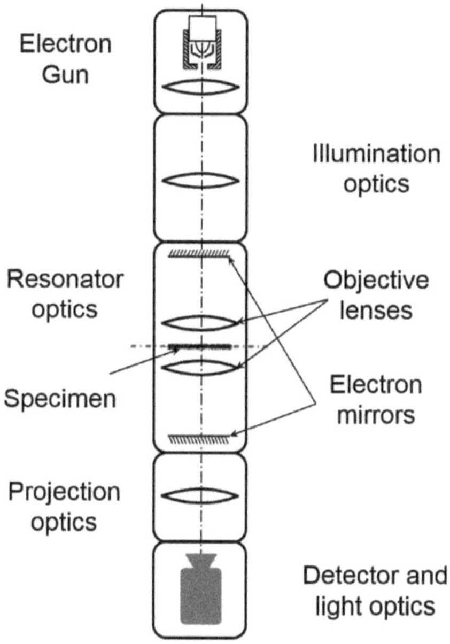

Figure 1 Schematic layout of the MPTEM column.

accumulated phase to proceed into the projection optics, which magnifies the image at the exit mirror onto the detector. Once the electron is detected, another pulse is triggered, and the whole process is repeated at high frequency (1-10 MHz) until a sufficient electron dose is accumulated by the detector. The pulsed electron gun determines the electron source parameters, including the pulse frequency, virtual source size, and the angular and energy distributions. The illumination optics provides the required illumination size and convergence angle at the specimen, while the spherical and chromatic aberrations of the objective lenses limit the ultimate achievable spatial resolution. The projection optics magnify the specimen image or diffraction pattern at the detection plane. Initially, both magnetic and electrostatic lenses were considered in the electron-optical design. While magnetic lenses provide superior spatial resolution, they introduce image rotations, which significantly increases the complexity of the resonator optics. The main goal of the prototype electron column is to demonstrate the multi-pass imaging approach. In order to minimize cost, development time and complexity in this phase of the development, an electrostatic column operating at a beam energy of 10 keV has been adopted. In this prototype, a resolution of approximately 5 nm is targeted, which is sufficient to verify the multi-pass concept. A higher resolution microscope can be developed in the future by implementing magnetic lenses and using higher beam energies, as well as by fully utilizing the aberration correction capabilities of the electron mirrors inherent in this approach. Specialized software packages developed by MEBS, Ltd. (SOFEM, SORAY, IMAGE and MIRROR DA) were used for the analysis of the MPTEM optics. The SOFEM software package computes the potential distributions of the electrostatic lenses using the Second-Order Finite Element Method. The computed axial lens field functions are subsequently used by the SORAY, IMAGE, and MIRROR DA packages, to compute its optical properties, including aberrations up to fifth-order. The SORAY software package computes and plots electron trajectories by direct ray-tracing through the computed lens fields. The IMAGE software package computes electron-optical properties by propagating bunches of particles through realistic electromagnetic fields, including the fields from the individual particles, by accurate direct ray-tracing to capture the combined effects of Coulomb interactions and aberrations. MIRROR DA is a differential algebra-based (DA) software package that computes the temporal and optical properties, including aberrations of any order, for electron mirrors with any symmetry and can handle combinations of electron mirrors and electron lenses in a unified way. Results computed with MIRROR DA were shown to be in good agreement with those extracted by direct ray tracing with relative deviations of less than 0.065% for all primary aberration coefficients (Munro, Rouse, Liu, & Wang, 2008).

2.1 Resonator optics

The resonator optics distinguishes the MPTEM from a conventional TEM. This optical element includes the objective and field lenses and electron mirrors, needed to establish the resonant, self-imaging beam path. The analysis that follows starts with the objective lens only and subsequently incorporates the field lenses and electron mirrors, which are used to re-image the specimen onto itself. The electron-optical properties of the objective lens determine the spatial resolution of the MPTEM column. Objective lenses are designed to generate a highly resolved, magnified image of the specimen. The resolution limit due to the spherical and chromatic aberration is only important for the objective lens because the magnification decreases the aperture angle in the following lenses. Even for moderate magnifications of 10-50×, the aperture angle at the subsequent lenses is so small that their spherical and chromatic aberrations can be neglected. For this prototype, a conventional Einzel lens operating in decelerating

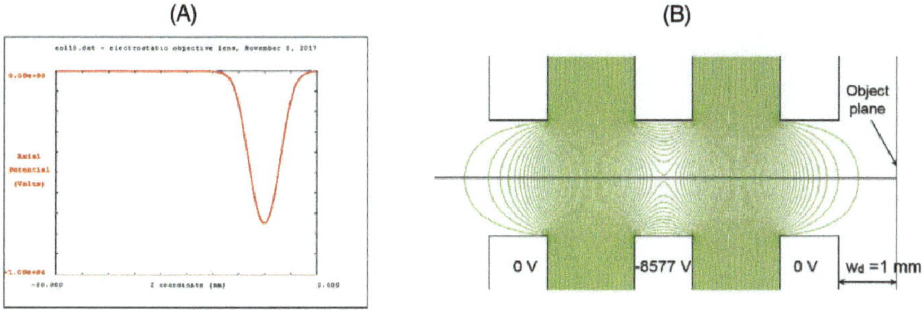

Figure 2 Axial distribution ((A), specimen location at z = 0 mm) and overall electrostatic potential distribution (B) of the MPTEM objective lens.

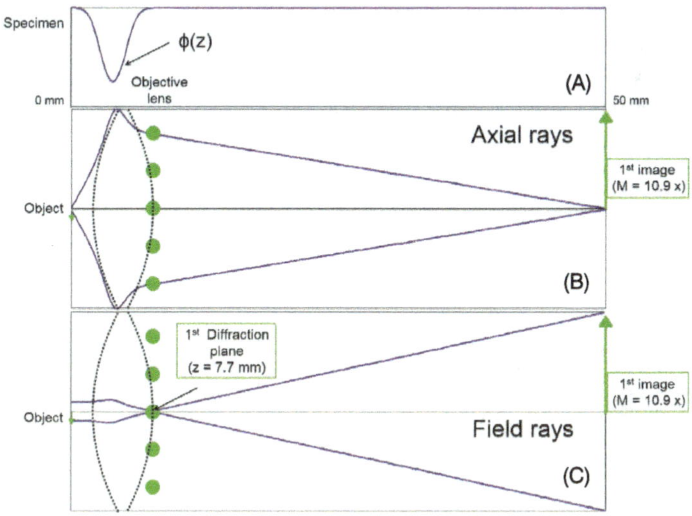

Figure 3 Axial distribution ((A), specimen location at z = 0 mm) and axial (B) and field rays (C) in the MPTEM objective lens.

mode has been selected. The geometry of the lens electrodes has been optimized for operation at 10 keV in order to achieve short focal lengths and thereby small chromatic and spherical aberration coefficients while maintaining peak electrostatic fields below 10 kV mm^{-1} in order to minimize arcing. The geometry of the objective lens and its electrostatic potential distribution is shown in Fig. 2. The ray diagram showing the impact of the objective lens is shown in Fig. 3. Here, the focusing voltage is approximately −8.6 kV for a working distance $w_d = 1$ mm, where the specimen is located 1 mm behind the final electrode face. The specimen is in a field-

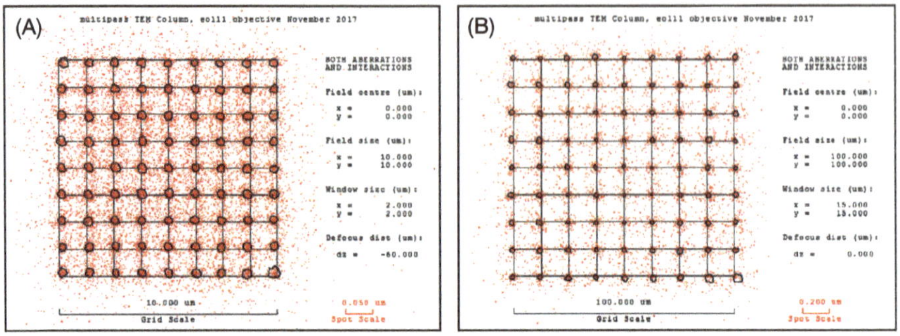

Figure 4 Optical blur introduced by the objective lens in the first image plane for a field of view of 1 and 10 μm and an object angle of 3 mrad (at specimen plane).

free region, sandwiched symmetrically from both sides by the electrodes of the two objective lenses. The objective lens magnifies the image, located at $z = 50$ mm, by a factor of approximately 10.9×. Further, the back-focal (diffraction) plane is conveniently located in a field-free region, just outside the last ground electrode of the Einzel lens. With this geometry and stated working distance, the spherical aberration coefficient (referred to the object) is 78.4 mm, while the chromatic aberration coefficient is 17.3 mm. The combined effect of axial (chromatic and spherical) as well as off-axis aberrations (coma, astigmatism, field curvature and distortion) has been evaluated for fields of view of 1 and 10 μm (at the specimen), and for a range of object angles. Fig. 4 shows the blur in the first image plane, after the initial magnification by the objective lens. As illustrated in Fig. 4A, the blur is dominated by axial aberrations for a 1 μm field of view (at the specimen) with an object angle of 3 mrad. Here, the field aberrations of the objective lens do not appreciably deteriorate the spatial resolution across the field of view. The contour plots show an average 20-80% blur of less than 20 nm and a distortion of less than 10 nm at the corner of the field of view, which corresponds to a blur of 2 nm at the specimen plane. For an object angle of 3 mrad, the diffraction blur is 2.2 nm and comparable to the blur of the geometric and chromatic aberrations, yielding an optimum total blur of approximately 3 nm. Even for the larger field of view of 10 μm (at the specimen) shown in Fig. 4B, the field aberrations of the objective lens are still negligible. The contour plots show an average 20-80% blur of less than 30 nm and a distortion of less than 10 nm at the corner of the field of view, which corresponds to a blur of 3 nm at the specimen plane, yielding an optimum total blur of less than 4 nm.

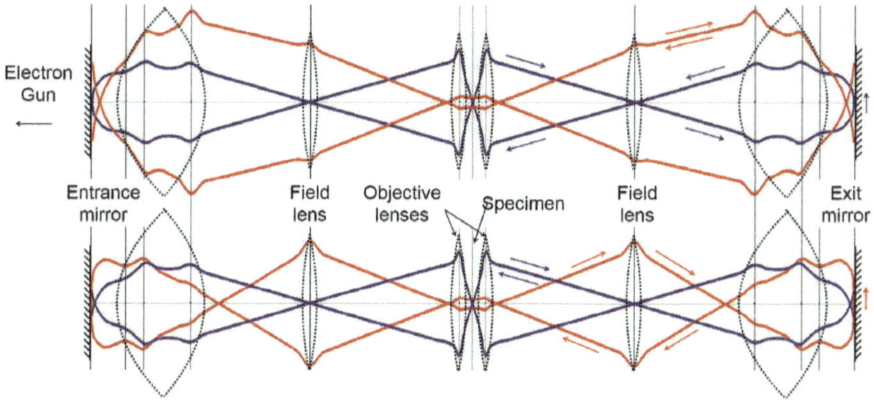

Figure 5 Ray diagram of the full resonator optics for 2 imaging modes, with a real image (top) or a crossover (bottom) at the mirror reflection plane. Axial rays are shown in blue, and field rays are shown in red.

A detailed layout of the resonator optics tuned for multi-passing is shown in Fig. 5. Here, the electron mirrors are biased for reflection. As shown, the optical layout is symmetric about the specimen plane with each side containing an objective lens, a field lens, and a mirror. In order to accumulate the change in the phase imparted to the electrons by the sample, the electrons must impact the sample at the same location with each reflection. To this end, the optics of the resonator must image the specimen onto itself with a $+1\times$ magnification. The multi-pass imaging condition also implies that the back-focal plane of the objective lens must also be imaged onto itself in order to maintain collimated illumination. This desirable imaging condition is illustrated in the upper ray diagram in Fig. 5, where the path of the fundamental rays are shown over one full period. Here, the axial rays (shown in blue) are defined to emanate from the sample plane on-axis with a unit slope, while the field rays (shown in red) are defined to start off-axis at a unit distance with zero slope. The axial ray is tuned by the objective lens to pass through the center of the field lens to make the ray practically insensitive to the action of this lens. The mirror is tuned to place the axial ray on axis at the reflection plane. This condition guarantees that the axial ray traces a path back onto the sample that is symmetric about the optical axis, thereby imaging the sample onto itself. The field lens is then tuned to make the field ray retrace its path upon reflection. Another possible imaging condition of the resonator is illustrated in the lower ray diagram in Fig. 5. As in the previous case, the objective

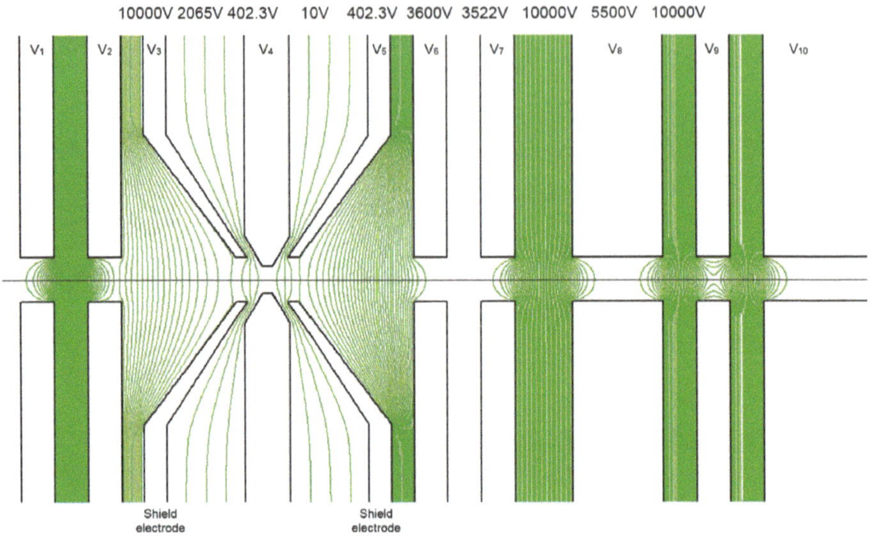

Figure 6 Overall electrostatic potential distribution of the MPTEM electron mirror.

lens is tuned to place the axial ray at the center of a field lens. Here, the mirror is tuned to make the axial ray retrace its path upon reflection, and the field lens is tuned to place the field ray on axis at the reflection plane, thereby reversing the role of the two fundamental rays. These conditions, however, yield a net $-1\times$ magnification. To recover a $+1\times$ magnification, either magnetic lenses can be used to rotate the image by 180°, or the specimen can be restricted to reside in a half-plane, as it is common in electron holography. The advantage of this imaging condition is the relative ease with which the electron mirrors can be tuned to correct both the spherical and chromatic aberrations of the objective lens to significantly decrease the round-trip aberrations. In the first mode, only the chromatic aberration of the objective lens has been successfully corrected so far. A more detailed discussion of the aberration correction properties of the resonator optics is described in a separate paper (Koppell et al., 2019). The geometry of the entrance electron mirror, which is composed of rotationally-symmetric electrodes, is shown in Fig. 6, together with the equipotential distribution for correcting the chromatic aberration of the objective lens. The entrance and exit electrodes are grounded (V1 = V10 = 0 keV) while the thick, V-shaped gated electrode (V4 = −10.09 keV) is biased at a potential that is slightly more negative than the electron emitter to reflect the electron beam. The two slanted electrodes surrounding the gated electrode are biased a few hundred volts more positive (V3 = V5 = −9.6 keV) than the gated electrode. These electrodes effectively shield the gated electrode to minimize the extent of the transient fields when the mirror is pulsed to transmit the electron beam. The next two electrodes provide the two potentials (V6 = −6.4 keV, V7 = −6.5 keV) needed to tune the spherical and chromatic aberration coefficients of the objective lens and to focus the mirror in conjunction with the potential of the gated and shielding electrodes (V4 and V5). The next electrode is grounded (V8 = 0 keV), and the remaining electrode (V9 = −6.4 keV) forms an auxiliary lens that provides an additional focusing degree of freedom for the mirror. The axial distribution of the electrostatic potential in the entrance mirror for the two distinct states (open and closed) is shown in Fig. 7. To change states, the gate electrode is pulsed with the remaining electrodes held constant. When the gate electrode is set to V4 = −10.09 keV, the electrons cannot penetrate through the mirror potential (green in Fig. 7). When an additional +100 V bias is applied to the gate electrode (V4 = −9.01 keV), electrons have sufficient kinetic energy to penetrate the mirror and enter or exit the resonator. Transient effects caused by the ramp-up and ramp-down of the gating voltage pulse are not considered here; a detailed discussion can be found here (Koppell et al., 2019). At a 10 keV beam energy, the period of one full cycle lasts approximately 10–15 ns, which for an optimum number of 10 passes (Juffmann et al., 2017) implies one electron can be detected in about 100 ns. Hence, with a 10 MHz repetition rate, a total of 10^7 electrons can be detected in 1 s.

2.2 Illumination optics

In MPTEM, the illumination optics must transmit the electron beam from the electron emitter through the entrance electron mirror, rendered transparent by a voltage pulse synchronized with the laser beam, and flood-illuminate the desired imaging area with a nearly uniform, parallel electron beam. This task is accomplished by focusing the virtual electron source into the back focal plane of the first objective lens. The illumination path includes the first half of the resonator optics, i.e. the strongly focusing mirror lens, as well as the field lens. Further, the area illuminated at the specimen is required to vary from about 1 to 50 μm, without the use of an aperture so that no beam current is lost. To cover this range, the illumination optics must vary the magnification of the virtual source by 50–100× in order to vary the convergence angle at the back focal plane, which in turn varies the illuminated area at the sample. The illumination optics must perform

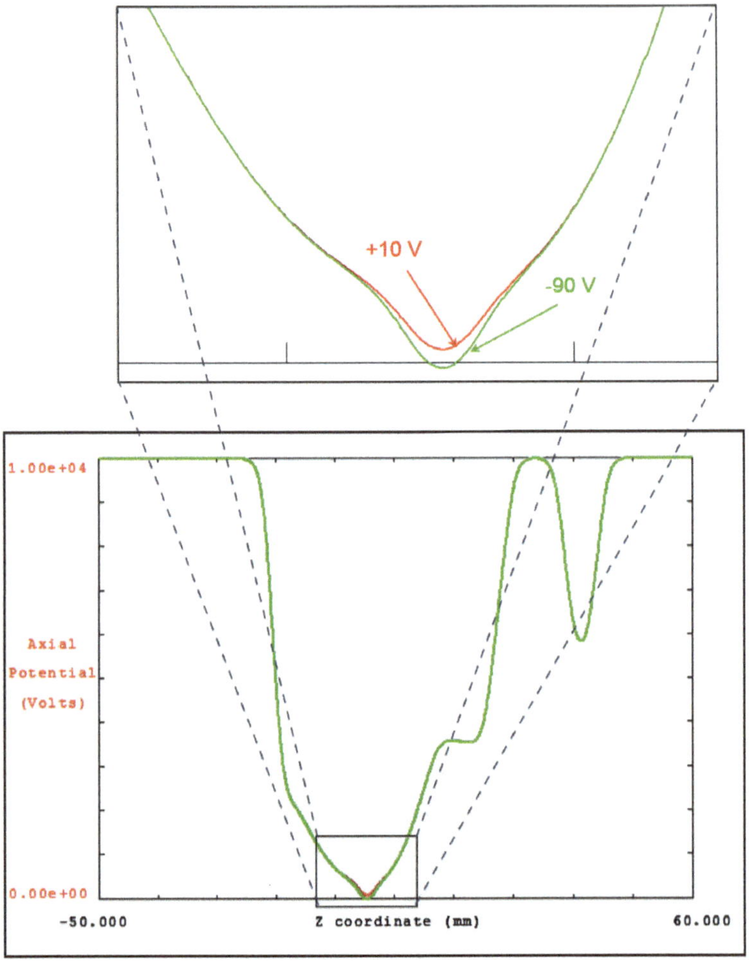

Figure 7 Axial electrostatic potential distribution of the MPTEM electron mirror The zoomed-in region marked by the box covers the vicinity of the gated and shield electrodes.

the above functions without altering any of the settings in the resonator optics (with the exception of the mirror electrode bias of 100 V required to pass the electrons through). In the MPTEM optics, the change in the illumination area is accomplished by utilizing the (magnetic) gun lens and one additional electrostatic condenser lens, located between the gun lens and the mirror. A detailed layout of the illumination optics including the entrance part of the resonator optics is shown in Fig. 8, with the gated entrance mirror electrode now biased for in-coupling of the electron beam.

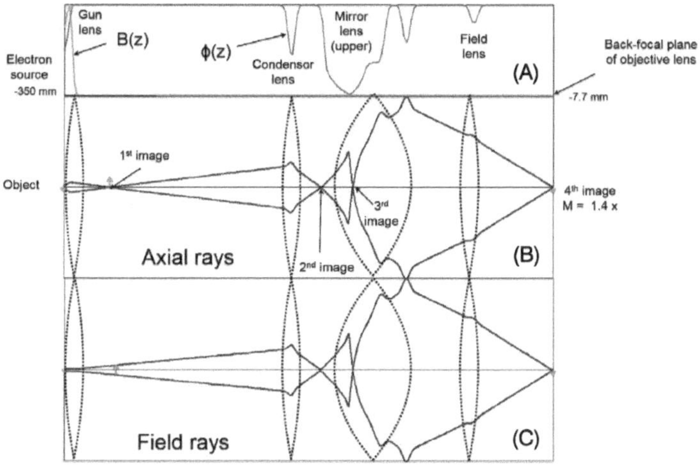

Figure 8 Axial distribution ((A), specimen location at z = 0 mm) and axial (B) and field rays (C) in the MPTEM illumination optics.

The electron emitter is a Schottky electron gun manufactured by Delong Instruments (2019). The gun was modified to allow for pulsed operation via laser illumination of the Schottky electron emitter. The electron gun includes a magnetic immersion lens that is used to focus the exiting beam. Note that since the magnetic lens is not part of the resonator optics, it does not introduce any image rotation inside the resonator. The gun lens forms a first image of the virtual source in front of the condenser lens, which in turn forms the second image near the entrance of the electron mirror. The strong negative bias of the gated electrode, just below reflection, produces a strong focusing action to yield a third image inside the mirror, in the vicinity of the gated electrode. The combined focusing action of the preset mirror and field lens then results in a fourth image at the back focal plane of the objective lens. The strength of the gun and condenser lens can be varied to adjust the magnification between the electron source and the back-focal plane of the objective lens to vary the area illuminated at the sample over a large range from approximately 0.5 to 50 μm. The ray diagram shown in Fig. 8 illustrates a case with low net magnification of 1.4×, resulting in a large convergence angle and thus a relatively large illuminated specimen area of 26 μm. The resulting spot shape generated by the illumination optics in the back-focal plane of the MPTEM objective lens is shown in Fig. 9A. Here the spot size is limited by spherical aberration accumulated in the mirror lens, resulting in a 12-88 blur of approximately 35 μm, while

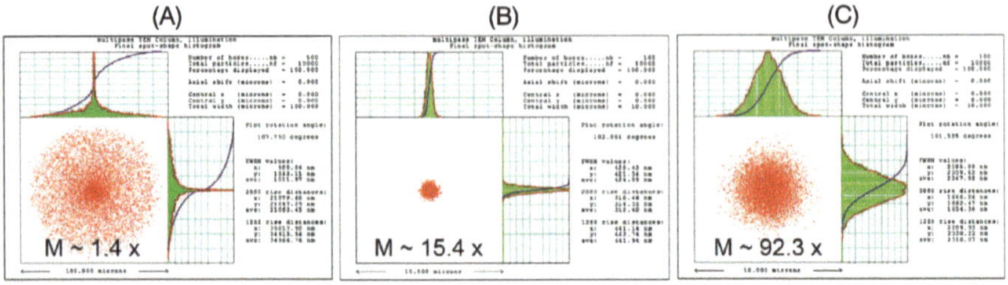

Figure 9 Spot shape generated by the illumination optics in the back-focal plane of the MPTEM objective lens for a net magnification of 1.4× (A) that corresponds to a large field of view of 26 μm, a net magnification of 15.4× (B) that corresponds to an optimum field of view of 26 μm, and for a net magnification of 92.3× (C) that corresponds to a small field of view of 0.5 μm.

the FWHM is approximately 1 μm. The resulting spot shape generated by the illumination optics with a large net magnification of 92.3× with a field of view of approximately 0.5 μm at the specimen is shown in Fig. 9C. In this case, the spot size profile is nearly Gaussian and now limited by the magnification of the virtual source, resulting in a 12-88 blur of approximately 2.3 μm, and a FWHM of approximately 2.2 μm. The optimum field of view, determined by the expected optimum resolution and our 2 kpixel × 2 kpixel detector, is approximately 2.5 μm. This field of view is obtained with a magnification of 15.4×. In this case, the spot size is near its minimum and limited by the magnification of the virtual source, resulting in a 12-88 blur of approximately 0.44 μm, and a FWHM of approximately 0.42 μm. This ensures that the specimen illumination is collimated to better than 0.1 mrad.

2.3 Projection optics

In MPTEM, the electron beam must be transmitted from the resonator optics into the projection optics, through the exit electron mirror, rendered transparent by a voltage pulse synchronized with the laser beam. The MPTEM projection optics follows the resonator optics and consists of an intermediate lens and two projector lenses. The intermediate lens focuses either the specimen image or its diffraction pattern at the entrance plane of the projection optics, and the projection lenses are then used to achieve the desired magnification on the viewing screen. The design of intermediate and projector lenses is distinctly different from those corresponding to high quality objective or condenser lenses. In the projector, the aperture-dependent (axial) aberrations are negligible because the in-

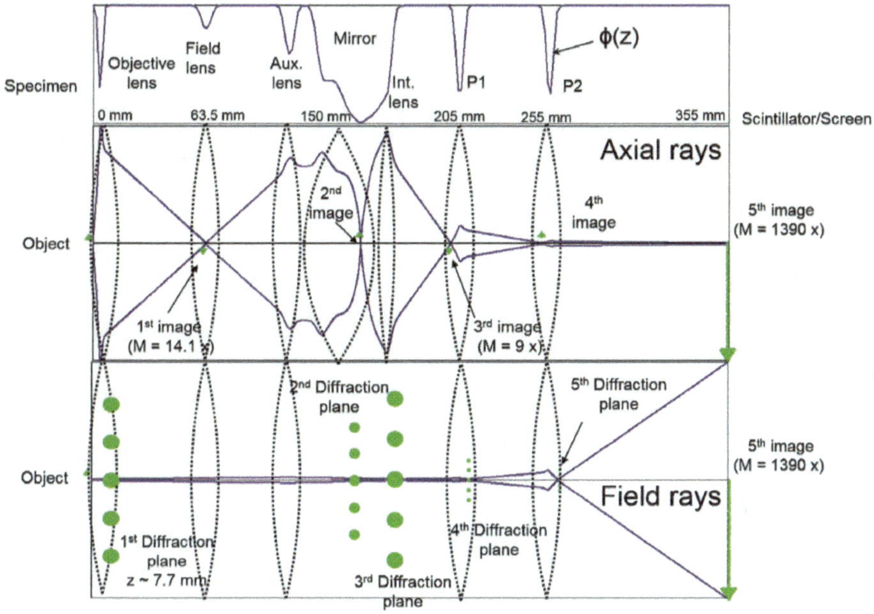

Figure 10 Axial distribution ((A), specimen location at z = 0 mm) and axial (B) and field rays (C) in the MPTEM projection optics for a net magnification of 1390×.

termediate images at the object plane of the projector lenses are already at considerable magnification, rendering the corresponding aperture angles small in comparison to the aperture angle present in the objective lens. Hence, there is no benefit in optimizing these lenses to reduce the spherical and chromatic aberration coefficients. On the other hand, projector lenses must admit relatively large image fields. Hence, the aberrations that dominate in these lenses are the so-called field aberrations, including distortion, coma, field curvature, and the chromatic aberration of magnification and rotation. A detailed layout of the projection optics including the exit part of the resonator optics is shown in Fig. 10, with the gated exit mirror electrode now biased for out-coupling of the electron beam. Starting the analysis from the specimen plane, the objective lens forms a first image of the specimen at the center of the field lens, and the mirror forms a second image near its gated electrode. The electrode following the shield electrode forms the intermediate lens, which focuses a third image in front of the first projector lens (P1). The combined focusing action of the objective, mirror and intermediate lens then results in a net magnification of approximately 9×. The strength of the projector lenses P1 and P2 can be varied

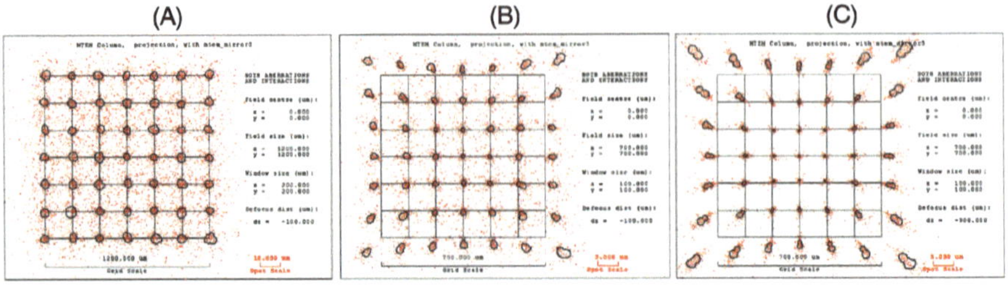

Figure 11 Optical blur introduced by the projection optics in the detector plane for three net electron-optical magnification settings and fields of view of 1, 3 and 10 μm and an object angle of 3 mrad at the specimen plane: (A) – 1390×, (B) – 246.5×, (C) – 79.8×.

to adjust the magnification between the specimen plane and the detector plane from approximately 50 to 5000×. The ray diagram shown in Fig. 10 illustrates a case with a net electron-optical magnification of 1390×. The electron image formed on the scintillator in the detection plane can then be further magnified by an optical zoom lens by up to 20× onto a CCD camera. The resulting spot diagrams generated by the projection optics at the detector plane of the MPTEM are shown in Fig. 11 for three magnification settings with an object angle of 3 mrad. The results show that a magnified image of the specimen can be transported through the mirror into the projection optics with high resolution for practical fields of view ranging from 1 to 10 μm. For the very high magnification case (Fig. 11A, 1390×) corresponding to a small field of view of 1 μm at the specimen, the blur is dominated by the axial aberrations of the objective lens: the field aberrations of the complete projection optics including the transfer through the exit mirror are negligible. The contour plots show an average 20-80% blur of less than 3.5 μm and a distortion of less than 1 μm at the corner of the field of view, at the detector plane, which corresponds to a blur of approximately 2.5 nm and a distortion of less than 1 nm at the specimen plane. For the high magnification case (Fig. 11B, 246.5×) corresponding to a medium field of view of 3 μm at the specimen, the blur is beginning to be affected by the field aberrations of the projection optics. Still, the deterioration in the spatial resolution across the field of view remains small. Near the axis, the contour plots show an average 20-80% blur of less than 0.7 μm, which corresponds to a blur of less than 3 nm at the specimen plane. At the corner of the field of view, field aberrations increase the average 20-80% blur to approximately 1.5 μm and introduce a distortion of approximately 4 μm, which corresponds to a blur of approximately 6 nm and a maximum distortion of 16 nm at the specimen plane. For the low magnification case (Fig. 11B, C, 79.8×) corresponding to a large field of view of 10 μm at the specimen, the blur is appreciably affected by the field aberrations of the projection optics. However, the deterioration in the spatial resolution across the field of view is still tolerable, as this image setting is meant for orientation of the specimen and for locating areas of interest. Near the axis, the contour plots show an average 20-80% blur of less than 0.5 μm, which corresponds to a blur of less than 6 nm at the specimen plane. At the corner of the field of view, field aberrations increase the average 20-80% blur to approximately 4 μm and introduce a distortion of approximately 10 μm, which corresponds to a tolerable blur of approximately 50 nm and distortion of 125 nm at the specimen plane.

3. Summary

The electron-optical concept of a novel multi-pass transmission electron microscope has been presented. In this imaging technique, the electron beam interacts elastically with the specimen multiple times so that the change in the phase accumulates before reaching the detector. The novel electron-optical column utilizes an electron-optical resonator formed by two electron mirrors sandwiching the sample and objective lenses. Detailed electron-optical analysis of the MPTEM and its key components were performed using state-of-the-art simulation software, yielding an optical design that is suitable for a prototype operating at a 10 keV beam energy. A gated electrostatic mirror and complementary illumination and projection optics has been proposed to achieve efficient in- and out-coupling of the electron beam. Simulations of the electron-optical properties of the MPTEM show that this microscope can achieve a spatial resolution of approximately 5 nm for a practical range of illumination and image magnifications.

Acknowledgments

This work was done as part of the Quantum Electron Microscope collaboration funded by the Gordon and Betty Moore foundation. Brannon Klopfer acknowledges support from the Stanford Graduate Fellowship. Thomas Juffmann acknowledges support from the ERC Micromoupe Grant 758752.

References

Agard, D., Cheng, Y., Glaeser, R. M., & Subramaniam, S. (2014). Chapter two – Single-particle cryo-electron microscopy (cryo-EM): Progress, challenges, and perspectives for further improvement. In *Advances in imaging and electron physics: Vol. 185* (pp. 113–137). Elsevier.

Delong Instruments (2019). http://www.delong.cz.

Giovannetti, V., Lloyd, S., & Maccone, L. (2006). Quantum metrology. *Physical Review Letters, 96*, 010401.

Giovannetti, V., Lloyd, S., & Maccone, L. (2011). Advances in quantum metrology. *Nature Photonics, 5*, 222–229.

Glaeser, R. M. (2015). How good can cryo-EM become? *Nature Methods, 13*, 28.

Juffmann, T., Klopfer, B. B., Frankort, T. L., Haslinger, P., & Kasevich, M. A. (2016). Multi-pass microscopy. *Nature Communications*, 7, 12858.

Juffmann, T., Koppell, S. A., Klopfer, B. B., Ophus, C., Glaeser, R. M., & Kasevich, M. A. (2017). Multi-pass transmission electron microscopy. *Scientific Reports*, 7(1), 1699.

Khoshouei, M., Radjainia, M., Baumeister, W., & Danev, R. (2017). Cryo-EM structure of haemoglobin at 3.2 A determined with the Volta phase plate. *Nature Communications, 8*, 16099.

Koppell, S. A., Mankos, M., Bowman, A. J., Israel, Y., Juffmann, T., Klopfer, B. B., & Kasevich, M. A. (2019). Design for a 10 keV multi-pass transmission electron microscope. *Ultramicroscopy*. Submitted for publication.

Kruit, P., Hobbs, R. G., Kim, C-S., Yang, Y., Manfrinato, V. R., Hammer, J., . . . Berggren, K. K. (2016). Designs for a quantum electron microscope. *Ultramicroscopy, 164*, 3–45.

Munro, E., Rouse, J., Liu, H., & Wang, L. (2008). Aberration analysis of electron mirrors by a differential algebraic method. *Optik, 119*(2), 90–96.

Ono, T., Okamoto, R., & Takeuchi, S. (2013). An entanglement-enhanced microscope. *Nature Communications, 4*.

Putnam, W. P., & Yanik, M. F. (2009). Noninvasive electron microscopy with interaction-free quantum measurements. *Physical Review A, 80*.

Reimer, L. (1984). *Transmission electron microscopy. Springer series in optical sciences: Vol. 36*. Springer-Verlag Berlin Heidelberg.

CHAPTER FIVE

A simulation program for electron mirrors using Boundary Element Method

Eric Munro*, Haoning Liu, Catherine Rouse, John Rouse
Munro's Electron Beam Software Ltd, London, United Kingdom
*Corresponding author. e-mail address: eric@mebs.co.uk

Contents

Advances in Imaging and Electron Physics, Volume 212
ISSN 1076-5670
https://doi.org/10.1016/bs.aiep.2019.08.006

1. Introduction

Electron mirrors have many very important applications in charged particle optics, including aberration correctors (Preikszas & Rose, 1997), time-of-flight mass spectrometers (Schury et al., 2014), monochromators (Mankos, Shadman, & Kolarik, 2016) and multi-pass mirror systems for reducing specimen damage in electron microscopes (Mankos et al., 2019). The design and optimization of electron mirrors requires very high accuracy computation of fields, trajectories, focal properties and aberrations. The simulations are harder than for electron lenses, because ray slopes are infinite at the reflection point, so the trajectories must be computed as functions of time instead of axial position. A new software package has been developed for designing electron mirrors using Boundary Element Method ("BEM"). (See Bannerjee & Butterfield, 1981, for a general description of BEM; or Read & Bowring, 2011, for applications specifically in charged particle optics.)

Our new BEM program computes the surface charge density generated on the electrodes by the applied voltages. The potential and fields at any point inside the mirror are then computed by numerical evaluation of elliptic integrals of the surface charge distribution. This program has several advantageous features: (1) Data input is very simple, because only the electrode surfaces need to be discretized, and not the intervening space; (2) Numerical accuracy can be quantified, by comparing computed potentials at auxiliary points on the electrodes with their known values; (3) The axial potential and its derivatives can be computed with great accuracy; (4) All the optical properties, including geometrical and chromatic aberrations, can be computed by a Differential Algebraic ("DA") Method (Berz, 1989; Wang, Tang, & Cheng, 2000); (5) Potentials and fields at off-axis points can also be computed with great accuracy, so the program could also be used for designing wide-angle lenses and mirrors, or "zonal" electron lenses and mirrors, which may have electrodes on the axis.

2. Important applications of electron mirrors

Electron mirrors have many important applications. For example:

2.1 Aberration correctors

This is probably the most important application of electron mirrors. As is well-known, all conventional rotationally-symmetric electron lenses have positive spherical and chromatic aberration (C_s and C_c) (Scherzer, 1936). This imposes fundamental limitations on the resolving power of conventional electron microscopes and electron beam lithography and inspection equipment for the semiconductor manufacturing industry. Scherzer (1947) also showed that electron mirrors can have negative C_s and C_c, so they can be used to correct these aberrations in electron lenses. A practical example is the Low Energy Electron Microscope ("LEEM") designed and built by Tromp et al. (2010), which uses an electron mirror to correct both C_s and C_c.

2.2 Multi-pass TOF mass spectrometers

A pair of mirrors can be incorporated in a Time-of-Flight ("TOF") mass spectrometer, to make the particles pass back and forth through the spectrometer several times, thereby enhancing its mass resolution capability (Schury et al., 2014).

2.3 Monochromators

A novel type of monochromator has been designed using an electron mirror (Mankos et al., 2016). It contains an electron mirror, a 90° prism, and a knife edge. Electrons of nominal energy are bent through 90° by the prism, enter the mirror along the mirror axis and are reflected back and through the output side of the prism. These electrons just miss the knife-edge, which is located near the axis between prism and mirror. Faster electrons are refracted less than 90° and are stopped by the knife-edge before reaching the mirror. Slower electrons are refracted more than 90° and after being reflected by the mirror, they are stopped by the knife-edge on their way back to the prism. Only electrons close to the nominal energy emerge from the monochromator.

2.4 Multi-pass mirrors for reducing specimen damage in electron microscopes

This type of electron microscope (Mankos et al., 2019) uses two electron mirrors, one on each side of the specimen, and the electrons pass back and forth through the sample several times. This improves signal-to-noise ratio and reduces specimen damage, and looks promising for high-resolution imaging of complex molecules in TEM.

3. Simulation methods for electron mirrors

Simulation methods for electron mirrors are quite similar to those for electron lenses:

First, the electrostatic potential distribution $\Phi(z, r)$ is computed, using either a partial differential equation method such as Finite Difference Methods (e.g. Forsythe & Wasow, 2013), or Finite Element Method (e.g. Munro, 1973), or an integral equation method such as Boundary Element Method (e.g. Read & Bowring, 2011). From these calculations we obtain the axial potential distribution $\varphi(z)$.

Second, the optical properties (paraxial rays, focusing properties and aberrations) are computed by expanding the axial potential $\varphi(z)$ off-axis in a power series of the off-axis distance r, and computing the paraxial rays and aberrations by successive approximations in increasing powers of r.

The main difficulty in simulating optical properties of electron mirrors arises because the ray slopes become infinite at the reflection point. In electron lens simulations, trajectories are generally computed using axial coordinate z as independent variable, and the dependent variables are radial off-axis distance $r(z)$ and radial slope $r'(z)$.

Unfortunately, this method is unsuitable for ray-tracing in electron mirrors, because the ray slope $r'(z)$ becomes infinite at the reflection point. To overcome this difficulty, we must use the time t as the independent variable, and compute the electron's position $(z(t), r(t))$ and velocity components $(\dot{z}(t), \dot{r}(t))$ as functions of t.

To achieve this, Rose and Preikzsas (1995) developed a "time-dependent perturbation formalism", with t as independent variable. This method proved very successful for computing the primary and secondary aberrations of electron mirrors.

Our group at MEBS Ltd subsequently developed an analogous method for simulating the optics of electron mirrors, up to high order aberrations, using time t as independent variable. However, we computed the aberrations of electron mirrors directly, using the Differential Algebra ("DA") Method, a powerful technique originally pioneered and developed mainly by Berz (1989). This method obviates the need for complicated aberration formulae. We published a description of the application of this method to the aberration analysis of electron mirrors (Munro, Wang, Rouse, & Liu, 2006; Wang, Rouse, Munro, Liu, & Zhu, 2008). We incorporated this method into a MEBS Software Package, "MIRROR_DA".

4. Advantages of our new BEM software

The main modification we have made in this new version of our mirror simulation software is to compute the potential distribution using the Boundary Element Method ("BEM") instead of the Finite Element Method. The Boundary Element Method has several advantages in this application, including:

- Data input is very simple.
- The accuracy of the computed potential distribution can be quantified.
- Axial potential and its derivatives can be computed with great accuracy.
- Optical properties can then be computed very accurately by DA Method.
- Fields at off-axis points can also be computed accurately, which allows accurate simulation of "zonal" mirrors or lenses, with electrodes on the axis.

5. Theory of our new BEM software

5.1 The basic principle of the Boundary Element Method

The basic idea of BEM is to assign a set of n electrostatic surface charge densities σ_i $(i = 1, n)$ to points on the electrode surfaces, and set up a matrix equation relating these surface charge densities σ_i to the known voltages V_i on the electrodes. This gives a matrix equation of the form

$$\left[A_{ij}\right][\sigma_i] = [V_i] \quad \text{for } i = 1, \ldots, n; \; j = 1, \ldots, n \tag{1}$$

where $\left[A_{ij}\right]$ is an $n \times n$ matrix whose coefficients represent the relationship between the unknown surface charge densities σ_i and the known voltages V_i on the electrodes.

So if we can compute the matrix elements A_{ij}, we can then solve the matrix equation (1) for the surface charge densities σ_i. We can then use the known values of σ_i to compute the potential at any point in space, and in particular we can calculate the axial potential distribution.

By considering Eq. (1), we can understand how to compute the matrix elements A_{ij}. Line 1 of the matrix equation (1) states that

$$\sum_{j=1}^{j=n} A_{ij}\sigma_j = V_i \tag{2}$$

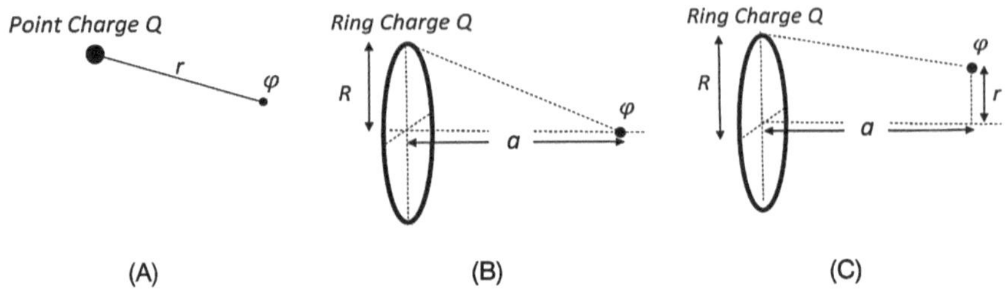

Figure 1 Illustrating formulae for the potential generated by point and ring charges. (A) Potential φ for point charge Q, at distance r from charge Q. (B) Potential φ at distance a along the axis of a ring with charge Q. (C) Potential φ at axial distance a and radius r off the axis of a ring with charge Q.

Now if we set $\sigma_j = 1$ and all the other values of $\sigma = 0$, then Eq. (1) becomes simply

$$A_{ij} = V_i \tag{3}$$

Therefore, to compute A_{ij}, we simply set the surface charge density at sample point j to $\sigma_j = 1$, and set the values of σ at all the other sample points to zero, and compute the resulting potential V_i at sample point i.

5.2 Formulae for the potential generated by a point charge and a ring charge

Some basic formulae for the potentials generated by point charges and ring charges are illustrated in Fig. 1.

The starting point for the theory is the well-known formula for the potential φ due a point charge Q, at distance r from the charge (see Fig. 1A):

$$\varphi = \frac{Q}{4\pi\varepsilon_0} \cdot \frac{1}{r} \tag{4}$$

where ε_0 is the permittivity of free space. This formula is easily generalized for the potential on the axis of a circular ring charge with a total charge Q and ring radius R, at distance z along the axis, see Fig. 1B:

$$\varphi = \frac{Q}{4\pi\varepsilon_0} \cdot \frac{1}{\sqrt{a^2 + R^2}} \tag{5}$$

For the general case of a ring with total charge Q, at axial distance a from the ring and radius distance r off the ring axis, as in Fig. 1, the potential φ

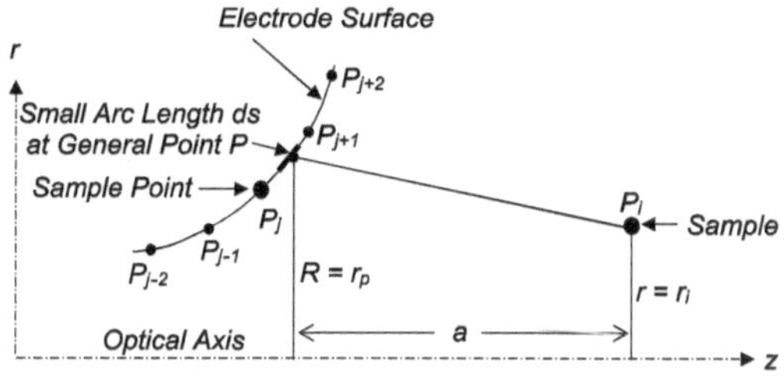

Figure 2 Illustrating computation of the coefficients A_{ij} of the BEM Equations.

is given by

$$\varphi(z, r) = \frac{Q}{4\pi\varepsilon_0} \cdot \frac{2}{\pi} \cdot \frac{K(k)}{\sqrt{q}} \tag{6}$$

where

$$q = a^2 + (r + R)^2 \quad k = \sqrt{4rR/q} \tag{7}$$

and

$$K(k) = \int_0^{\pi/2} \frac{d\phi}{\sqrt{1 - k^2 \sin^2 \phi}} \tag{8}$$

Eqs. (6)-(8) are derived by several authors, e.g. Mandre (2007).

The function $K(k)$ in Eq. (8) is called the "*complete elliptic integral of the first kind*". This is a standard transcendental mathematical function, described in textbooks on higher mathematics (e.g. Boas, 1983). This function can be evaluated with standard numerical algorithms (e.g. Press, Teukolsky, Vetterling, & Flannery, 2007), so the elliptic integral $K(k)$ can be treated like an elementary function such as $\sin\theta$ or e^x, giving exact values to the rounding error of the computer.

5.3 Computing the coefficients A_{ij} of the BEM equations

As explained in Section 5.1, the coefficient A_{ij} represents the potential V_i generated at sample point i by unit surface charge density σ_j applied at mesh-point j, with all other surface charge densities set to zero ($\sigma = 0$). We consider this case as shown in Fig. 2.

As will be explained in the section on Data Specification, the electrode surfaces can be defined either by straight lines or circular arcs. In Fig. 2

we have illustrated the case of an electrode surface with a circular arc. In general, in the Data Specification, we automatically arrange the layout of sample points in groups of 5 points to allow for a smooth variation of surface charge density. For example, on the curved electrode surface shown in Fig. 2, we have shown 5 adjacent charge density computation points $(P_{j-2}, P_{j-1}, P_j, P_{j+1}, P_{j+2})$. The sample point P_i, where the potential V_i is to be calculated, can be either on the same electrode as the sample point P_j, or it can be on a different electrode, as shown in Fig. 2. The program can handle both cases.

In Fig. 2, we have shown an elemental arc length ds in the (z, r) plane on the electrode surface, at the point P, at radius $R = r_p$. This generates an elemental surface area

$$dA_p = 2\pi r_p ds \tag{9}$$

with a surface charge density

$$\sigma_p \tag{10}$$

This gives an elemental ring charge

$$dQ_p = \sigma_p dA_p = \sigma_p . 2\pi r_p ds \tag{11}$$

By analogy with Eqs. (6)-(8), we can now write the following formulae for the elemental potential dV_i at the sample point i generated by the ring charge dQ:

$$dV_i = \frac{\sigma_p . r_p ds}{\pi \varepsilon_0} . \frac{K(k)}{\sqrt{q}} \tag{12}$$

where

$$q = (z_i - z_p)^2 + (r_i + r_p)^2 \tag{13}$$

$$k = \sqrt{4 r_i r_p / q} \tag{14}$$

$$K(k) = \int_0^{\pi/2} \frac{d\phi}{\sqrt{1 - k^2 \sin^2 \phi}} \tag{15}$$

These formulae must now be integrated over the 5 sample points P_{j-2} to P_{j+2}. In these numerical integrations, we use 5-point Lagrange multipliers (Rockafellar, 1993) to set the surface charge density to 1 at the sample point being evaluated, and to 0 at each of the other 4 sample points. The numerical integrations are performed using 32-point Gauss-Legendre quadrature formulae (Abramowitz & Stegun, 1972, page 917).

5.4 Solving the BEM equations

After computing the coefficients A_{ij} of the BEM matrix equation (1),

$$\left[A_{ij}\right][\sigma_i] = [V_i] \quad \text{for } i = 1, \ldots, n;\ j = 1, \ldots, n \tag{16}$$

this equation can be solved to obtain the surface charge density vector $[\sigma_i]$ for an given voltage vector $[V_i]$ on the electrodes. This equation can readily be solved by Gaussian elimination and backward substitution, using any standard algorithm (Press et al., 2007). In general, the mirror will have several electrodes (typically about 4), whose voltages can be set independently, so we need to solve Eq. (16) for several "unit potential distributions on the electrodes", i.e. with 1 Volt applied to each electrode in turn, with the other electrodes grounded. This can be done most efficiently by doing the Gaussian elimination of the matrix $\left[A_{ij}\right]$ once only, and then doing the backward substitution on each unit potential distribution on the electrodes sequentially. This algorithm can easily solve 1000 simultaneous equations or more on a modern laptop computer in a matter of seconds, regardless of the number of electrodes in the mirror.

5.5 Computing the unit axial potential distributions

Once the surface charge vectors $[\sigma_i]$ at each Lagrange integration point have been computed, the corresponding vector of ring charges $\left[Q_i\right]$ can be obtained, and these can be used to compute the unit axial potential distributions, using the simple formula in Eq. (5), summed up over all the ring charges, at each required point on the optical axis.

6. Data specification

A useful feature of our Boundary Element Program is that the Data Specification is quick and simple, because we only have to specify the shapes of the electrode surfaces themselves and discretize the surfaces into "Boundary Elements". This is much simpler than with our Finite Element Software, in which we also have to define a 2D Finite Element Mesh in the space between the electrodes.

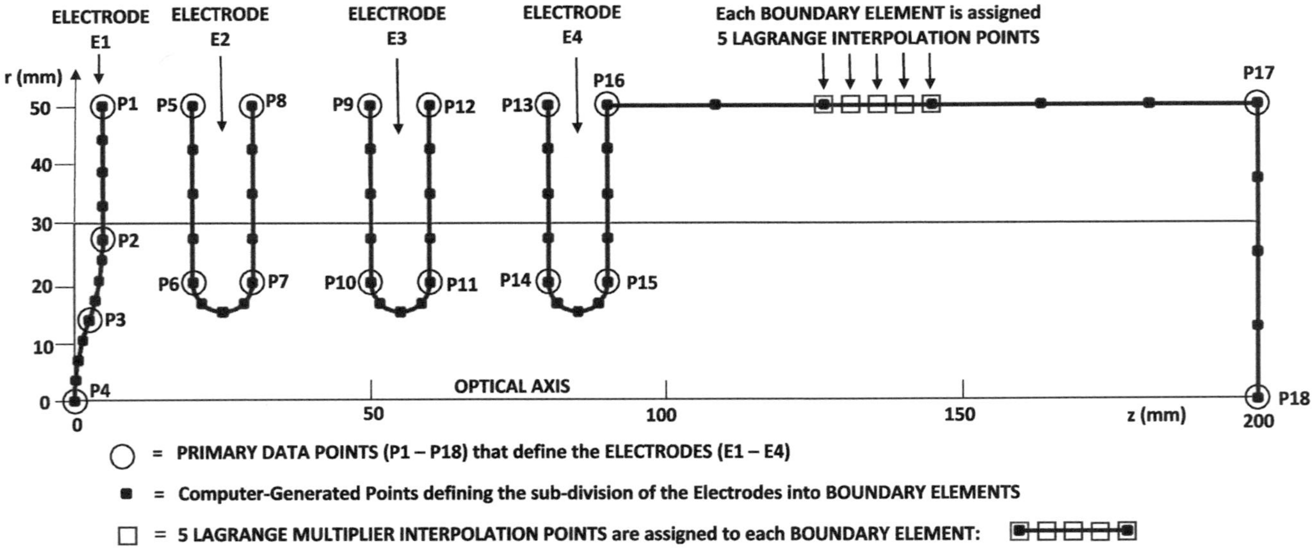

Figure 3 The electron mirror corresponding to the Data Specification in Table 1.

Table 1 Data specification for simulating the electron mirror in Fig. 3.

Electrode number	Point	z (mm)	r (mm)	Radius of curvature (mm)	Number of boundary elements
E1	P1	4.8246	50	–	–
	P2	4.8246	27.3616	0	4
	P3	2.4123	13.6808	−40	4
	P4	0	0	40	4
E2	P5	20	50	–	–
	P6	20	20	0	4
	P7	30	20	5	4
	P8	30	50	0	4
E3	P9	50	50	–	–
	P10	50	20	0	4
	P11	60–	20	5	4
	P12	60	50	0	4
E4	P13	80	50	–	–
	P14	80	20	0	4
	P15	90	20	5	4
	P16	90	50	0	4
	P17	200	50	0	6
	P18	200	0	0	4

As an example, Fig. 3 shows a typical 4-electrode mirror, and Table 1 shows the complete Data Specification for simulating the potential distribution in this mirror.

Fig. 3 shows the mirror cross-section in the (z, r) plane, with the optical axis (z-axis) horizontally at the bottom, and the radial axis (r-axis) vertically on the left. Electrons enter the mirror from the right-hand side, and travel towards the Reflection Electrode E1 at the left-hand side and are reflected close to the surface of E1. Electrodes E2, E3 and E4 are focusing electrodes. The right-hand boundary of E4 should be chosen far enough away for the stray field on the axis to be negligible.

Having 4 electrodes allows several degrees of freedom in adjusting the operating conditions, and the aim is to adjust the electrode voltages to create a mirror with negative spherical and chromatic aberration (C_s and C_c), which can cancel the C_s and C_c of a conventional round lens. Adjusting the curvature and shape of the Reflection Electrode E1 in the design can also play an important role in correcting the aberrations.

The Data Specification rules are simple, as can be seen from Fig. 3 and Table 1:

1. The electrode positions are defined by the **PRIMARY DATA POINTS (P1-P18)**, with their (z, r) coordinates listed in Table 1.
2. Each electrode is divided into **SEGMENTS**, with each **SEGMENT** separated by two consecutive **PRIMARY DATA POINTS**.
3. Each **SEGMENT** can be either a straight line or a circular arc. For circular arcs, the radius of curvature is specified. For straight lines, the curvature is specified as zero.
4. Each **SEGMENT** is divided into a number of **BOUNDARY ELEMENTS**. The number of **BOUNDARY ELEMENTS** in each segment is listed in the final column of Table 1.
5. Finally, each **BOUNDARY ELEMENT** is assigned 5 equally-spaced **LAGRANGE INTEGRATION POINTS** (see Fig. 3).

The **LAGRANGE INTEGRATION POINTS** form the points where the Surface Charge Density σ_i is computed in the numerical computation. In this example, the total number of Lagrange Integration Points is 236, requiring the solution of 236 simultaneous equations to compute the Surface Charge Density Distribution.

7. A worked example

As a typical example, we have simulated the properties of a 4-electrode mirror, using the mirror design described in Section 6, with reference to Fig. 3 and Table 1.

7.1 Computation of potential distribution and unit axial potential functions

With this simple data, we ran our new BEM Software and computed the potential distribution $\Phi(z, r)$ in the (z, r) plane, with the following voltages applied to the 4 electrodes simultaneously: $V_1 = 1$ Volt, $V_2 = 2$ Volts, $V_3 = 3$ Volts, and $V_4 = 4$ Volts. The computed equipotentials are plotted in Fig. 4A.

We also computed the unit axial potential functions $\varphi(z)$, for 1 Volt applied to each of the 4 electrodes separately, and the results for this are plotted in Fig. 4B.

For comparison, Fig. 5 shows the corresponding results obtained using our MEBS "SOFEM" Software, which uses Second Order Finite Element Method (Zhu & Munro, 1995). It can be seen that the equipotentials and unit axial potential functions computed with our new BEM Software agree almost exactly with the results of our SOFEM Software.

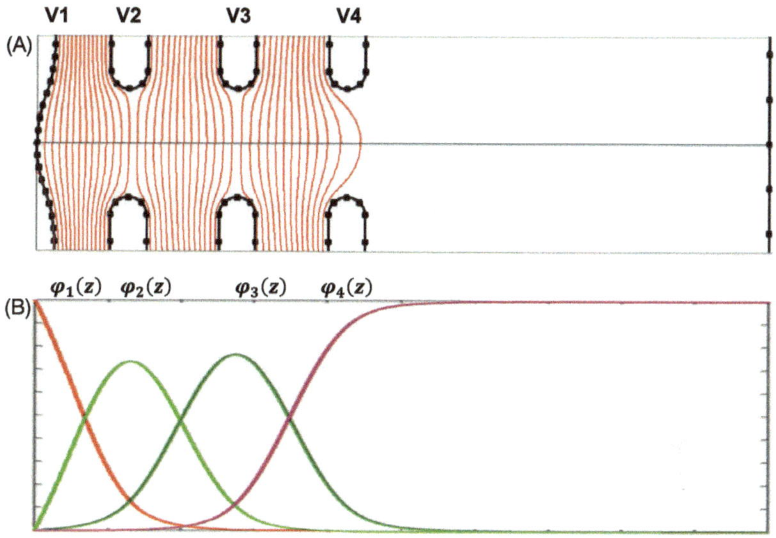

Figure 4 Results computed with new BEM Software. (A) BEM equipotentials. (B) BEM unit axial potentials.

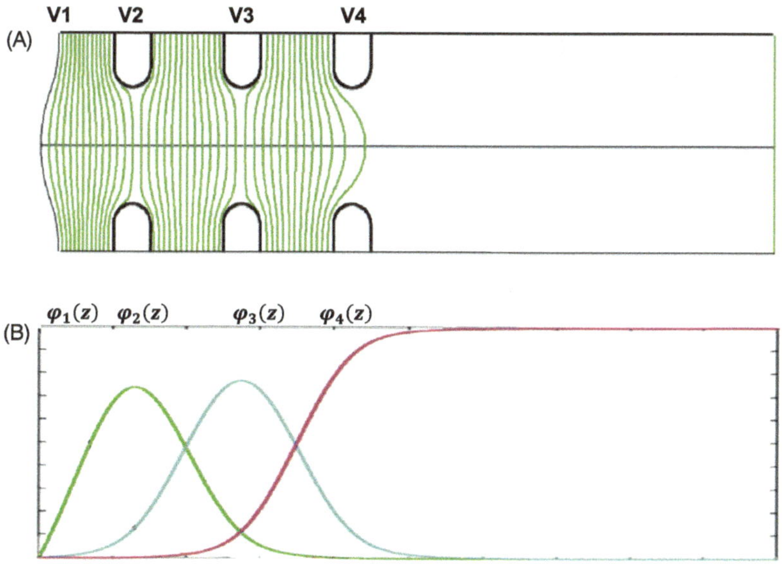

Figure 5 Results computed with SOFEM Software. (A) SOFEM equipotentials. (B) SOFEM unit axial potentials.

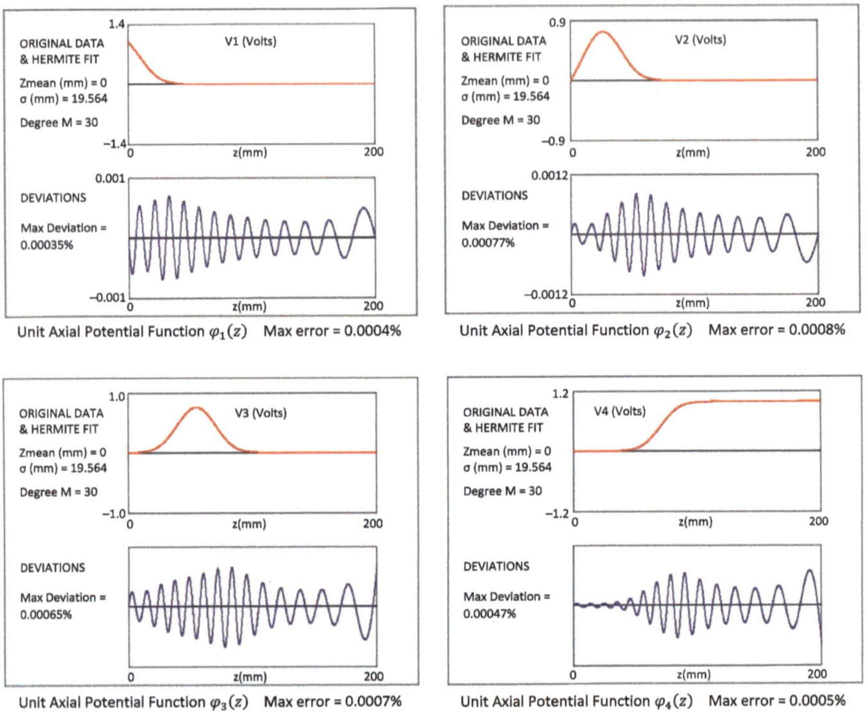

Figure 6 Fitting the Unit Axial Potential functions with series of 30 Hermite Functions. Maximum fitting error = 0.0005%.

7.2 Computation of the optical properties

After computing the Unit Axial Potential Functions with the BEM Software (Fig. 4B), we computed the optical properties of the mirror, using our MEBS MIRROR_DA Software (Munro et al., 2006; Wang et al., 2008).

The first step is to fit each Unit Axial Potential Function in Fig. 4B with a series of Hermite Functions (Boas, 1983). This is illustrated in Fig. 6. We used Hermite Series Fits with 30 terms, which gave analytics fits, accurate to within 0.0005% of the unit axial potential functions generated by our BEM Software.

These Hermite Series Fits were then used as the axial functions required for computing the optical properties of the mirror. The optical properties were computed by the Differential Algebra ("DA") Method, using the principles pioneered by Berz (1989). The results are summarized in Fig. 7 and Table 2.

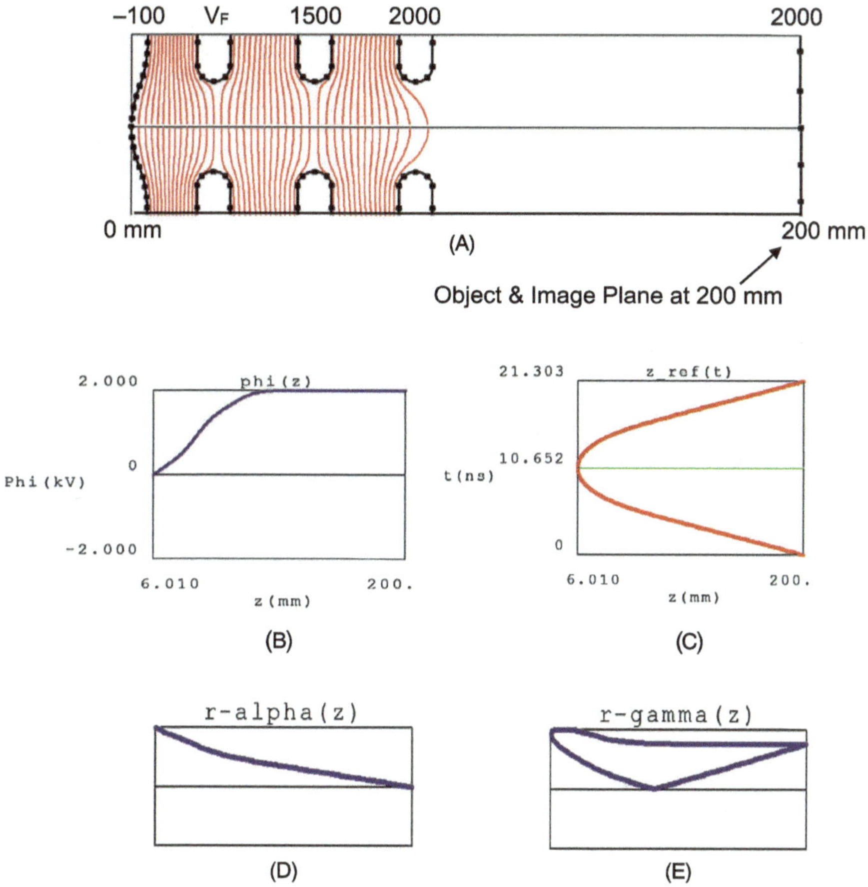

Figure 7 Mirror optics by Differential Algebra with BEM Software. (A) Electron mirror. (B) Axial potential. (C) Reference particle vs time. (D) Focusing ray. (E) Field ray.

The 4 electrodes of the mirror were set to the following potentials:

$$V_1 = -100 \text{ Volts}, \quad V_2 = V_f, \quad V_3 = 1500 \text{ Volts}, \quad V_4 = 2000 \text{ Volts} \tag{17}$$

The mirror was used in unity magnification mode, with object and image planes at the same location, at $z_o = z_i = 200$ mm. The electrons enter the mirror traveling in the negative z direction with initial kinetic energy of 2000 eV, towards the reflection electrode, which is biased to $V_1 = -100$ Volts. The Focus Electrode has a variable focus voltage, $V_2 = V_f$, and Electrode 3 has a fixed voltage $V_3 = 1500$ Volts.

Table 2 Comparison of mirror properties computed with BEM and SOFEM.

Optical property	BEM	SOFEM
Focus voltage V_F (Volts)	333.629	335.98
C_s (m)	−72.323	−69.201
C_c (m)	−6.132	−6.198

In the MIRROR_DA Software, the focus voltage V_F is automatically adjusted to focus the return beam at the specified image plane ($z_i = 200$ mm in this example).

Fig. 7A shows the electrode potentials used in the computation, and the program automatically adjusted the focus voltage to $V_F = 333.629$ Volts. Fig. 7B shows the computed axial potential function, with reflection point at $z_r = 6.01$ mm in front of the reflection electrode. Fig. 7C shows the reference particle position $z(t)$ along the axis, as a function of the time t; the total transit time from object plane to reflection point and back to the image plane is $t = 21.303$ ns. The paraxial focusing ray r_α and field ray r_γ are also plotted in Figs. 7D and 7E respectively.

Table 2 shows the focusing conditions and optical properties computed by MIRROR_DA, using Differential Algebra Method, starting from the Unit Axial Potential Functions computed with our new BEM Software. We have listed computed values of:

- V_f (Focus Voltage)
- C_s (Primary Spherical Aberration Coefficient)
- C_c (Primary Chromatic Aberration Coefficient)

In the final column of Table 2, we have listed the computed values of the same optical parameters, computed with our MIRROR_DA Software, starting from the Unit Axial Potential Functions computed with our SOFEM Software.

By comparing the two columns of results in Table 2, we can conclude that:

1. The optical properties of electron mirrors, computed using our new BEM Software, yield results very similar to those of our older SOFEM Software. This gives us confidence that the new BEM Software is working correctly.
2. In this electron mirror, the primary spherical and chromatic aberration coefficients, C_s and C_c, are both **negative**. This type of mirror should therefore be able to eliminate the positive C_s and C_c (Scherzer, 1936) of conventional rotationally symmetric electron lenses.

8. Plans for future work

Plans for our future work on our new BEM Software for Mirrors include:

8.1 Computing higher order spherical and chromatic aberration coefficients

Since our program uses Differential Algebraic ("DA") Method, we can also compute the higher order aberrations of electron mirrors, for example the 5th order spherical aberration C_{s5} and the higher-order chromatic aberration coefficients, e.g. C_{c3} and also chromatic terms of higher rank (involving different powers of the energy spread ΔV). To correct these higher order terms, we would like several degrees of freedom, e.g. perhaps using additional electrodes and varying the shape of the reflection electrode.

8.2 Simulation of multi-pass mirror systems

Our new BEM Software can already be used to calculate the optics of Multi-Pass Mirror Systems, such as Multi-Pass Time-Of-Flight Mass Spectrometers (Schury et al., 2014), and Multi-Pass Mirrors for reducing specimen damage in Transmission Electron Microscopes (Mankos et al., 2019). The main modification for this purpose has been achieved by adding the ability to apply pulses to the strength of the input and output mirrors as a function of time, to inject and recover electrons after a selected number of passes between the mirrors.

Our BEM Software can simulate multiple passes between a pair of mirrors, which can each be "Gated", i.e. switched between acting as a LENS and acting as a MIRROR, simply by switching the potentials on the relevant electrodes. Figs. 8–13 illustrate the operation of this "Gated Mirror" capability.

Fig. 8 shows a simple arrangement with two electrostatic "Einzel Elements", each containing three electrodes, with the two outer electrodes set at a potential V_O and the central electrode set at a potential V_C. At high settings of the central potential V_C, the Einzel Element acts as a Lens. On the other hand, at low settings of the potential V_C, the Einzel Element can act as a Mirror. In this example, the Einzel Elements are centered at $z_1 = -100$ mm and $z_2 = +100$ mm respectively.

Fig. 9 shows both Elements with V_C set so that they both act as Conventional Lenses, with an object plane at $z_o = -200$ mm and an image plane $z_i = +200$ mm respectively, with a parallel beam between the two Lenses.

System

Place two such lenses in a system

lenses at z=-100 mm and z=+100 mm
Object at z=-200 mm
Image at z=+200 mm

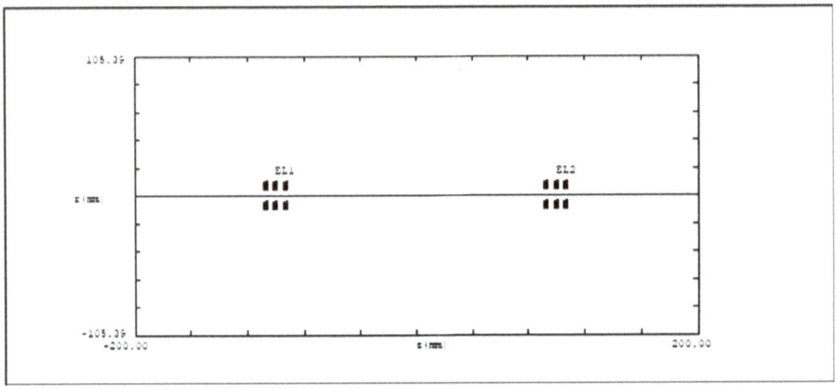

Figure 8 Two electrostatic "Einzel Elements" that can act as either Lenses or Mirrors.

Straight-through case

Adjust central voltage V2 in each lens symmetrically

Parallel beam in between lenses

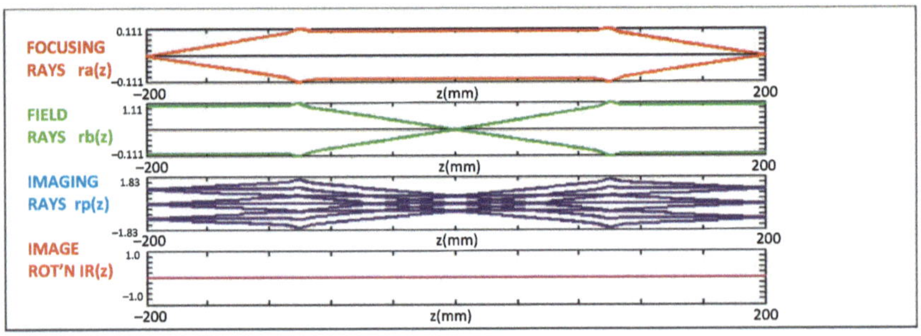

	Lens1	Lens2
V1	+1000	+1000
V2	+421	+421
V3	+1000	+1000

Cs = 108 m
Cc = 0.5 m

Positive Cs and Cc, as expected for regular round lenses

Figure 9 Both Element 1 and Element 2 acting as Lenses, with positive C_S and C_C.

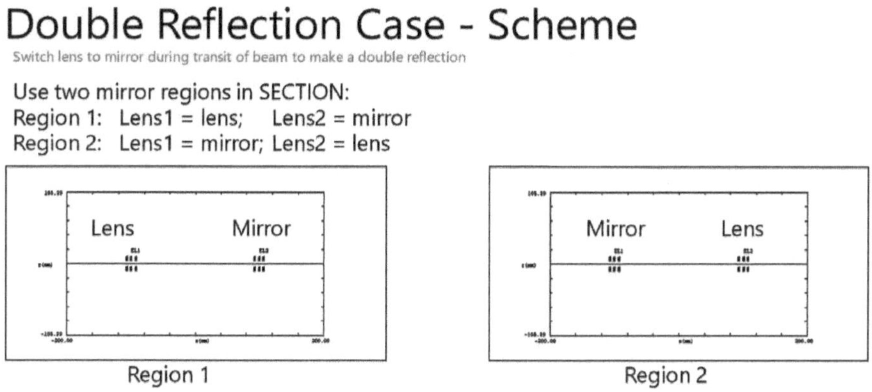

Figure 10 Scheme for a Multi-Pass Mirror System. The two "Einzel Elements" can be set as a lens followed by a Mirror (Region 1) or as a Mirror followed by a Lens (Region 2).

In this example, the beam passes straight through the system with no reflection and C_s and C_c are both positive as is always the case for ordinary Electron Lenses.

Fig. 10 shows a scheme for a Multi-Pass Mirror System. The two "Einzel Elements" can be switched to act as a Lens Followed by a Mirror (Region 1), or as a Mirror followed by a Lens (Region 2).

We then energize the system as a double mirror. This is done in two steps: An Injection Step (collimation-reflection-focusing); and an Ejection Step (reflection-collimation-focusing).

For the Injection Step, Fig. 11 shows Element 1 with its control voltage V_{C1} set to act as a Lens and Element 2 with its control voltage V_{C2} set to act as a Mirror. In this case, a small object at z_o will be collimated to a parallel beam by Element 1 (acting as a Lens) and will be reflected by Element 2 (acting as a Mirror), to form a unity magnification image at $z_i = 0$ mm, midway between Element 1 and Element 2. Fig. 11 shows the paraxial ray paths for this situation.

For the Ejection Step, Fig. 12 shows how, with Element 1 set to act as a Mirror, Element 2 can now be switched to act as a Lens, which then releases the electrons to form a unity magnification image on the exit side of the system, at the image plane z_i.

The software computes the path of the electron beam, from object to final image plane, including the reflections and specified image planes and shows the paths of the principal paraxial rays along an unfolded axis (Fig. 13). As can be seen from the raytrace diagrams, the software predicts

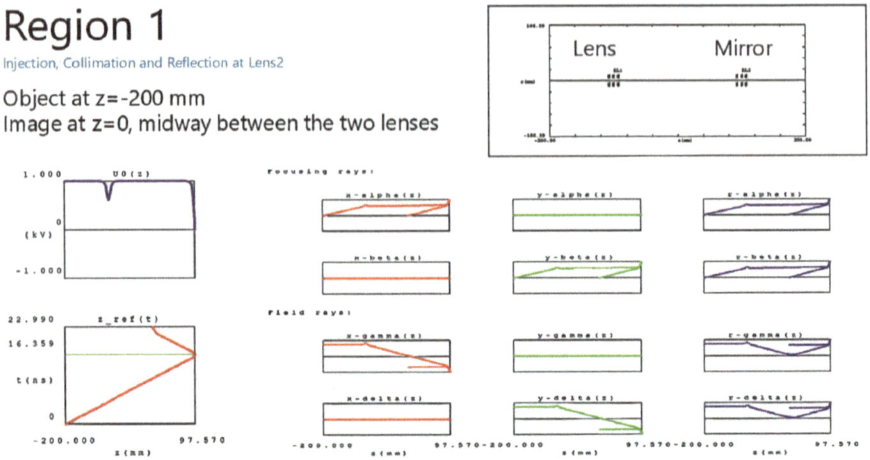

Region 1	Lens1	Lens2
V1	+1000	+1000
V2	+421	-801
V3	+1000	+1000

Figure 11 Injection Step: Element 1 acts as a Lens and Element 2 acts as a Mirror.

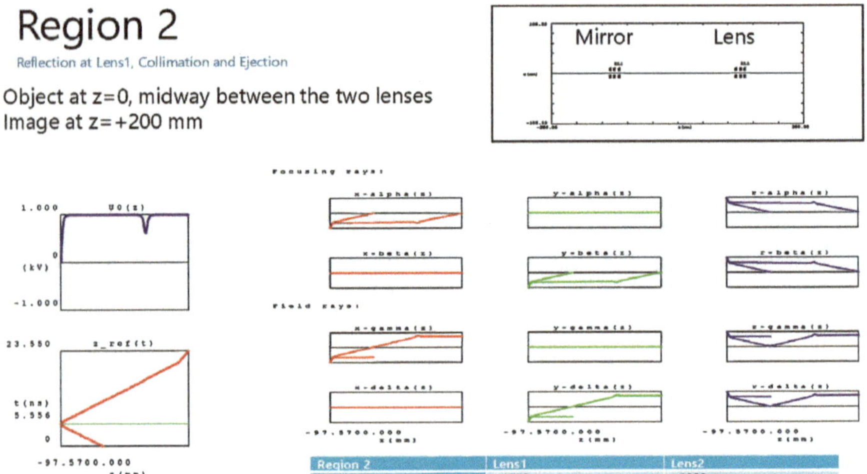

Region 2	Lens1	Lens2
V1	+1000	+1000
V2	-801	+421
V3	+1000	+1000

Figure 12 Ejection Step: Element 1 acts as a Mirror and Element 2 acts as a Lens to eject the beam.

the required voltages to achieve unity magnification focusing and computes the primary C_s and C_c values, which are now both negative.

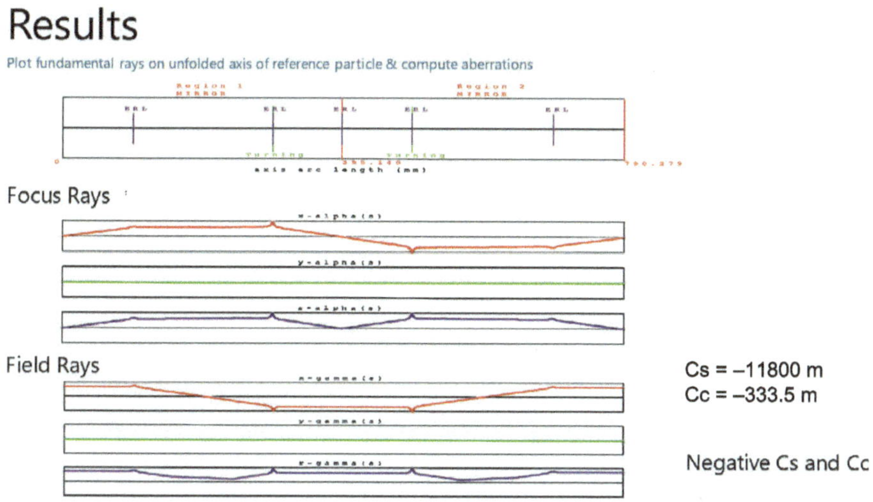

Figure 13 Plot of Fundamental Paraxial Rays on an unfolded axis of the reference particle, including computed values of the primary C_S and C_C, which are now both negative.

By suitable tuning of the system using the software, it should be possible to design a Double-Mirror System to cancel the primary aberrations of a round-lens system.

In addition, once the primary aberrations have been zeroed, the software can also predict the secondary axial aberrations of the system (fifth-order spherical aberration and higher rank axial chromatic aberrations), as these will then become the performance-limiting ideal aberrations.

If more than one reflection pass is required (as required, for example, by the Multi-Pass Electron Microscope of Mankos et al., 2019), then it is a simple matter to add pairs of lenses, both acting as mirrors, into the system specification between the injection and ejection sections already shown.

8.3 Adding direct ray-tracing capabilities in the BEM software

A key advantage of our new BEM Software is that the potential and electric field components due to any set of ring charges should obey Laplace's equation exactly, to the rounding error accuracy of the computer, because the formulae are all based ultimately on the simple formula in Eq. (6) for the potential generated by a rotationally symmetric ring charge, which gives a physically accurate potential distribution that is an exact solution of Laplace's equation. This makes the BEM Software a good candidate for highly accurate ray-tracing at large off-axis distances. The analytic formulae

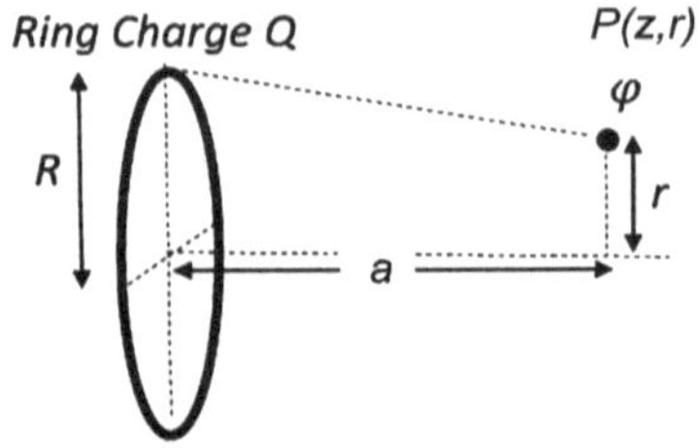

Figure 14 Illustrating formulae for field components E_z and E_r generated at the point $P(z, r)$ by a ring charge.

for the electrostatic field components, $E_z(z, r)$ and $E_r(z, r)$, generated by a ring charge, at a general off-axis point, $P(z, r)$, are illustrated in Fig. 14.

The field components $E_z(z, r)$ and $E_r(z, r)$ are obtained by differentiating the formulae for the potential φ in Eqs. (6)-(8), and the resulting formulae (Mandre, 2007) are:

$$E_z(z, r) = \frac{Q}{4\pi\varepsilon_0} \cdot \frac{2}{\pi} \cdot \frac{1}{q^{3/2}(1-k^2)} . a . E(k) \tag{18}$$

$$E_r(z, r) = \frac{Q}{4\pi\varepsilon_0} \cdot \frac{2}{\pi} \cdot \frac{1}{q^{3/2}(1-k^2)k^2} . \{2RK(k).(1-k^2) - E(k).[2R - k^2(r+R)]\} \tag{19}$$

where

$$q = a^2 + (r+R)^2 \quad k = \sqrt{4rR/q} \tag{20}$$

$$K(k) = \int_0^{\pi/2} \frac{d\phi}{\sqrt{1 - k^2 \sin^2\phi}} \quad (= \text{Complete elliptic integral of first kind}) \tag{21}$$

$$E(k) = \int_0^{\pi/2} \sqrt{1 - k^2 \sin^2\phi}\, d\phi \quad (= \text{Complete elliptic integral of second kind}) \tag{22}$$

These formulae yield results for the field components which are accurate to the rounding error precision of the computer.

Having developed the capability to compute E_z and E_r extremely accurately, we can then write a very accurate direct ray-tracing algorithm, using a high-order Runge-Kutta algorithm with adaptive step-size control.

8.4 Application of the BEM program to design wide-angle imaging systems

Novel wide-angle rotationally symmetric electrostatic lenses have been proposed with a coaxial construction, with the inner electrode being a solid cylindrical "core electrode" on the optical axis, with the electrons traveling in a hollow beam surrounding this core electrode (e.g. Gabor, 1946; Dungey & Hull, 1947). This type of lens could be simulated with our new BEM Software, and might provide alternative designs for correcting C_s in conical hollow beams.

8.5 Use of BEM software for simulating point cathode TFE electron sources

Simulation of point cathode TFE Electron Sources using our SOFEM software is time-consuming, because of the complexity in generating a Finite Element Mesh near the surface of the sub-micron radius cathode. Since BEM only requires the discretization of the electrode surfaces, our new software will be very useful in this application.

9. Summary

A new Boundary Element Software has been developed for the simulation of electron mirrors. The computed results agree well with those of our existing Finite Element Software. The new program offers much simpler data input, which will make it easier to use. As summarized in Section 8, we have already enhanced the software to handle "Multi-Pass Mirror Systems". We have also described several further improvements which we plan to implement soon to enhance the capabilities of this new software.

Acknowledgments

We gratefully acknowledge helpful discussions with Prof. Hermann Wollnik on Multi-Pass TOF Spectrometers and with Dr. Marian Mankos on several novel applications of electron mirrors, including the use of Multi-Pass Mirrors in Electron Microscopes.

References

Abramowitz, M., & Stegun, I. (1972). *Handbook of mathematical functions*. New York: Dover Publications.

Bannerjee, P. K., & Butterfield, R. (1981). *Boundary element methods in engineering science*. UK: McGraw-Hill.

Berz, M. (1989). Differential algebra. *Particle Accelerators*, *24*, 109–124.

Boas, M. (1983). *Mathematical methods in the physical sciences* (2nd edition) (pp. 474–477, 530–531). John Wiley and Sons.

Dungey, J. W., & Hull, C. R. (1947). Coaxial electron lenses. *Proceedings of the Physical Society of London. Series B*, *59*, 828–843.

Forsythe, G. E., & Wasow, W. R. (2013). *Finite difference methods for partial differential equations: Applied mathematics series paperback*.

Gabor, D. (1946). A zonally corrected electron lens. *Nature*, *158*, 198.

Mandre, I. (2007). Off-axis electric field of a ring of charge. http://www.mare.ee/indrek.

Mankos, M., Koppell, S. A., Klopfer, B. B., Juffmann, T., Kolarik, V., Shadman, K., & Kasevich, M. (2019). Electron optics for a multi-pass transmission electron microscope. In *Advances in imaging and electron physics: Vol. 212* (CPO-10 proceedings volume). In press.

Mankos, M., Shadman, K., & Kolarik, V. (2016). Novel electron monochromator for high resolution imaging and spectroscopy. *Journal of Vacuum Science & Technology. B*, *34*.

Munro, E. (1973). Computer-aided design of electron lenses by the finite element method. In P. W. Hawkes (Ed.), *Image processing and computer-aided design in electron optics* (pp. 284–323). Academic Press.

Munro, E., Wang, L., Rouse, J., & Liu, H. (2006). Differential algebraic method for electron optical design. In *Recent trends in charged particle optics 2006* (pp. 53–54). Czech Republic: Skalsky Dvur.

Preikszas, D., & Rose, H. (1997). Correction properties of electron mirrors. *Journal of Electron Microscopy*, *46*, 1–9.

Press, W. H., Teukolsky, S. A., Vetterling, W. T., & Flannery, B. P. (2007). *Numerical recipes* (3rd edition). Cambridge University Press.

Read, F. H., & Bowring, N. J. (2011). *Nuclear Instruments & Methods in Physics Research. Section A*, *645*, 273–277.

Rockafellar, R. T. (1993). Lagrange multipliers and optimality. *SIAM Review*, *35*, 183–238.

Rose, H., & Preikzsas, D. (1995). Time-dependent perturbation formalism for calculating the aberrations of systems with large ray gradients. *Nuclear Instruments & Methods in Physics Research. Section A*, *363*, 301–315.

Scherzer, O. (1936). Über einige Fehler von Elektronenlinsen. *Zeitschrift für Physik*, *101*, 593–603.

Scherzer, O. (1947). Spharische und chromatische Korrektur von Elektronen-Linsen. *Optik*, *2*, 114–132.

Schury, P., Wada, M., Ito, Y., Arai, F., Naimi, S., Sonoda, T., ... Shchepunov, V. A. (2014). A high resolution multi-reflection time-of-flight mass spectrograph for precision mass measurements at RIKEN/SLOWRI. *Nuclear Instruments & Methods in Physics Research. Section B*, *335*, 39–53.

Tromp, R. M., Hannon, J. B., Ellis, A. W., Wan, W., Berghaus, A., & Schaff, O. (2010). A new aberration-corrected, energy-filtered LEEM/PEEM instrument. I. Principles and design. *Ultramicroscopy*, *110*, 852–861.

Wang, L., Rouse, J., Munro, E., Liu, H., & Zhu, X. (2008). Aberration analysis of electron mirrors by a differential algebra method. *Optik*, *119*(2), 90–96.

Wang, L., Tang, T., & Cheng, B. (2000). Differential algebraic theory and method for arbitrary high order aberrations of electron optics. *Optik*, *111*, 285–289.

Zhu, X., & Munro, E. (1995). Second-order finite element method and its practical application in electron optics. *Journal of Microscopy*, *179*, 161–169.

CHAPTER SIX

An algorithm for simulating the geometric optics of charged particle instruments

Khashayar Shadman
Electron Optica Inc., Palo Alto, CA, United States of America
e-mail address: kashadman@gmail.com

Contents

Advances in Imaging and Electron Physics, Volume 212
ISSN 1076-5670
https://doi.org/10.1016/bs.aiep.2019.08.004

List of abbreviations

ODE Ordinary differential equation
DA Differential algebraic
SA Standard algebraic

1. Introduction

Charged particle instruments, such as electron microscopes and spectrometers, are essential for characterizing materials at the nanoscale. These particles can probe samples with high resolution, owing to their subnanometer wavelengths. To achieve this resolution, the beam of particles has to be steered and focused by electromagnetic fields in precise ways. Hence, the design of these instruments not only requires precision in the construction of the optical elements that produce these fields, it also requires precision in the simulation of the particle trajectories within them.

The particle trajectories are governed by the relativistic equation of motion, which relates the force imparted by the electric and magnetic flux densities, $\vec{E}$ and $\vec{B}$, to the particle acceleration. The equation of motion is described by a second order, nonlinear, ordinary differential equation (ODE). This equation may be recast as a set of coupled, first order, nonlinear ODEs for the position $\vec{r}$ and momentum $\vec{p}$ parametrized in time t. In relativistic mechanics, the operative coordinates are the proper time and velocity, here reduced by the speed of light c: that is,

$$d\tau \equiv cdt/\gamma$$
$$\vec{v} \equiv \frac{1}{mc}\vec{p}$$

where m represents the rest-mass and γ describes the relativistic mass increase,

$$\gamma \equiv \sqrt{1+v^2}$$

In these coordinates, the equation of motion is expressed concisely as

$$\frac{d\vec{r}}{d\tau} = \vec{v}$$
$$\frac{d\vec{v}}{d\tau} = \vec{e} + \vec{v} \times \vec{b}$$

with $\vec{e}$ and $\vec{b}$ describing the reduced electric and magnetic flux densities

$$(\vec{e}, \vec{b}) \equiv \frac{q}{mc^2}(\gamma\vec{E}, c\vec{B})$$

with q defining the particle charge.

For static fields, time may be eliminated in favor of the spatial coordinate along the optical axis, in which case the equations evolve the two dimensional transverse position and slope of the particle's ray with the conservation of energy closing the set. The rays describe the geometric optics of the system, which yields an accurate description of the system's optical performance for low values of the beam brightness and beam current. As the beam brightness increases, the numerical aperture of the system may be reduced to lessen the impact of the aberrations. Here, the impact of diffraction from the system apertures can limit the optical resolution. As the beam current increases, the impact of Coulomb interactions between the particles degrades the beam emittance and thereby the optical resolution. Hence, in general, the geometric optics model provides an upper bound on the system's resolving power. However, the utility of this model extends beyond this approximation. The geometric optics solution defines the path integral in Fourier optics, which incorporates the wave nature of the particles and thereby the impact of diffraction. Furthermore, the particle trajectories resulting from the bulk electromagnetic fields can serve as a starting point for a statistical-mechanical treatment of the Coulomb interactions to model the degradation in the beam emittance.

For the analysis to follow, the ray trajectory is represented generically by

$$\boldsymbol{q} = (\boldsymbol{p}; \boldsymbol{p}')$$

where $\boldsymbol{p}$ describes the I components of the particle position,

$$\boldsymbol{p} = (p_1; p_2; \ldots p_I)$$

and $\boldsymbol{p}'$ describes the derivatives of these components with respect to the independent variable,

$$\boldsymbol{p}' = (p_1'; p_2'; \ldots p_I')$$

Labeling the independent variable z, the equation of motion is expressed as the set of coupled, first order differential equations

$$\frac{d\boldsymbol{q}}{dz} = (\boldsymbol{p}'; \boldsymbol{p}'')$$

with the second derivative of $\boldsymbol{p}$ derived from the forcing functions,

$$\boldsymbol{p}'' = \boldsymbol{f}(z, \boldsymbol{q}) : \boldsymbol{f} = (f_1; f_2; \ldots f_I)$$

where, depending on the nature of the independent variable, f_i may either represent the particle acceleration or the ray curvature in the i'th dimension.

An optical system is characterized by examining the deviation of the particle coordinates from the optical axis as a function of the initial conditions. The trajectories may be derived from a numerical solution of the differential equation; however, this method is inefficient as it requires launching particles with numerous combinations of initial conditions. The elegant approach directly characterizes the optical system, not the particles.

To this end, the forcing functions in the differential equation are approximated by their values near the optical axis via a Taylor expansion. Here, the transverse and axial derivatives of the electromagnetic fields are related by Maxwell's equations. For practical cases, where the fields can be viewed as static during the particle transit time, the only equation needed is Poisson's equation, which simplifies to the Laplace equation if the effects of the beam space charge and current can be neglected. The result is a series expansion of $\boldsymbol{f}$ in the coordinates $\boldsymbol{q}$ with series coefficients that are expressed solely as a function of the axial derivatives of the electromagnetic potentials along the optical axis.

In the paraxial approximation, the dominant contributions to the forcing functions yield a linear differential equation for the fundamental ray trajectories, which can be readily solved. The next step is traditionally performed using perturbation theory (Hawkes & Kasper, 2018, chapter 22). Here, the next order contributions to the forcing functions are combined with the solutions for the fundamental rays to produce integral expressions for the primary aberration coefficients. Although illuminating, the derivation of these expressions becomes intractable for the higher order coefficients.

What is desired is a procedure for deriving these coefficients in an automated manner. One such procedure is provided by the differential algebraic (DA) method (Berz, 1989, 1999; Berz, Makino, & Wan, 2015). In the

DA method, use is made of a structure from nonstandard analysis with properties that enable the direct computations of the series expansions and derivatives of functions: calculations that are necessary for extracting the lens coefficients up to any order in an automated manner.

From the viewpoint of truncated power series algebra, the structure of the coordinates in DA are arrays that contain the coefficients of their polynomial expansions in the initial conditions up to a desired order. Binary operations such as addition and multiplication of the coordinates are extended to their arrays; in particular, multiplication of the coordinates translates to a convolution of their arrays, here denoted by $\times$. Functions of the coordinates are translated to function-arrays that are derived from a Taylor series in the monomials of the coordinate-arrays whose powers are calculated by a repeated application of the $\times$ operator. In this manner, the differential equation for the coordinates translates to one for their arrays, which may likewise be integrated via a finite difference method like that of Runge-Kutta (Munro, Wang, Rouse, & Liu, 2006).

The algorithm that is being presented here also constructs a differential equation for the coordinate-arrays, but it does so without invoking any results from nonstandard analysis. This algorithm uses standard algebraic (SA) techniques, namely a combination of the binomial and Taylor expansions, to manipulate functions of coordinates that are represented by a truncated power series. In essence, the operator in SA associated with a multinomial expansion, $\otimes$, consolidates the repeated application of the $\times$ operator in DA.

2. The standard algebraic algorithm

For the purpose of introducing the algorithm, let $\boldsymbol{p}$ be composed of a single variable p. In the standard approach, the forcing function f is formally expressed as a power series in the coordinates

$$\boldsymbol{q} = (p; p')$$

truncated to a desired order L. The power series is represented as a dot product of the coordinates with the array of series coefficients $\underline{f}$:

$$\begin{aligned} f(z, \boldsymbol{q}) &= \boldsymbol{q} \cdot \underline{f}(z) \\ &\equiv \sum_{l_1 l_2} (p)^{l_1} (p')^{l_2} f_{l_1 l_2}(z) : l_1 + l_2 \leq L \end{aligned}$$

As described in Section 3.1, the coordinates $\boldsymbol{q}$ are decomposed into a principal component $\bar{\boldsymbol{q}}(z)$, which is independent of the initial conditions, and the remainder $\tilde{\boldsymbol{q}}(z, \boldsymbol{q}_0)$, which houses the dependencies on the initial coordinates $\boldsymbol{q}_0 \equiv \boldsymbol{q}(z_0)$: that is,

$$\boldsymbol{q} = \bar{\boldsymbol{q}}(z) + \tilde{\boldsymbol{q}}(z, \boldsymbol{q}_0)$$

The principal component, $\bar{\boldsymbol{q}}$, describes a deviation in the coordinates from the optical axis that is independent of the initial conditions, impacting the beam of particles uniformly.

By applying a binomial expansion, the power series of f in $\boldsymbol{q}$ is regrouped into a series in $\tilde{\boldsymbol{q}}$ with the principal component, $\bar{\boldsymbol{q}}$, folded into the series coefficients. Stated mathematically,

$$f = (\bar{\boldsymbol{q}} + \tilde{\boldsymbol{q}}) \cdot \underline{f} \equiv \tilde{\boldsymbol{q}} \cdot (\bar{\boldsymbol{q}} \odot \underline{f})$$

where the circle-dot operator $\odot$ represents the new series coefficients as a convolution of $\bar{\boldsymbol{q}}$ with the array $\underline{f}$. As illustrated in Section 3.1, this convolution yields a power series in $\tilde{\boldsymbol{q}}$.

The coordinates $\tilde{\boldsymbol{q}}$ are then expressed as a power series in the initial conditions $\boldsymbol{q}_0$:

$$\tilde{q}_i(z, \boldsymbol{q}_0) = \boldsymbol{q}_0 \cdot \underline{c_i}(z) : i \in 1, 2$$

Note that by design, the array of coefficients

$$\underline{c_i} = [c_{i,00}, c_{i,10}, c_{i,01} \ldots]$$

has no principal contribution: that is, $c_{i,00} = 0$, which makes $\tilde{q}_i(z, \mathbf{0}) = 0$. This property, which is related to nilpotency (Berz, 1999, chapter 2), guarantees that only the coefficients in the truncated series of f in $\boldsymbol{q}$ contribute to those in the truncated series of f in $\boldsymbol{q}_0$.

The truncated series in $\boldsymbol{q}_0$ is derived from two dot products because it is produced by a series of monomials that are, themselves, represented by a power series. The compound dot product generates a multinomial expansion. The resulting series coefficients are derived from a convolution of the two arrays $\underline{\boldsymbol{c}}$ and $\bar{\boldsymbol{q}} \odot \underline{f}$. This convolution is symbolized by the circled-times operator $\otimes$ to make

$$f = (\boldsymbol{q}_0 \cdot \underline{\boldsymbol{c}}) \cdot (\bar{\boldsymbol{q}} \odot \underline{f}) \equiv \boldsymbol{q}_0 \cdot \left(\underline{\boldsymbol{c}} \otimes (\bar{\boldsymbol{q}} \odot \underline{f})\right)$$

The expressions represented by the $\odot$ and $\otimes$ operators are introduced in Section 3.1 and expounded in Appendix B.

What remains is a derivation of the array $\underline{f}$ for the series representation of f in the local coordinates $\boldsymbol{q}$. The translation of a function to its array is described in Section 3.2. The only intricate case is that of a composite function:

$$f = f(g(\boldsymbol{q}))$$

where the z dependence has been suppressed. In this case, the function g is decomposed into a principal component, which is independent of $\boldsymbol{q}$, and the remainder,

$$g(\boldsymbol{q}) = \bar{g} + \tilde{g}(\boldsymbol{q})$$

and the function f is Taylor expanded about $\bar{g}$ as

$$f = \tilde{g} \cdot \underline{\mathcal{D}} f(\bar{g}) : \mathcal{D}_n \equiv \frac{1}{n!} \frac{d^n}{d\bar{g}^n}$$

Introducing the series expansion

$$\tilde{g}(\boldsymbol{q}) = \boldsymbol{q} \cdot \underline{\tilde{g}}$$

into the Taylor series produces a multinomial expansion for the array $\underline{f}$.

As described in Section 4, the translations of the coordinates and the functions of coordinates to their respective arrays translate the differential equation for the coordinates $\boldsymbol{q}$ to two differential equations: one for the deviation of the beam from the optical axis, $\bar{\boldsymbol{q}}$, which would arise from a deflection; and one for the array of coefficients $\underline{c}$, where the linear coefficients describe the fundamental rays, and the higher order coefficients describe the aberrations of the instrument when evaluated at the image plane.

These coefficients are accumulated using two complementary, independent methods dubbed serial and parallel integration. These methods are described in Sections 4.1 and 4.2. In serial integration, the coefficients up to a desired order are accumulated incrementally along the independent variable z. In parallel integration, the coefficients over all z are accumulated by ascending order in the indices. The method of parallel integration is illustrated for a round magnetic lens in Section 5.

A comparison of the results from the two methods serves as a check on the accuracy of the calculation; that is, the convergence to the true solution. The precision in the calculation or the degree of convergence is determined by the discretization of the optical axis. As described in Appendix C, the trapezoid rule is used to make the precision quadratic in the segment lengths, thereby producing the following scaling for the trade-off between the precision in the calculation, ϵ, and the computation time T:

$$T \propto \epsilon^{-1/2}$$

The precision and the computation time for the calculation of the series coefficients for a round magnetic lens are demonstrated in Section 6.

Once the coefficients are determined up to a desired order, the initial coordinates of a particle, $\boldsymbol{q}_0$, are readily mapped to the coordinates at any value of z, $\boldsymbol{q}(z)$, with an accuracy that is determined by the order of the expansion. This accuracy does not account for any errors in the calculation of the axial derivatives of f. These derivatives can be calculated exactly if the electromagnetic fields can be described analytically. If, on the other hand, the fields are calculated numerically, the resulting profiles must be fitted to analytic functions in order to calculate their derivatives. To this end, the field profiles may be represented as a series in orthonormal functions (Hawkes & Kasper, 2018, chapter 34) with the fit to the profiles providing the series coefficients. Although the resulting approximations to the fields can be made to high precision, the approximations to their higher order axial derivatives eventually break down not just because of noise in the data, but because of the inherent ill-posed nature of differentiation. Consequently, the calculation of the aberration coefficients from numerically derived fields is generally limited to a finite order. As stated in (Berz, 1999, chapter 3), the growing errors in the derivatives may be alleviated via the use of the Gaussian Wavelet.

Although the analysis is centered on the solution to a second order nonlinear ODE for characterizing charged particle instruments, it is by no means limited to this order. The generalization of the algorithm to an N'th order, nonlinear ODE in a system of I dependent variables

$$\boldsymbol{p} = (p_1; p_2; \dots p_I)$$

is described in Appendix A. In short, the generalized algorithm gives the solution to the equation

$$\frac{d^N p_i}{dz^N} = f_i(\boldsymbol{q}) : i \in 1 \dots I$$

as a power series in the initial condition of $\boldsymbol{q}(z)$, which comprises the $N-1$ lower order derivatives of $\boldsymbol{p}$ in the independent variable z:

$$\boldsymbol{q} = (\boldsymbol{p}^{(0)}; \boldsymbol{p}^{(1)}; \ldots \boldsymbol{p}^{(N-1)}) : \boldsymbol{p}^{(n)} \equiv \frac{d^n \boldsymbol{p}}{dz^n}$$

3. Power series representation of a function

To facilitate the description of the algorithm, a formalism is developed in this section to concisely express the complex operations involving a power series. The power series expansion of a function f in a set of I coordinates

$$\boldsymbol{r} = (r_1; r_2; \ldots r_I)$$

is described here as the dot product

$$f(\boldsymbol{r}) = \boldsymbol{r} \cdot \underline{f} \equiv \sum_{\mathrm{k}} \boldsymbol{r}^{\mathrm{k}} f_{\mathrm{k}}$$

where $\underline{f}$ describes the array of coefficients

$$\underline{f} \equiv [f_{00\ldots0}, f_{10\ldots0}, f_{010\ldots0}, f_{0010\ldots0}, \ldots] \equiv [\ldots f_{\mathrm{k}} \ldots]$$

with indices

$$\mathrm{k} \equiv (k_1 k_2 \ldots k_I)$$

defining the collection of exponents to which the individual coordinates in $\boldsymbol{r}$ are raised; that is

$$\boldsymbol{r}^{\mathrm{k}} \equiv \prod_i (r_i)^{k_i} = (r_1)^{k_1} (r_2)^{k_2} \ldots (r_I)^{k_I}$$

For a series in two coordinates, for example, the series is expressed as

$$f = f_{00} + r_1 f_{10} + r_2 f_{01} + (r_1)^2 f_{20} + r_1 r_2 f_{11} + (r_2)^2 f_{02} + \ldots$$

In practice, the series is truncated to order K such that

$$\sum_i a_i k_i = K$$

where a_i is the weight of coordinate i – a positive integer. If

$$\boldsymbol{r} = \boldsymbol{0} = (0; 0; \ldots 0)$$

the series reduces to the zeroth-order or principal coefficient:

$$\bar{f} \equiv \mathbf{0} \cdot \underline{f} = f_{\mathbb{O}} : \mathbb{O} \equiv (00\ldots0)$$

The *bar* accent and the $\mathbb{O}$ subscript will be used interchangeably to connote the principal coefficient.

The dot product is linear in the array of coefficients; that is,

$$\boldsymbol{r} \cdot (c_1\underline{f_1} + c_2\underline{f_2}) = c_1(\boldsymbol{r} \cdot \underline{f_1}) + c_2(\boldsymbol{r} \cdot \underline{f_2})$$

For a set of J functions,

$$\boldsymbol{f} \equiv (f_1; f_2; \ldots f_J)$$

the definition of the dot product is extended to apply to each entry in the set of arrays

$$\underline{\boldsymbol{f}} \equiv (\underline{f_1}; \underline{f_2}; \ldots \underline{f_J})$$

That is,

$$\boldsymbol{f} = (\boldsymbol{r} \cdot \underline{f_1}; \boldsymbol{r} \cdot \underline{f_2}; \ldots \boldsymbol{r} \cdot \underline{f_J}) \equiv \boldsymbol{r} \cdot \underline{\boldsymbol{f}}$$

If $J = I$ and the functions $\boldsymbol{f}$ are set equal to the coordinates $\boldsymbol{r}$, then

$$\boldsymbol{r} = \boldsymbol{r} \cdot \underline{\boldsymbol{I}} \leftrightarrow r_i = \boldsymbol{r} \cdot \underline{I_i} \qquad (3.1)$$

with $\underline{\boldsymbol{I}}$ representing the I arrays of identity coefficients for the first order monomials: that is,

$$I_{i,\mathrm{k}} \equiv \delta_{k_i 1} \prod_{j \neq i} \delta_{k_j 0}$$

where δ_{kl} represents the Kronecker delta, equal to 1 if $k = l$ and to 0 otherwise. For $I = 2$,

$$\underline{I_1} = [0, 1, 0, 0, \ldots 0]$$
$$\underline{I_2} = [0, 0, 1, 0, \ldots 0]$$

which makes

$$r_1 = 0(r_1)^0(r_2)^0 + 1(r_1)^1(r_2)^0 + 0(r_1)^0(r_2)^1 + 0\ldots$$
$$r_2 = 0(r_1)^0(r_2)^0 + 0(r_1)^1(r_2)^0 + 1(r_1)^0(r_2)^1 + 0\ldots$$

as required. The product of the Kronecker delta's is recast in the concise form

$$I_{i,\mathrm{k}} = \delta_{\mathrm{k},\mathbb{1}(i)}$$

where $\mathbb{1}(i)$ defines the i'th index of first order:

$$\mathbb{1}(i) \equiv (k_1 k_2 \ldots k_I) : k_j = \delta_{ji} \tag{3.2}$$

For $I = 2$,

$$\mathbb{1}(1) = (10), \mathbb{1}(2) = (01)$$

3.1 Translation of coordinates

The I coordinates represented by $\boldsymbol{r}$ may themselves be functions of another set of J coordinates

$$\boldsymbol{q} = (q_1; q_2; \ldots q_J)$$

The operative example is the one where $\boldsymbol{r}$ represents the local particle coordinates and $\boldsymbol{q}$ its initial conditions. In this case, the function $f(\boldsymbol{r})$ is represented as a power series in $\boldsymbol{q}$ by decomposing $\boldsymbol{r}$ as

$$\boldsymbol{r} = \bar{\boldsymbol{r}} + \tilde{\boldsymbol{r}}(\boldsymbol{q})$$

with $\bar{\boldsymbol{r}}$ describing the principal contribution and with $\tilde{\boldsymbol{r}}$ expressed as the power series

$$\tilde{\boldsymbol{r}}(\boldsymbol{q}) = \boldsymbol{q} \cdot \underline{\boldsymbol{c}} \leftrightarrow \tilde{r}_i = \boldsymbol{q} \cdot \underline{c_i}$$

where $\underline{\boldsymbol{c}}$ houses the I arrays of coefficients

$$\underline{\boldsymbol{c}} \equiv (\underline{c_1}; \underline{c_2}; \ldots \underline{c_I}) : \underline{c_i} = [\ldots c_{i,\mathrm{l}} \ldots]$$

with each coefficient $c_{i,\mathrm{l}}$ corresponding to the powers

$$\mathrm{l} = (l_1 l_2 \ldots l_J)$$

to which the J coordinates in $\boldsymbol{q}$ are raised to give $\tilde{r}_i$. By design, the zeroth-order contribution has been attributed to $\bar{\boldsymbol{r}}$ to make

$$\bar{c}_i = c_{i,\mathbb{0}} = 0 \ \forall i$$

This property will limit the contributions to the truncated series in $\boldsymbol{q}$ to the coefficients in the truncated series in $\boldsymbol{r}$. If $\boldsymbol{r}$ were not decomposed, it would not be possible to derive a relation between the coefficients in the two power series.

The decomposition of $\boldsymbol{r}$ is inserted into the series representation of f to yield a binomial expansion,

$$f(\bar{\boldsymbol{r}}+\tilde{\boldsymbol{r}}) = (\bar{\boldsymbol{r}}+\tilde{\boldsymbol{r}}) \cdot \underline{f} \equiv \tilde{\boldsymbol{r}} \cdot (\bar{\boldsymbol{r}} \odot \underline{f})$$

where the circled-dot operator $\odot$ encapsulates a regrouping of the old series in $\boldsymbol{r}$ into a new series in $\tilde{\boldsymbol{r}}$ with the new coefficients given as a series in $\bar{\boldsymbol{r}}$. For $I=2$, the binomial expansion is expressed as

$$\begin{aligned} f &= \sum_{k_1 k_2} (\bar{r}_1+\tilde{r}_1)^{k_1} (\bar{r}_2+\tilde{r}_2)^{k_2} f_{k_1 k_2} \\ &= \sum_{k_1 k_2} (\tilde{r}_1)^{k_1} (\tilde{r}_2)^{k_2} (\bar{\boldsymbol{r}} \odot \underline{f})_{k_1 k_2} \end{aligned}$$

with

$$\{\bar{\boldsymbol{r}} \odot \underline{f}\}_{k_1 k_2} = \sum_{j_1 j_2} \frac{1}{j_1! j_2!} (\bar{r}_1)^{j_1} (\bar{r}_2)^{j_2} \{f\}_{j_1+k_1, j_2+k_2}$$

where the curly brackets denote the Taylor coefficients:

$$\{f\}_{k_1 k_2} \equiv k_1! k_2! f_{k_1 k_2}$$

By inspection, it can be seen that the principal coefficient resulting from the $\odot$ product reduces to the dot-product:

$$(\bar{\boldsymbol{r}} \odot \underline{f})_{\mathbf{0}} = \bar{\boldsymbol{r}} \cdot \underline{f}$$

The transformation is completed by inserting the power series for $\tilde{\boldsymbol{r}}(\boldsymbol{q})$ into the series for f. Defining the array

$$\underline{d} \equiv (\bar{\boldsymbol{r}} \odot \underline{f})$$

the dependence of f on the coordinate $\boldsymbol{q}$ is expressed as

$$g(\boldsymbol{q}) = f(\bar{\boldsymbol{r}} + \boldsymbol{q} \cdot \underline{c}) = (\boldsymbol{q} \cdot \underline{c}) \cdot \underline{d}$$

As is evident, the series in $\boldsymbol{q}$ is derived from a composition of two dot products because the initial series is a function of coordinates that, themselves, are represented as a power series. This composition is derived from a repeated application of the binomial expansion to give a multinomial expansion for the coefficients in the series in $\boldsymbol{q}$. The new series coefficients are derived from a convolution of the old. Labeling this convolution by the circled-times operator $\otimes$ makes

$$g(\boldsymbol{q}) = \boldsymbol{q} \cdot \underline{g} : \underline{g} \equiv (\underline{c} \otimes \underline{d})$$

For $I = J = 2$, the multinomial expansion is expressed as

$$\begin{aligned} g &= \sum_{k_1 k_2} \prod_{i=1}^{2} \left(\sum_{l_1 l_2} (q_1)^{l_1} (q_2)^{l_2} c_{i, l_1 l_2} \right)^{k_i} d_{k_1 k_2} \\ &= \sum_{m_1 m_2} (q_1)^{m_1} (q_2)^{m_2} (\underline{c} \otimes \underline{d})_{m_1 m_2} \end{aligned}$$

The coefficients in the series are derived from a sum over the set of indices $k_{i,\mathrm{l}}$, parametrized by $i \in 1, 2$ and $\mathrm{l} = (l_1 l_2)$: that is,

$$(\underline{c} \otimes d)_{m_1 m_2} \equiv \sum_{k_{1,\mathrm{l}}, k_{2,\mathrm{l}}} \prod_{i=1}^{2} \frac{(c_{i,\mathrm{l}})^{k_{i,\mathrm{l}}}}{k_{i,\mathrm{l}}!} \{d\}_{k_1 k_2}$$

where

$$k_i \equiv \sum_{l_1 l_2} k_{i,\mathrm{l}} : i \in 1, 2$$

and the indices $\mathrm{m} = (m_1 m_2)$, $\mathrm{l} = (l_1 l_2)$, and $k_{i,\mathrm{l}}$ are related by the following constraint:

$$\mathrm{m} = \sum_{l_1 l_2} \mathrm{l}(k_{1,\mathrm{l}} + k_{2,\mathrm{l}})$$

The relation among the indices shows that for $\mathrm{m} = \mathbb{0}$, l must also be $\mathbb{0}$. Given that $c_{i,\mathbb{0}} = 0 \, \forall i$, the only contribution to the principal coefficient arises from $k_{i,\mathbb{0}} = 0 \, \forall i$, which makes

$$(\underline{c} \otimes \underline{d})_{\mathbb{0}} = d_{\mathbb{0}}$$

There are no constraints on the values of $k_{i,\mathbb{0}}$. Hence, if $\boldsymbol{r}$ had not been decomposed and the binomial expansion not applied, then $c_{i,\mathbb{0}}$ would have remained finite, equal to $\bar{r}_i$, and the above relation would have broken down.

The relation among the indices determines which coefficients in the arrays $\underline{c}$ and $\underline{d}$ contribute to a particular coefficient in the array $\underline{g}$. This exercise is too elaborate to be addressed here. Instead, a derivation is presented in Appendix B.1 that sidesteps this task via a recursive application of the binomial expansion.

The $\odot$ and $\otimes$ products should be recognized as the workhorse of this algorithm. Like the dot operation, these two operations are linear in the second array of coefficients and are extended in definition to apply to a set of such arrays:

$$(\bar{r}+\tilde{r})\cdot\underline{f} \equiv \tilde{r}\cdot(\bar{r}\odot\underline{f}) \tag{3.3a}$$

$$(q\cdot\underline{c})\cdot\underline{d} \equiv q\cdot(\underline{c}\otimes\underline{d}) \tag{3.3b}$$

3.2 Translation of functions

The expression for the ODE will contain sums and products of functions of the coordinates as well as composite functions; that is, functions of functions of the coordinates. In order to represent these expressions as a power series in the coordinates, the composition of functions must be translated to a composition of their series coefficients.

Sums and products of the functions $f(q)$ and $g(q)$ are translated readily as

$$f+g = q\cdot(\underline{f}+\underline{g})$$
$$fg = q\cdot(\underline{f}\times\underline{g})$$

where the commutative and associative, times operator $\times$ defines the coefficients of the latter as

$$(\underline{f}\times\underline{g})_{\mathrm{l}} \equiv \sum_{\mathrm{k}} f_{\mathrm{k}} g_{\mathrm{l-k}}$$

Derivatives of a function $f(q)$,

$$\nabla^{\mathrm{k}} f = q\cdot\underline{\nabla^{\mathrm{k}} f}$$
$$\nabla^{\mathrm{k}} \equiv (\partial_1)^{k_1}(\partial_2)^{k_2}\ldots(\partial_J)^{k_J} : \partial_j \equiv \frac{\partial}{\partial q_j}$$

translate to a shift in the index of its Taylor coefficients; that is,

$$\{\nabla^{\mathrm{k}} f\}_{\mathrm{l}} = \{f\}_{\mathrm{l+k}} \tag{3.4a}$$

where

$$\{f\}_{\mathrm{l}} \equiv \mathrm{l}!f_{\mathrm{l}} : \mathrm{l}! \equiv l_1!l_2!\ldots l_J! \tag{3.4b}$$

Integrals of the function are given by the same relation, but with negative entries in the index k.

For functions whose arguments are themselves functions of the coordinates, the transformation is more involved. To extract the array of coefficients associated with the composite function $f(g(\boldsymbol{q}))$, the function $g(\boldsymbol{q})$ is first decomposed as

$$g(\boldsymbol{q}) = \bar{g} + \tilde{g}(\boldsymbol{q})$$

with $\bar{g}$ defining its principal contribution. The function $f(g)$ is then Taylor expanded about $\bar{g}$. Defining $\underline{\mathcal{D}}$ as the collection of derivatives that comprise a Taylor expansion,

$$\underline{\mathcal{D}} = [\mathcal{D}_0, \mathcal{D}_1 \ldots] : \mathcal{D}_n \equiv \frac{1}{n!}\frac{d^n}{d\bar{g}^n}$$

makes

$$f(g) = (\tilde{g} \cdot \underline{\mathcal{D}})f(\bar{g}) \equiv \sum_n \tilde{g}^n \mathcal{D}_n f(\bar{g})$$

This Taylor series is compounded with the series expansion for $\tilde{g}$,

$$\tilde{g}(\boldsymbol{q}) = \boldsymbol{q} \cdot \underline{\tilde{g}}$$

to make

$$f(g(\boldsymbol{q})) = ((\boldsymbol{q} \cdot \underline{\tilde{g}}) \cdot \underline{\mathcal{D}})f(\bar{g}) = \boldsymbol{q} \cdot (\underline{\tilde{g}} \otimes \underline{\mathcal{D}})f(\bar{g})$$

A notable function of practical interest is

$$f(g) = g^\nu$$

where ν is a rational number. For this function, the derivatives yield the relation

$$\mathcal{D}_n \bar{g}^\nu = \bar{g}^{\nu-n} C_n^\nu$$

with C_n^ν describing the combination formula:

$$\{C^\nu\}_n = n! C_n^\nu \equiv \frac{\nu!}{(\nu-n)!} = \prod_{m=0}^{n-1} (\nu - m) : C_0^\nu \equiv 1$$

Consequently, the series coefficients for this function are found as

$$\underline{g^\nu} = \bar{g}^\nu \left((\bar{g}^{-1} \underline{\tilde{g}}) \otimes \underline{C^\nu} \right) : \underline{C^\nu} \equiv [C_0^\nu, C_1^\nu, \ldots] \tag{3.5}$$

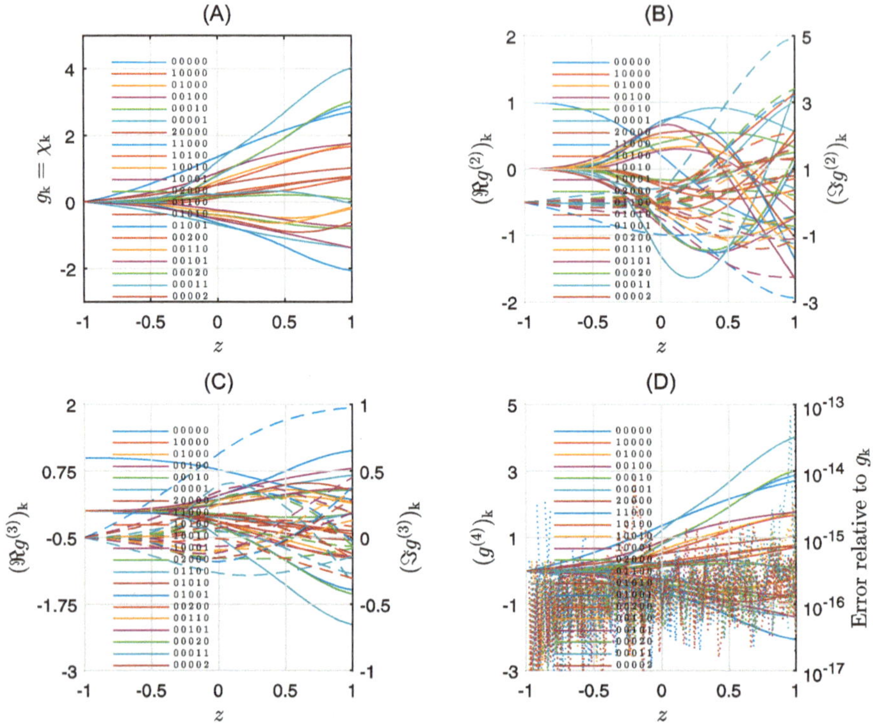

Figure 1 Subplot (A): the series coefficients of the gyro-phase, $g = \chi$, in the coordinates $\boldsymbol{R}_0$ through order 2. Subplot (B): the series coefficients of the real (left side, solid line) and the imaginary (right side, dashed line) parts of $g^{(2)} = \exp(-ig)$. Subplot (C): the series coefficients of the real (left side, solid line) and the imaginary (right side, dashed line) parts of $g^{(3)} = (g^{(2)})^{-1/2}$. Subplot (D): (left side, solid line) the series coefficients of $g^{(4)} = -2i \ln g^{(3)}$; (right side, dotted line) the relative error between the coefficients of $g^{(4)}$ and g.

The derivation of the series coefficients for a second composite function of practical importance, the exponential function,

$$f(g) = e^g$$

is simpler. Its array of coefficients is found as

$$\begin{aligned} \underline{e^g} &= e^{\bar{g}} \underline{\tilde{g}} \otimes \underline{e} : \underline{e} = [e_0, e_1, \ldots] \\ \{e\}_n &= n! e_n \equiv 1 \end{aligned} \tag{3.6}$$

These formulas are demonstrated in Fig. 1. Subplot (A) shows the axial profiles of the series coefficients for a function that happens to describe the gyro-phase of an electron, $g = \chi$, in a round magnetic lens. The series is in the initial values of the five coordinates in $\boldsymbol{R}$ (Eq. (5.24a)), which describes the quartet of quadratics (products of the transverse position ρ and slope ρ' of the ray) plus the relative energy spread $\tilde{u}/\bar{u}$. The coefficients are shown through order 2. Order 2 in the quadratics corresponds to 5th order in $(\rho; \rho')$. Of interest are the series coefficients of the exponential, $g^{(2)} = \exp(-i\chi)$. These coefficients are derived from the application of Eq. (3.6). Their real and imaginary parts are shown in subplot (B). To demonstrate the validity of the formulas in Eqs. (3.5) and (3.6), the array for $g^{(2)}$ is raised to some power, here chosen as $\nu = -1/2$, using Eq. (3.5) to make the array for $g^{(3)} = (g^{(2)})^{-1/2}$. The real and imaginary parts of the resulting coefficients are shown in subplot (C). Then, the natural-logarithm of $g^{(3)}$ is taken to give back the original function; that is, $g^{(4)} = -2i\ln g^{(3)}$. The formula for the array of coefficients for the natural logarithm of a function is found as

$$\underline{\ln g} = \underline{1}\ln \bar{g} + (\bar{g}^{-1}\underline{\tilde{g}}) \otimes \underline{e^{-1}} : \underline{e^{-1}} = [0, e_1^{-1}, \ldots]$$
$$\{e^{-1}\}_{n+1} = (n+1)! e_{n+1}^{-1} \equiv (-1)^n n!$$

As shown in subplot (D), the gyro-phase χ is recovered. The relative error between $g^{(4)}$ and g is shown to be of the order of the machine precision for most of the coefficients, with the maximum error less than 10^{-13}.

4. The array equation

As discussed in the introduction, the equation of motion for a charged particle can be cast as an ODE for the $J = 2I$ particle coordinates

$$\boldsymbol{q} \equiv (\boldsymbol{p}; \boldsymbol{p}')$$

comprising the I particle positions $\boldsymbol{p}$ and their first derivatives $\boldsymbol{p}'$ in the independent variable z:

$$\boldsymbol{p} = (p_1; p_2; \ldots p_I)$$
$$\boldsymbol{p}' = (p_1'; p_2'; \ldots p_I')$$

The first derivatives may represent particle speeds if z represents the travel time; or they may represent the slopes of the trajectories if z represents the position along the optical axis. The ODE is expressed as

$$\frac{d\boldsymbol{q}}{dz} = \boldsymbol{q}' : \boldsymbol{q}' = (\boldsymbol{p}'; \boldsymbol{f}) \tag{4.1a}$$

where the second derivative of the positions, $\boldsymbol{p}''$, has been set equal to the I components of acceleration or ray curvature, $\boldsymbol{f}$, depending on the nature of the independent variable.

The coordinates $\boldsymbol{q}$ are formally represented as a series in the initial coordinates $\boldsymbol{q}_0$:

$$\boldsymbol{q} \equiv \bar{\boldsymbol{q}} + \boldsymbol{q}_0 \cdot \underline{\boldsymbol{c}} : \boldsymbol{c}_{\mathbb{0}} \equiv \mathbf{0} \tag{4.1b}$$

The component $\bar{\boldsymbol{q}}$ describes a deviation from the optical axis experienced by all particles, independent of their initial position and slope; and the array $\underline{\boldsymbol{c}}$ houses the linear and nonlinear optical characteristics. The component $\bar{\boldsymbol{q}}$ would become active if, for example, the optical element possesses deflectors or the lens contains imperfections that produce a dipole field.

To translate Eq. (4.1a) into one for the evolution of $\bar{\boldsymbol{q}}$ and the set of arrays $\underline{\boldsymbol{c}}$, each f_i is expanded as a power series in $\boldsymbol{q}$,

$$f_i(z, \boldsymbol{q}) = \boldsymbol{q} \cdot \underline{f_i}(z) : \underline{f_i} = [\ldots f_{i,1} \ldots]$$

to make

$$\frac{d\boldsymbol{q}}{dz} = \boldsymbol{q} \cdot \underline{\boldsymbol{q}'}(z) : \underline{\boldsymbol{q}'} = (\underline{\boldsymbol{p}'}; \underline{\boldsymbol{f}}) \tag{4.1c}$$

The i'th array of coefficients in the set $\underline{\boldsymbol{p}'}$,

$$\underline{p_i'} = [\ldots p_{i,1}' \ldots]$$

has only one non-zero component as it represents one of the coordinates: q_{i+I}. Hence, the entries in its array are expressed as

$$p_{i,1}' = \delta_{1,\mathbb{1}(i+I)}$$

with $\mathbb{1}(i+I)$ representing the $(i+I)$'th index of first order.

Introducing Eq. (4.1b) into Eq. (4.1c) and applying the binomial and multinomial expansions, Eqs. (3.3a) and (3.3b), the array equation is derived as

$$\frac{d\bar{\boldsymbol{q}}}{dz} + \boldsymbol{q}_0 \cdot \frac{d\underline{\boldsymbol{c}}}{dz} = \boldsymbol{q}_0 \cdot (\underline{\boldsymbol{c}} \otimes (\bar{\boldsymbol{q}} \odot \underline{\boldsymbol{q}'})) \tag{4.2a}$$

The principal contribution to the driving force,

$$(\underline{\boldsymbol{c}} \otimes (\bar{\boldsymbol{q}} \odot \underline{\boldsymbol{q}'}))_{\mathbb{0}} = (\bar{\boldsymbol{q}} \odot \underline{\boldsymbol{q}'})_{\mathbb{0}} = \bar{\boldsymbol{q}} \cdot \underline{\boldsymbol{q}'} \tag{4.2b}$$

evolves $\bar{q}$ for the deviation in the beam from the optical axis:

$$\frac{d\bar{q}}{dz} = \bar{q} \cdot \underline{q}' \tag{4.2c}$$

The remainder evolves $\underline{c}$ for the lens characteristics. Defining the circled-ring operator $\circledcirc$ as the operator $\odot$ with the principal coefficient removed,

$$(\bar{q} \circledcirc \underline{q}')_1 \equiv (1 - \delta_{1 0})(\bar{q} \odot \underline{q}')_1 \tag{4.2d}$$

the differential equation for the array $\underline{c}$ is expressed as

$$\frac{d\underline{c}}{dz} = \underline{c} \otimes (\bar{q} \circledcirc \underline{q}') \tag{4.2e}$$

As shown next, the differential equations for $\bar{q}$ and $\underline{c}$ can be integrated in one of two ways via serial or parallel integration.

4.1 Serial integration

In serial integration, the independent variable is first discretized by adequately subdividing the range of interest $[z_0, z_1]$. Starting with Eq. (4.1c), the ODE is integrated across a segment, say segment m:

$$q_{m+1} = q_m + \Delta q(z_m) \tag{4.3a}$$

with

$$\Delta q(z) \equiv \int_z^{z+h} dz\, q \cdot \underline{q}' \tag{4.3b}$$

where h defines the segment length. The evolution of the coordinates across the segment is determined by the difference scheme, the simplest of which is given by Euler:

$$\Delta q(z_m) = q_m \cdot \underline{\Delta q}(z_m) : \underline{\Delta q}(z) \approx h\underline{q}'(z) \tag{4.3c}$$

As described in Appendix C.2, the segment lengths h are calculated from the local behavior of the functions f in order to maintain the discretization errors among the segments comparable.

For the Euler scheme, the discretization error is quadratic in h, making the error over all segments linear in h. Consequently, the number of required segments for this scheme becomes inversely proportional to the desired precision. The number of segments may be reduced by employing a discretization scheme that is higher order in h. A scheme with an overall error that is quadratic in h is described in Appendix C.1. In this scheme, the number of required segments becomes inversely proportional to the square-root of the precision. As the demand for precision increases, the reduction in the number of segments provides a speed advantage despite the extra cost in the calculation of the higher order discretization. The prescription for the 2nd order discretization scheme that is given in Appendix C.1 may be extended to a higher order scheme like that of Runge-Kutta.

To derive the evolution of $\boldsymbol{q}$ over the many differential segments comprising the optical element, the evolution of $\boldsymbol{q}_m$ over all prior segments $(1, 2, \ldots m-1)$ is formally decomposed into the principal component, which is independent of the initial conditions, and the remainder, which is represented as a power series in the initial conditions $\boldsymbol{q}_0$: that is,

$$\boldsymbol{q}_m \equiv \bar{\boldsymbol{q}}^{(m)} + \boldsymbol{q}_0 \cdot \underline{\boldsymbol{c}}^{(m)} : \boldsymbol{c}_{\mathbb{0}}^{(m)} \equiv \boldsymbol{0} \tag{4.4a}$$

Eq. (4.4a) is inserted into Eq. (4.3c) and the binomial and multinomial expansions, Eqs. (3.3a) and (3.3b), are performed to make

$$\Delta \boldsymbol{q}(z_m) = \boldsymbol{q}_0 \cdot \left(\underline{\boldsymbol{c}}^{(m)} \otimes \left(\bar{\boldsymbol{q}}^{(m)} \odot \underline{\Delta \boldsymbol{q}}(z_m) \right) \right) \tag{4.4b}$$

Eqs. (4.4a) and (4.4b) are then inserted into Eq. (4.3a) and the result is separated into a principal contribution and a remainder to give the evolution of the composites as

$$\bar{\boldsymbol{q}}^{(m+1)} = \bar{\boldsymbol{q}}^{(m)} + \bar{\boldsymbol{q}}^{(m)} \cdot \underline{\Delta \boldsymbol{q}}(z_m) \tag{4.5a}$$

$$\underline{\boldsymbol{c}}^{(m+1)} = \underline{\boldsymbol{c}}^{(m)} + \underline{\boldsymbol{c}}^{(m)} \otimes \left(\bar{\boldsymbol{q}}^{(m)} \circledcirc \underline{\Delta \boldsymbol{q}}(z_m) \right) \tag{4.5b}$$

with the starting conditions

$$\bar{\boldsymbol{q}}^{(0)} = \boldsymbol{0} \tag{4.5c}$$

$$\underline{\boldsymbol{c}}^{(0)} = \underline{\boldsymbol{I}} \tag{4.5d}$$

Note that $\boldsymbol{0} \cdot \underline{\Delta \boldsymbol{q}} = (\Delta \boldsymbol{q})_{\mathbb{0}}$; hence, if $\underline{\Delta \boldsymbol{q}}$ does not have a principal component, there will be no deviation in the beam from the optical axis: that is, $\bar{\boldsymbol{q}}^{(m)} = 0 \, \forall m$.

4.2 Parallel integration

Instead of accumulating the series coefficients for all indices of interest from one segment of the element to the next, the series coefficients for all segments may be accumulated from one index to the next. This parallel integration of the coefficients requires an integral solution to the ray equation. This method provides an independent check on the solution that is derived via serial integration. Furthermore, it improves the speed of the calculation by eliminating the repeated application of the ⊗ operation, which is computationally expensive.

To start, the set of coordinates $\boldsymbol{p}$ is limited to a single entry p. The differential equation for this coordinate,

$$p'' = f(\boldsymbol{q}) : \boldsymbol{q} \equiv (p; p')$$

is solved by separating the contributions to f that are linear in p and p': that is,

$$f = \boldsymbol{q} \cdot \underline{f} = f_{00} + pf_{10} + p'f_{01} + \boldsymbol{q} \cdot \underline{f_2}$$

where f_{00} defines the zeroth-order contribution and the last term contains all non-linear contributions,

$$\underline{f_2} \equiv [\ldots f_{l_1 l_2} \ldots]|_{l_1+l_2>1}$$

The linear contributions enter into the homogeneous equation

$$p'' = f_{10}p + f_{01}p'$$

whose solution is expressed as a linear combination of the initial conditions. Interpreting $\boldsymbol{q}$ as a column vector, the homogeneous solution is expressed as

$$\boldsymbol{q}(z, \boldsymbol{q}_0) = \mathrm{C}(z)\boldsymbol{q}_0$$

with

$$\mathrm{C} \equiv \begin{bmatrix} c_{10} & c_{01} \\ c'_{10} & c'_{01} \end{bmatrix}$$

The matrix C is derived from the solution to

$$\mathrm{C}' = \mathrm{AC} : \mathrm{A} \equiv \begin{bmatrix} 0 & 1 \\ f_{10} & f_{01} \end{bmatrix} \tag{4.6}$$

with the initial condition $\mathrm{C}(z_0) = \mathrm{I}$, where I is the identity matrix. As described in Appendix C, the evolution of the matrix C is derived by discretizing the differential equation and by evolving the matrix from one segment to the next.

The overall solution is found from the method of variation of parameters, established by Lagrange. In this method, the particle coordinate p is expressed as a linear combination of the homogeneous solutions with variable coefficients $\theta_n(z)$; that is,

$$p = \sum_n C_{1n}(z)\theta_n(z) : n \in 1, 2$$

As there are two unknowns and one differential equation, the following judiciously chosen constraint is placed on the derivatives of the coefficients,

$$\sum_n C_{1n}\theta'_n = 0$$

This constraint makes $\boldsymbol{\theta}$ the back-projection of the coordinates $\boldsymbol{q}$ onto the initial plane, $z = z_0$:

$$\boldsymbol{q} = \mathrm{C}\boldsymbol{\theta} : \boldsymbol{\theta} \equiv (\theta_1; \theta_2) \tag{4.7a}$$

More significantly, the constraint simplifies the differential equation for $\boldsymbol{\theta}$ to

$$\begin{aligned} \mathrm{C}\boldsymbol{\theta}' &= \boldsymbol{g} \\ \boldsymbol{g} &= (f_{00} + \boldsymbol{q} \cdot \underline{f_2})(0; 1) \end{aligned} \tag{4.7b}$$

where $\boldsymbol{g}$ is interpreted as a 2×1 column vector with a nontrivial second component.

This first order ODE is formally integrated as

$$\boldsymbol{\theta} = \boldsymbol{\theta}(z_0) + \int dz\, \boldsymbol{\theta}' \tag{4.8a}$$

with the initial conditions $\boldsymbol{\theta}(z_0) = \boldsymbol{q}_0$. Expressing $\boldsymbol{\theta}'$ as a power series in the coordinates $\boldsymbol{q}$,

$$\boldsymbol{\theta}'(z, \boldsymbol{q}) = \boldsymbol{q} \cdot \underline{\boldsymbol{\theta}'}(z) : \underline{\boldsymbol{\theta}'} = \mathrm{C}^{-1}\underline{\boldsymbol{g}} \tag{4.8b}$$

and decomposing $\boldsymbol{q}$ into its principal component and the remainder as described by Eq. (4.1b), the integral equation for $\boldsymbol{\theta}$, Eq. (4.8a), is transformed to one for the series coefficients:

$$\mathrm{C}^{-1}\bar{\boldsymbol{q}} = \int dz\, \bar{\boldsymbol{q}} \cdot \underline{\boldsymbol{\theta}'} \tag{4.9a}$$

$$\mathrm{C}^{-1}\underline{\boldsymbol{c}} = \underline{\boldsymbol{I}} + \int dz\, \underline{\boldsymbol{c}} \otimes (\bar{\boldsymbol{q}} \circledcirc \underline{\boldsymbol{\theta}'}) \tag{4.9b}$$

Introducing Eq. (4.8b) for $\underline{\boldsymbol{\theta}'}$, using the linearity of the operators, and noting that the $\circledcirc$ operator (Eq. (4.2d)) removes the principal contribution, the integral equations are expressed as

$$\mathrm{C}^{-1}\bar{\boldsymbol{q}} = \int dz\, \mathrm{C}^{-1}(\boldsymbol{g}_{00} + \bar{\boldsymbol{q}} \cdot \underline{\boldsymbol{g}_2}) \tag{4.9c}$$

$$\mathrm{C}^{-1}\underline{\boldsymbol{c}} = \underline{\boldsymbol{I}} + \int dz\, \mathrm{C}^{-1}(\underline{\boldsymbol{c}} \otimes (\bar{\boldsymbol{q}} \circledcirc \underline{\boldsymbol{g}_2})) \tag{4.9d}$$

Eq. (4.9c) is solved iteratively for the deviation in the beam from the optical axis, starting with $\bar{\boldsymbol{q}}^{(0)} = \mathbf{0}$ in the integral. Hence, on the first iteration, $\bar{\boldsymbol{q}}$ is evaluated as

$$\bar{\boldsymbol{q}}^{(1)} = \mathrm{C} \int dz\, \mathrm{C}^{-1}\boldsymbol{g}_{00} \tag{4.10a}$$

If $f_{00} = 0$, the beam remains on the optical axis. If f_{00} is finite, the calculation of $\bar{\boldsymbol{q}}$ is refined on the second iteration, which begins to incorporate the nonlinear contributions to f. In general,

$$\bar{\boldsymbol{q}}^{(m+1)} = \bar{\boldsymbol{q}}^{(1)} + \mathrm{C} \int dz\, \mathrm{C}^{-1}(\bar{\boldsymbol{q}}^{(m)} \cdot \underline{\boldsymbol{g}_2}) \tag{4.10b}$$

The iterative process converges if the beam does not deviate far from the optical axis such that $|\bar{\boldsymbol{q}}^{(1)} \cdot \underline{f_2}|$, whose dominant term is quadratic in $\bar{\boldsymbol{q}}$, remains small compared to $|f_{00}|$. The convergence is assessed by examining the evolution of $\bar{\boldsymbol{q}}^{(m)}$ with each iteration.

The solution for $\bar{\boldsymbol{q}}$ is then inserted into Eq. (4.9d) for the array $\underline{\boldsymbol{c}}$. This equation is also solved iteratively, starting with $\underline{\boldsymbol{c}}^{(0)} = \mathbf{0}$ in the integral. Noting that $\underline{\mathbf{0}} \otimes \underline{\boldsymbol{d}} = \underline{1}\boldsymbol{d}_{00}$ (Eq. (B.5b)), the integral in Eq. (4.9d) vanishes on the first iteration to make

$$\underline{\boldsymbol{c}}^{(1)} = \mathrm{C}\underline{\boldsymbol{I}} \tag{4.11a}$$

Hence, the first iteration just recovers the homogeneous solution for the fundamental rays. On the next iteration, the primary nonlinear coefficients are evaluated. In general,

$$\underline{\boldsymbol{c}}^{(m+1)} = \underline{\boldsymbol{c}}^{(1)} + \mathrm{C} \int dz\, \mathrm{C}^{-1}(\underline{\boldsymbol{c}}^{(m)} \otimes (\bar{\boldsymbol{q}} \circledcirc \underline{\boldsymbol{g}_2})) \tag{4.11b}$$

The number of iterations increases with the order of the coefficients being calculated with the total number given by the order of the truncated series for f.

The integrals are calculated as a discretized sum. For an overall precision that improves quadratically with the number of segments, the matrix A in the homogeneous equation, Eq. (4.6), and the array of coefficients $\underline{f_{\mathbb{2}}}$ in Eq. (4.9d) are evaluated by the trapezoid rule as described in Appendix C.

The above solution may be readily extended to a set of variables $\boldsymbol{p}$. For I such variables, the solution is expressed with the matrix C extended from size 2 to size $J = 2I$ for the J coordinates and first order indices. Concurrently, the 0 and 1 entries in the matrix A are extended to the zero and identity matrices of size I, respectively, and the bottom row of A is extended to an $I \times J$ matrix with the entries $f_{i,\mathbb{1}(j)}$. Last, the column vector $\boldsymbol{g}$ is generalized to a column vector of size J with I leading 0 entries followed by the I entries $f_{i,00} + \boldsymbol{q} \cdot \underline{f_{i,\mathbb{2}}}$.

4.3 Composition of optical elements

A system may comprise regions with electromagnetic fields that are separated by field free regions. These isolated regions form distinct optical elements, which can be modeled independently. The model for the system would then require the evolution of the coordinates from one element to another. This evolution is derived in the same manner as the serial evolution of the coordinates over the differential segments within an element.

To start, the coordinates $\mathbf{Q}$ at the two ends of element $n+1$ are related by

$$\mathbf{Q}_{n+1} = \bar{\mathbf{Q}}_{n+1} + \mathbf{Q}_n \cdot \underline{\boldsymbol{C}}_{n+1} : \boldsymbol{C}_{n+1,\mathbb{0}} \equiv \mathbf{0} \tag{4.12a}$$

with $\bar{\mathbf{Q}}_{n+1}$ describing the deviation in the beam from the optical axis in this region and $\underline{\boldsymbol{C}}_{n+1}$ describing the properties of the lens in this region. Next, the evolution of $\mathbf{Q}_n$ over all prior elements is formally decomposed into a principal component and the remainder with the latter represented as a power series in $\mathbf{Q}_0$, the initial conditions for the collection of all elements:

$$\mathbf{Q}_n \equiv \bar{\mathbf{Q}}^{(n)} + \mathbf{Q}_0 \cdot \underline{\boldsymbol{C}}^{(n)} : \boldsymbol{C}_{\mathbb{0}}^{(n)} \equiv \mathbf{0} \tag{4.12b}$$

Eq. (4.12b) is inserted into Eq. (4.12a) and Eqs. (3.3a) and (3.3b) are applied to give the evolution of the composites as

$$\bar{\mathbf{Q}}^{(n+1)} = \bar{\mathbf{Q}}_{n+1} + \bar{\mathbf{Q}}^{(n)} \cdot \underline{\boldsymbol{C}}_{n+1} \quad : \bar{\mathbf{Q}}^{(0)} = \mathbf{0} \tag{4.13a}$$

$$\underline{C}^{(n+1)} = \underline{C}^{(n)} \otimes (\bar{\mathbf{Q}}^{(n)} \circledcirc \underline{C}_{n+1}) \quad : \underline{C}^{(0)} = \underline{I} \tag{4.13b}$$

Eqs. (4.13a) and (4.13b) give the composite deflection in the beam and the composite magnification and aberrations in a system of optical elements.

5. Application of the algorithm to a round lens

The derivation of the ray equation for a round lens is described in great detail in (Hawkes & Kasper, 2018, chapter 15). Nevertheless, in order to present this equation in a form that is amenable to the SA algorithm, a brief derivation is given in Appendix D. In this appendix, a general ray equation is derived first, which is applicable to a multipole optical element. Then, the equation is specified to a round lens.

The general ray equation, Eq. (D.6), can also be applied specifically to a round lens. A comparison of the solution to Eq. (D.6) with the solution derived from Eq. (D.7), which is specific to the round lens, provides a check on the implementation of the ray equations as well as on the accuracy of the algorithm. Furthermore, the general ray equation is solved using both serial and parallel integration. A comparison of the solutions via the two integration methods will provide another layer of inspection of the algorithm.

For a few specific electromagnetic configurations, it is possible to derive analytic solutions for the fundamental rays, which in turn produce integral expressions for the primary aberration coefficients that can be solved in closed form or, more practically, calculated with adequate precision via numerical quadrature. Such a solution is provided by Glaser's bell-shaped magnetic field profile. A comparison of the analytic expressions for the primary aberration coefficients with the calculations derived from the SA algorithm will provide the most direct (albeit limited) inspection of the algorithm.

The algorithm is validated for the calculation of the higher order aberrations by comparing the solution to the array equation for the round lens with the solution to the array equation for the general, multipole lens. The latter is derived succinctly in Appendix D.2. The array equation for the round lens and its integral solution are described in depth in the following subsections. In particular, what is presented is a solution by parallel integration for the integral expressions of the monochromatic aberrations of any order, followed by a prescription for incorporating the chromatic aberrations.

The axi-symmetric nature of a round lens nullifies many of the coefficients when representing the transverse position and slope of the ray as a power series in their initial values. These trivial coefficients can be side-stepped by working in quadratic coordinates as described next. Furthermore, the canonical angular momentum is used as one such coordinate in order to exploit its constancy. These advantages are obtained by stepping in and out of the rotating frame of reference of a gyrating particle in a magnetic field. The associated coordinate transformations provide ample examples of the many manipulations of a power series as described in Section 3, making this demonstration quite illustrative of the SA algorithm.

5.1 Matrix representation in the rotating frame

To start, the ray equation for the round lens is described in the rotating frame of reference of a gyrating particle in a magnetic field. The resulting second order ODE, Eq. (D.7), is recast as a set of first order, coupled ODEs, which is encapsulated in matrix form as

$$\boldsymbol{q}' = \mathrm{A}\boldsymbol{q} : \boldsymbol{q} \equiv (\varrho; \varrho') \tag{5.1}$$

where $\boldsymbol{q}$ is interpreted as a 2×1 vector containing the transverse position ϱ and slope ϱ' in the rotating frame and

$$\mathrm{A} = \begin{bmatrix} 0 & 1 \\ A_{21} & A_{22} \end{bmatrix} \tag{5.2a}$$

with

$$\left(A_{21}; A_{22}\right) \equiv \frac{1 + |\varrho'|^2}{2\Gamma} \left(2\partial_{|\varrho|^2}; -\partial_z\right) \Gamma \tag{5.2b}$$

$$\Gamma \equiv \frac{1}{2}(\gamma^2 - 1) + P\dot{\chi} - \frac{1}{2}|\varrho|^2 \dot{\chi}^2 \tag{5.2c}$$

where γ is the relativistic mass increase, P is the canonical angular momentum, and $\dot{\chi}$ is the gyro-frequency.

The mass increase is related to the kinetic energy as

$$\gamma = 1 + u - \phi^e(z, |\varrho|^2) \tag{5.3}$$

with u and ϕ^e describing the initial kinetic energy and the electric potential energy, respectively, reduced by the mass energy. The gyro-frequency is proportional to the magnetic flux per radian ψ,

$$\dot{\chi} = |\varrho|^{-2}\psi(z, |\varrho|^2) \tag{5.4a}$$

$$\psi = -\frac{1}{2}\int d|\varrho|^2 \partial_z \phi^b(z, |\varrho|^2) \tag{5.4b}$$

where ϕ^b is the product of the speed of light with the magnetic potential energy, reduced by the mass energy.

As indicated by the expressions for γ and $\dot{\chi}$, the electromagnetic potentials in a round lens are functions of $|\varrho|^2$. Hence, the matrix entries A_{2j} are functions of coordinates that are quadratic in $\boldsymbol{q}$. To these coordinates, P is added to make the quartet

$$\mathbf{Q} = \left(|\varrho|^2; \Re(\varrho^*\varrho'); |\varrho'|^2; P\right) \tag{5.5}$$

Although the matrix entries do not depend explicitly on the local value of $\Re(\varrho^*\varrho')$, they will gain a dependence on its initial value through the other quadratic variables.

5.2 The matrix array

The matrix entries of A are expanded as a series in these coordinates:

$$\mathrm{A} = \mathbf{Q} \cdot \underline{\mathrm{A}} : \underline{\mathrm{A}} \equiv \begin{bmatrix} \underline{0} & \underline{1} \\ \underline{A_{21}} & \underline{A_{22}} \end{bmatrix} \tag{5.6}$$

where the dot-product is applied to each matrix entry. The arrays for the top two entries are trivial:

$$\underline{k} = [k, 0 \ldots 0] : k \in 0, 1$$

The arrays for the bottom two entries require a translation of the functions $A_{2j}(\mathbf{Q})$.

As described in Section 3.2, the translation of a function to its array is performed by transforming products to convolutions and composite functions to composite series using Taylor expansions. Specifically, the functions in Eq. (5.2b) are translated as

$$\underline{A_{21}} \equiv \frac{1}{2}(\underline{1} + \underline{I_3}) \times \underline{\Gamma^{-1}} \times \underline{2\partial_1\Gamma}$$

$$\underline{A_{22}} \equiv \frac{1}{2}(\underline{1} + \underline{I_3}) \times \underline{\Gamma^{-1}} \times \underline{-\partial_z\Gamma}$$

where $\underline{I_3}$ is the array of identity coefficients corresponding to $|\varrho'|^2$. The function Γ is translated as

$$\underline{\Gamma} = \frac{1}{2}(\underline{\gamma^2} - \underline{1}) + \underline{I_4} \times \underline{\dot{\chi}} - \frac{1}{2}\underline{I_1} \times \underline{\dot{\chi}^2}$$

with

$$\underline{\gamma} = (1+u)\underline{1} - \underline{\phi^e}$$
$$\underline{\dot{\chi}} = \underline{I_1^{-1}} \times \underline{\psi} : \underline{\psi} = -\frac{1}{2}\underline{\partial_1^{-1}\partial_z\phi^b}$$

As described by Eq. (D.10a), the electromagnetic potentials are expanded as a series in $|\varrho|^2$ (the magnitude of the transverse position is the same in the rotating and fixed frames). Their nontrivial coefficients in the power series in $\mathbf{Q}$ correspond to indices $\mathrm{l} = (l_1 l_2 l_3 l_4) = (l_1 000)$. These coefficients are related to the axial derivatives of the base potential $\bar{\phi} = \phi_{0000}$ as described by Eq. (D.10b).

The array for γ^2 can be derived from Eq. (3.5) with $g = \gamma$ and $\nu = 2$ or more readily by the convolution $\underline{\gamma} \times \underline{\gamma}$; the array for $\dot{\chi}^2$ is handled similarly.

As prescribed by Eq. (3.4a), the Taylor coefficients (Eq. (3.4b)) for the partial derivative of Γ with respect to Q_1 in the expression for $\underline{A_{21}}$ are given by the Taylor coefficients for Γ with the indices bumped by (1000): that is,

$$\{\partial_1\Gamma\}_{\mathrm{l}} = \{\Gamma\}_{\mathrm{l}+\mathbb{1}(1)} : \mathbb{1}(1) = (1000)$$

In the expression for $\underline{\dot{\chi}}$, the integral of ϕ^b with respect to Q_1 is handled similarly, but with the indices bumped by (-1000); that is,

$$\{\psi\}_{\mathrm{l}} = -\frac{1}{2}\{\partial_z\phi^b\}_{\mathrm{l}-\mathbb{1}(1)}$$

Here, the convolution with $\underline{I_1^{-1}}$ bumps the indices of the regular coefficients back by (1000) to make

$$\dot{\chi}_{\mathrm{l}} = \psi_{\mathrm{l}+\mathbb{1}(1)} = -\frac{\mathrm{l}!}{2(\mathrm{l}+\mathbb{1}(1))!}(\partial_z\phi^b)_{\mathrm{l}} = -\frac{1}{2(l_1+1)}(\partial_z\phi^b)_{\mathrm{l}}$$

Last, to obtain the array for Γ^{-1}, the principal contribution to the array $\underline{\Gamma}$ is extracted,

$$\underline{\Gamma} = \underline{1}\bar{\Gamma} + \underline{\tilde{\Gamma}}$$

$$\bar{\Gamma} = \frac{1}{2}(\bar{\gamma}^2 - 1) : \bar{\gamma} = 1 + u - \bar{\phi}^e$$

and the decomposition is introduced into Eq. (3.5) with $g = \Gamma$ and $\nu = -1$ to make

$$\underline{\Gamma^{-1}} = \bar{\Gamma}^{-1}((\bar{\Gamma}^{-1}\underline{\tilde{\Gamma}}) \otimes \underline{\mathcal{C}^{-1}})$$

5.3 Relation between the local and initial coordinates

The driving force in the ODE for the vector $\boldsymbol{q}$, Eq. (5.1), is described in matrix form with the matrix a function of the quadratics $\mathbf{Q}$ (Eq. (5.5)). As such, it is natural to represent the coordinates $\boldsymbol{q}$ as a product of a matrix C with the initial vector $\boldsymbol{q}_0$, with the matrix expanded as a series in the initial quadratics. Stated mathematically,

$$\boldsymbol{q}(z) = \mathrm{C}(z)\boldsymbol{q}_0 : \mathrm{C}(z) \equiv \mathbf{Q}_0 \cdot \underline{\mathrm{C}}(z) \tag{5.7}$$

where $\underline{\mathrm{C}}$ like $\underline{\mathrm{A}}$ represents a 2×2 matrix array of coefficients. Concomitantly, the local quadratic coordinates are themselves expressed formally as a power series in their initial values:

$$\mathbf{Q}(z) = \mathbf{Q}_0 \cdot \underline{\boldsymbol{c}}(z) : \underline{\boldsymbol{c}} = (\underline{c_1}; \underline{c_2}; \underline{c_3}; \underline{c_4}) \tag{5.8}$$

where it is reminded that the dot-product is applied individually to each of the four arrays in the set. The last array of coefficients, which represents a constant of motion, is equal to the array of identity coefficients for Q_4:

$$\underline{c_4} = \underline{I_4} \tag{5.9a}$$

making $P = P_0$ as required. The series coefficients for the remaining three quadratic variables are derived from products of the linear coordinates:

$$Q_{i+j-1} = \Re(q_i^* q_j) : i, j \in 1, 2$$

Introducing Eqs. (5.7) and (5.8) into this relation and noting that $\underline{\mathrm{C}}$ is real, the remaining arrays in $\underline{\boldsymbol{c}}$ are found as

$$\underline{c_{i+j-1}} = \sum_{kl} \underline{C_{ik}} \times \underline{C_{jl}} \times \underline{I_{k+l-1}} : i, j, k, l \in 1, 2 \tag{5.9b}$$

Eqs. (5.7) and (5.8) give the local, linear and quadratic coordinates as a series in the initial quadratics, paving the way for transforming the ODE, Eq. (5.1), to an ODE for the matrix array of coefficients $\underline{\mathrm{C}}$.

5.4 The integral equation in the rotating frame

Inserting Eq. (5.8) into Eq. (5.6) and applying the multinomial expansion, Eq. (3.3b), the series coefficients for A are transformed to coefficients in the initial quadratics:

$$\mathrm{A} = (\mathbf{Q}_0 \cdot \underline{c}) \cdot \underline{\mathrm{A}} = \mathbf{Q}_0 \cdot (\underline{c} \otimes \underline{\mathrm{A}})$$

Introducing this expression along with Eq. (5.7) into Eq. (5.1) gives the matrix differential equation

$$\underline{\mathrm{C}}' = (\underline{c} \otimes \underline{\mathrm{A}}) \times \underline{\mathrm{C}}$$

where the convolution is applied to each matrix entry; that is,

$$(\underline{\mathrm{A}} \times \underline{\mathrm{B}})_{ij} \equiv \sum_k \underline{\mathrm{A}_{ik}} \times \underline{\mathrm{B}_{kj}}$$

The matrix differential equation is decomposed by extracting the principal contributions, $\bar{\mathrm{A}}$ and $\bar{\mathrm{C}}$,

$$\underline{\mathrm{A}} \equiv \underline{1}\bar{\mathrm{A}} + \underline{\tilde{\mathrm{A}}}$$
$$\underline{\mathrm{C}} \equiv \underline{1}\bar{\mathrm{C}} + \underline{\tilde{\mathrm{C}}}$$

to make

$$\bar{\mathrm{C}}' = \bar{\mathrm{A}}\bar{\mathrm{C}} \tag{5.10a}$$
$$\underline{\tilde{\mathrm{C}}}' = \bar{\mathrm{A}}\underline{\tilde{\mathrm{C}}} + (\underline{c} \otimes \underline{\tilde{\mathrm{A}}}) \times \underline{\mathrm{C}} \tag{5.10b}$$

Here, use has been made of the relations

$$(\underline{c} \otimes \bar{\mathrm{A}}\underline{1}) \times \underline{\mathrm{C}} = \bar{\mathrm{A}}\underline{1} \times \underline{\mathrm{C}} = \bar{\mathrm{A}}\underline{\mathrm{C}}$$

The homogeneous equation, Eq. (5.10a), evolves $\bar{\mathrm{C}}$ to describe the fundamental rays. The differential equation is solved via serial integration by discretizing $\bar{\mathrm{C}}$ as described in Appendix C.2. By combining Eq. (5.10a) with the expression for the derivative of $\bar{\mathrm{C}}^{-1}$,

$$(\bar{\mathrm{C}}^{-1})' = \bar{\mathrm{C}}^{-1}\bar{\mathrm{C}}'\bar{\mathrm{C}}^{-1} = -\bar{\mathrm{C}}^{-1}\bar{\mathrm{A}}$$

the particular solution for the higher order matrices, for the lens aberrations, is transformed to an integral equation:

$$\bar{\mathrm{C}}^{-1}\underline{\tilde{\mathrm{C}}} = \int dz\, \bar{\mathrm{C}}^{-1}(\underline{c} \otimes \underline{\tilde{\mathrm{A}}}) \times \underline{\mathrm{C}} \tag{5.11}$$

Eq. (5.11) is a variant of the solution by parallel integration, Eq. (4.9d), tailored for the round lens. Like Eq. (4.9d), Eq. (5.11) is solved iteratively. The iteration starts with $\underline{\mathrm{C}} = \underline{1}\bar{\mathrm{C}}$ in the integrand, which includes evaluating Eq. (5.9b) for $\underline{c}$ with $C_{ij} = \bar{C}_{ij}$. There is no counterpart to Eq. (4.9c) for a round lens because $\bar{\boldsymbol{q}} = \mathbf{0}$ as there are no multipole fields to deflect the beam.

Eq. (5.11) is akin to the integral equation that is derived from perturbation theory for the primary aberration coefficients (Hawkes & Kasper, 2018, chapter 24). The notable difference here is that the iterative solution automates the calculation of the aberration coefficients to any desired order.

5.5 Transformation to the fixed coordinate system

Eq. (5.7) relates the local and initial coordinates in the rotating frame of reference. What is desired is an analogous relation in the fixed frame. The vector of coordinates in the fixed frame is given by

$$\boldsymbol{r} \equiv (\rho; \rho') : \rho \equiv \varrho e^{-i\chi} \tag{5.12a}$$

with χ defining the gyro-phase:

$$\chi = \int dz\, \chi' : \chi' = w^{-1}\dot{\chi} \tag{5.12b}$$

The axial speed in the expression for χ' is expressed in the rotating frame as

$$w = \sqrt{\frac{2\Gamma}{1 + |\varrho'|^2}} \tag{5.12c}$$

with Γ given by Eq. (5.2c).

The linear relation between the fixed and rotating coordinate systems is expressed in matrix form as

$$\boldsymbol{r} = \mathrm{B}\boldsymbol{q} \tag{5.13a}$$

$$\mathrm{B} = e^{-i\chi}(\mathrm{I} - i\chi'\mathrm{b}) \tag{5.13b}$$

with

$$\mathrm{I} = \begin{bmatrix} 1 & 0 \\ 0 & 1 \end{bmatrix} : \mathrm{b} = \begin{bmatrix} 0 & 0 \\ 1 & 0 \end{bmatrix}$$

Introducing Eq. (5.7) into Eq. (5.13a) gives the linear relation between the coordinates in the fixed frame:

$$\boldsymbol{r} = \mathrm{BCB}_0^{-1}\boldsymbol{r}_0 : \mathrm{B}^{-1} = \mathrm{B}^* \tag{5.14}$$

What remains is the expansion of the matrices as a series in the initial quadratics in the fixed frame.

First, the matrix B must be expanded as a series in the initial quadratics in the rotating frame, $\mathbf{Q}_0$. To this end, the function χ' is expressed as a series in the local quadratics,

$$\chi' = \mathbf{Q} \cdot \underline{\chi'} : \underline{\chi'} = \underline{w^{-1}} \times \underline{\dot{\chi}} \tag{5.15a}$$

with the first array in the convolution calculated from Eq. (3.5) with $g = w^{-2}$ and $\nu = 1/2$. Using Eq. (5.12c), the array for g in this calculation is expressed as

$$\underline{g} = \frac{1}{2}(\underline{1} + \underline{I_3}) \times \underline{\Gamma^{-1}} \tag{5.15b}$$

The second array in the convolution for $\underline{g}$ is itself found from Eq. (3.5) with $g = \Gamma$ and $\nu = -1$. Introducing Eq. (5.8) into Eq. (5.15a) and applying Eq. (3.3b) gives the series coefficients of χ' in the initial quadratics:

$$\chi' = \mathbf{Q}_0 \cdot (\underline{c} \otimes \underline{\chi'})$$

The series coefficients for the gyro-phase χ are found from the integral of this expression with respect to z.

Introducing the series expansions for χ and χ' into Eq. (5.13b) gives the sought after series representation:

$$\mathrm{B} = \mathbf{Q}_0 \cdot \underline{\mathrm{B}} : \underline{\mathrm{B}} = \underline{e^{-i\chi}} \times \left[\underline{1}\mathrm{I} - i(\underline{c} \otimes \underline{\chi'})\mathrm{b}\right] \tag{5.16a}$$

The array of coefficients for the exponential of the gyro-phase is derived via a Taylor expansion as prescribed by Eq. (3.6). Setting $g = -i\int dz\,\chi'$ in Eq. (3.6) and extracting the principal component of its array,

$$i\underline{g} = \underline{1}\bar{\chi} + \int dz\,\underline{c} \otimes \underline{\tilde{\chi}'} : \bar{\chi} = \int dz\,\bar{\chi}'$$

makes

$$\underline{e^{-i\chi}} = e^{-i\bar{\chi}}\left(\left(-i\int dz\,\underline{c} \otimes \underline{\tilde{\chi}'}\right) \otimes \underline{e}\right) \tag{5.16b}$$

The series expansions for B and C are introduced into Eq. (5.14) to make

$$r = \mathbf{Q}_0 \cdot \underline{\mathrm{D}} r_0 : \underline{\mathrm{D}} \equiv \underline{\mathrm{B}} \times \underline{\mathrm{C}} \times \underline{\mathrm{B}}_0^* \tag{5.17}$$

As it stands, Eq. (5.17) describes the relation between the local and initial coordinates in a hybrid of the rotating and fixed coordinate systems. To complete the transformation, the initial quadratics in the rotating frame, $\mathbf{Q}_0$, must be expressed as functions of their counterparts in the fixed frame. These relations are derived from the linear relation between the two coordinate systems, Eq. (5.13a), as

$$|\varrho|^2 = |\rho|^2$$
$$\Re(\varrho^* \varrho') = \Re(\rho^* \rho')$$
$$|\varrho'|^2 = |\rho'|^2 + (\chi')^2 |\rho|^2 + 2\chi' \Im(\rho^* \rho') : \chi' = w^{-1} \dot{\chi}$$

The fourth quadratic relation is given by the expression for the canonical angular momentum in the fixed frame:

$$P = w \Im(\rho^* \rho') - \psi$$

In these quadratic relations, the axial speed is evaluated in the fixed frame as

$$w = \sqrt{\frac{\gamma^2 - 1}{1 + |\rho'|^2}} \tag{5.18}$$

The quadratic relations are encapsulated as the series

$$\mathbf{Q} = \boldsymbol{R} \cdot \underline{\boldsymbol{d}} \tag{5.19a}$$

$$\boldsymbol{R} \equiv \left(|\rho|^2; \Re(\rho^* \rho'); |\rho'|^2; \Im(\rho^* \rho') \right) \tag{5.19b}$$

with

$$\underline{d_j} = \underline{I_j} : j \in 1, 2$$
$$\underline{d_3} = \underline{I_3} + \underline{(\chi')^2} \times \underline{I_1} + 2\underline{\chi'} \times \underline{I_4}$$
$$\underline{d_4} = \underline{w} \times \underline{I_4} - \underline{\psi} : \underline{\psi} = \underline{I_1} \times \underline{\dot{\chi}}$$

The series coefficients of χ' in the quadratics $\boldsymbol{R}$ are derived from Eqs. (5.15a) and (5.15b) with 2Γ replaced by $\gamma^2 - 1$. The array $\underline{w}$ is calculated from Eq. (3.5) with $g = w^2$ and $\nu = 1/2$ with the array for g in this calculation expressed as

$$\underline{g} = (\underline{\gamma} \times \underline{\gamma} - \underline{1}) \times (\underline{I_3} \otimes \underline{C^{-1}})$$

Evaluating Eq. (5.19a) in the initial plane, $z = z_0$, introducing the result into Eq. (5.17), and applying Eq. (3.3b) yields a relation between the local and initial coordinates that is described purely in the fixed frame:

$$\boldsymbol{r} = \boldsymbol{R}_0 \cdot (\underline{\boldsymbol{d}_0} \otimes \underline{\mathrm{D}})\boldsymbol{r}_0 \tag{5.20}$$

It is customary to transform the series in $\boldsymbol{R}$ to one in the quartet of linear coordinates:

$$\boldsymbol{R} = \boldsymbol{r}^{(2)} \cdot \underline{\boldsymbol{I}}^{(2)} : \boldsymbol{r}^{(2)} \equiv \left(\rho; \rho^*; \rho'; (\rho')^*\right)$$

where

$$\begin{aligned} |\rho|^2 &= \rho\rho^* & &\rightarrow \underline{I_1}^{(2)} = \underline{I_1} \times \underline{I_2} \\ \Re(\rho^*\rho') &= \frac{1}{2}\left(\rho^*\rho' + \rho(\rho')^*\right) & &\rightarrow \underline{I_2}^{(2)} = \frac{1}{2}(\underline{I_2} \times \underline{I_3} + \underline{I_1} \times \underline{I_4}) \\ |\rho'|^2 &= \rho'(\rho')^* & &\rightarrow \underline{I_3}^{(2)} = \underline{I_3} \times \underline{I_4} \\ \Im(\rho^*\rho') &= \frac{1}{2}\left(\rho^*\rho' - \rho(\rho')^*\right) & &\rightarrow \underline{I_4}^{(2)} = \frac{1}{2}(\underline{I_2} \times \underline{I_3} - \underline{I_1} \times \underline{I_4}) \end{aligned}$$

Defining

$$\underline{\mathrm{D}}^{(2)} \equiv \underline{\boldsymbol{I}}^{(2)} \otimes (\underline{\boldsymbol{d}_0} \otimes \underline{\mathrm{D}})$$

and performing the matrix-vector products, the final forms for the series expansions of the position and slope become

$$\begin{aligned} \rho &= \boldsymbol{r}_0^{(2)} \cdot \underline{c_1}^{(2)} : \underline{c_1}^{(2)} = \underline{\mathrm{D}_{11}}^{(2)} \times \underline{I_1} + \underline{\mathrm{D}_{12}}^{(2)} \times \underline{I_3} \\ \rho' &= \boldsymbol{r}_0^{(2)} \cdot \underline{c_3}^{(2)} : \underline{c_3}^{(2)} = \underline{\mathrm{D}_{21}}^{(2)} \times \underline{I_1} + \underline{\mathrm{D}_{22}}^{(2)} \times \underline{I_3} \end{aligned} \tag{5.21a}$$

Taking the complex conjugates of the position and slope and using the relation

$$\left(\boldsymbol{r}^{(2)}\right)^* = (\rho^*; \rho; (\rho')^*; \rho') \equiv \boldsymbol{r}^{(2)} \cdot \underline{\boldsymbol{I}}^*$$

where

$$(\underline{I_1}^*; \underline{I_2}^*; \underline{I_3}^*; \underline{I_4}^*) = (\underline{I_2}; \underline{I_1}; \underline{I_4}; \underline{I_3})$$

the coefficients for the remaining two coordinates are found as

$$\begin{aligned} \rho^* &= \boldsymbol{r}_0^{(2)} \cdot \underline{c_2}^{(2)} : \underline{c_2}^{(2)} = \underline{\boldsymbol{I}}^* \otimes \underline{c_1}^* \\ (\rho')^* &= \boldsymbol{r}_0^{(2)} \cdot \underline{c_4}^{(2)} : \underline{c_4}^{(2)} = \underline{\boldsymbol{I}}^* \otimes \underline{c_3}^* \end{aligned} \tag{5.21b}$$

The series expansions for the position, the slope, and their complex conjugates are encapsulated as

$$\boldsymbol{r}^{(2)} = \boldsymbol{r}_0^{(2)} \cdot \underline{\boldsymbol{c}}^{(2)} : \underline{\boldsymbol{c}}^{(2)} \equiv (\underline{c_1}; \underline{c_2}; \underline{c_3}; \underline{c_4})^{(2)} \tag{5.22}$$

5.6 Calculation of the chromatic aberrations

To incorporate the chromatic aberrations, the initial kinetic energy u, which appears only in the function γ (Eq. (5.3)), is decomposed into its mean value $\bar{u}$ and the spread about the mean, $\tilde{u}$; and the dependence on the relative energy spread $\tilde{u}/\bar{u}$, which happens to be a constant of motion, is incorporated in the series expansion of the matrix A at the start of the analysis by appending $\tilde{u}/\bar{u}$ to the quartet of quadratics $\mathbf{Q}$ in Eq. (5.5): that is,

$$\gamma = \mathbf{Q} \cdot \underline{\gamma} : \underline{\gamma} = (1 + \bar{u})\underline{1} - \underline{\phi^e} + \bar{u}\underline{I_5} \tag{5.23a}$$

$$\mathbf{Q} = \left(|\varrho|^2; \Re(\varrho^* \varrho'); |\varrho'|^2; P; \tilde{u}/\bar{u}\right) \tag{5.23b}$$

The amendment to $\mathbf{Q}$ is accompanied by an analogous one to $\boldsymbol{R}$ and to $\boldsymbol{r}^{(2)}$:

$$\boldsymbol{R} = \left(|\rho|^2; \Re(\rho^* \rho'); |\rho'|^2; \Im(\rho^* \rho'); \tilde{u}/\bar{u}\right) \tag{5.24a}$$

$$\boldsymbol{r}^{(2)} = \left(\rho; \rho^*; \rho'; (\rho')^*; \tilde{u}/\bar{u}\right) \tag{5.24b}$$

A similar prescription may be applied for examining the fluctuations in the applied currents and voltages that produce the electromagnetic fields.

6. Validation of the algorithm

As alluded to in the introduction to Section 5, the SA algorithm is validated in a few steps. First, the primary nonlinear coefficients of a round lens, as represented in the quadratic coordinates $\boldsymbol{R}$ (Eq. (5.24a)), are calculated using Eq. (5.20) and compared to the analytic results derived from Eq. (E.6) for an electromagnetic configuration studied by Glaser: one with no electric field and a bell-shaped magnetic field profile. Then, continuing with Glaser's configuration, the primary (3rd order) and secondary (5th order) nonlinear coefficients of a round lens, as represented in the coordinates $\boldsymbol{r}^{(2)}$ (Eq. (5.24b)), are calculated via Eq. (5.22) and compared to the results derived from the array equation that is general to a multipole optical element, Eq. (D.8), as described in Appendix D.2. The coefficients derived from this array equation are calculated using both serial (Eq. (4.5b)) and parallel (Eq. (4.9d)) integration.

The SA calculations are performed by an object-oriented code that has been developed in MATLAB®. All of the functionalities required to manipulate a power series, including the operators for the binomial and multinomial expansions, are conveniently codified in the PowerSeries object.

The calculations to follow are performed for the specific case where the rays start at $z = -1$ and are focused at $z = +1$ in units of λ, the scale length for the bell-shaped magnetic field. The analysis in Appendix E.1 shows that this condition can be achieved with many combinations of kinetic energies, $\bar{u}$, and magnetic field strengths, b_0:

$$\frac{b_0 \lambda}{2\bar{w}} = \sqrt{3} : \bar{w} = \sqrt{2\bar{u} + \bar{u}^2}$$

Here, the kinetic energy is set to $\bar{u} = 10$ keV.

Fig. 2 shows the pertinent parameters that enter into the calculation. Subplot (A) shows the axial profiles of the gyro-phase and its first three derivatives with the first derivative reflecting the bell-shaped profile of the magnetic field. Starting with $b^{(0)} = (1 + z^2)^{-1}$ and $b^{(1)} = -2z(1 + z^2)^{-2}$, higher order derivatives of the bell-shaped profile can be derived exactly from the following recursion relation:

$$b^{(n+2)} = -b^{(0)}\left(2(n+2)zb^{(n+1)} + c_{n+1}b^{(n)}\right)$$
$$c_{n+1} = c_n + 2(n+1) : c_0 = 0$$

Subplot (B) depicts the initial and tailored grids. The initial grid is composed of 100 uniform segments on which the exact values of the derivatives of the gyro-phase are evaluated. These values enter into the matrix entries A_{2j}. As described in the next subsection, the grid is tailored to place an upper bound ϵ on the relative error in the calculation of $\|\bar{\mathrm{D}}\|$, the determinant of the principal matrix. For this calculation, the values of the matrix A on the uniform grid are interpolated onto the tailored grid using cubic-spline. Eqs. (5.10a) and (5.10b) are then solved on the tailored grid for the matrix C. The values of the matrix C on this grid are then interpolated back onto the uniform grid using cubic-spline once more.

The relative error in $\|\bar{\mathrm{D}}\|$ for an upper bound or precision of $\epsilon = 10^{-6}$ is illustrated in subplot (C). The bound on the error in $\|\bar{\mathrm{D}}\|$ also limits the errors in the matrix entries as shown in subplot (D). The metric ϵ is

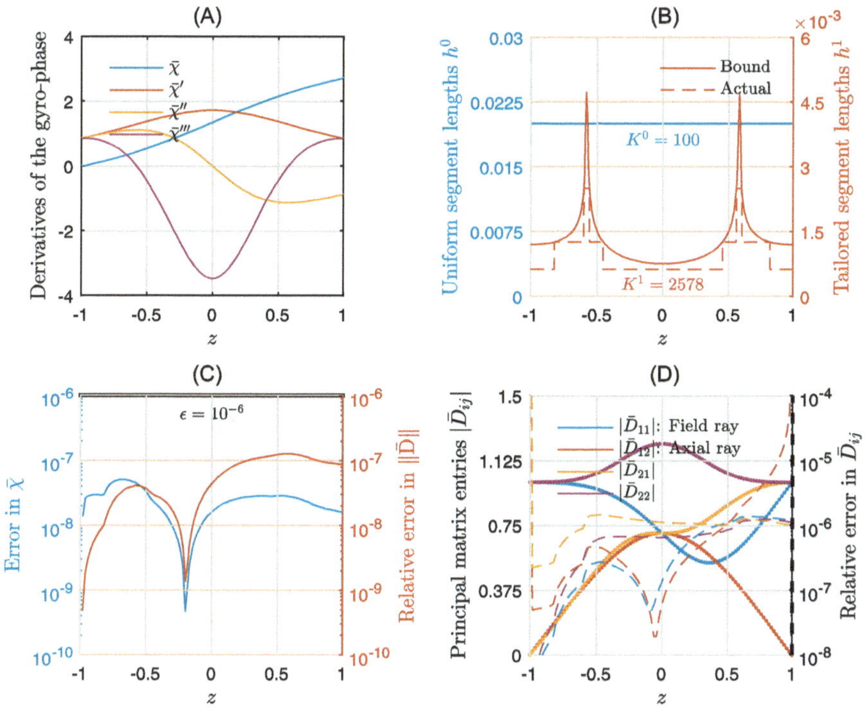

Figure 2 Subplot (A): the on-axis derivatives of the gyro-phase that impact the series coefficients to first order in the quadratic coordinates. Subplot (B): the segment lengths for the uniform grid (left side) and the tailored grid (right side) for a precision of $\epsilon = 10^{-6}$. Subplot (C): the error in the calculation of the on-axis gyro-phase (left side) and the relative error in the calculation of the matrix determinant (right side) for $\epsilon = 10^{-6}$. Subplot (D): left side, a comparison of the analytic (—) and the algorithmic (bullet) calculations of the principal matrix entries (magnitudes); right side, the relative error between the two calculations for $\epsilon = 10^{-6}$ (dashed line).

effective in curtailing the errors because it is derived from a quantity that is conserved by the ODE: the Helmholtz-Lagrange invariant.

6.1 The tailored grid

Using Eq. (5.10a) and Jacobi's formula (Eq. (C.1a)), the evolution of $\|\bar{\text{C}}\|$ is derived exactly to give the invariant

$$\frac{\|\bar{\text{C}}(z)\|}{\|\bar{\text{C}}(z_0)\|} e^{-\int dz \, \text{tr}\,\bar{\text{A}}} = 1 \tag{6.1a}$$

By extracting the principal contributions in Eqs. (5.2b) and (5.3) and inserting the results into $\bar{\mathrm{A}}$, the exponential term reduces to the square-root of the ratio of the axial speeds at the two locations:

$$e^{-\int dz\ \mathrm{tr}\,\bar{\mathrm{A}}} = e^{\frac{1}{2}(\ln \bar{w}(z) - \ln \bar{w}(z_0))} = \sqrt{\frac{\bar{w}(z)}{\bar{w}(z_0)}} \tag{6.1b}$$

If the ratio of $\|\bar{\mathrm{C}}\|$ were evaluated at the conjugate planes of the lens, it would reduce to the product of the linear and angular magnifications to give the Abbe sine law in the rotating frame of reference.

The determinant of the principal matrix in the fixed frame is derived from Eq. (E.4). Introducing Eq. (6.1a) into the expression for $\|\bar{\mathrm{D}}\|$, the exact relation for its evolution across a segment is found as

$$\|\bar{\mathrm{D}}(z+h)\| = \exp(\varepsilon^0)\|\bar{\mathrm{D}}(z)\|$$

$$\varepsilon^0 = \int_z^{z+h} dz\,(\mathrm{tr}\,\bar{\mathrm{A}} - 2i\bar{\chi}')$$

This exact relation is then compared to the approximate relation

$$\|\bar{\mathrm{D}}(z+h)\| \approx \exp(\varepsilon^1)\|\bar{\mathrm{D}}(z)\|$$

derived from the discrete evolution of $\bar{\mathrm{C}}$ and from the discretization of the integral for $\bar{\chi}$. The former is described by Eq. (C.1b). The latter is given by the Trapezoid rule as

$$\int_z^{z+h} dz\,\bar{\chi}' \approx \frac{1}{2}h\big(\bar{\chi}'(z) + \bar{\chi}'(z+h)\big) \equiv h\langle\bar{\chi}'\rangle$$

For Glaser's configuration, where there is no electric field, the axial speed is a constant. As a result, the contribution from $\mathrm{tr}\,\bar{\mathrm{A}}$ vanishes to make $\|\bar{\mathrm{C}}(z+h)\| = \|\bar{\mathrm{C}}(z)\|$. An evaluation of Eq. (C.3a) shows that this equality also holds to order h^3 for the discretized evolution of $\|\bar{\mathrm{C}}(z)\|$. Hence, to this order, the relative error between the exact and discretized relations,

$$\big|\exp(\varepsilon^1 - \varepsilon^0) - 1\big| \approx |\varepsilon^1 - \varepsilon^0|$$

is reduced to

$$|\varepsilon^1 - \varepsilon^0| = 2\left|h\langle\bar{\chi}'\rangle - \int_z^{z+h} dz\,\bar{\chi}'\right|$$

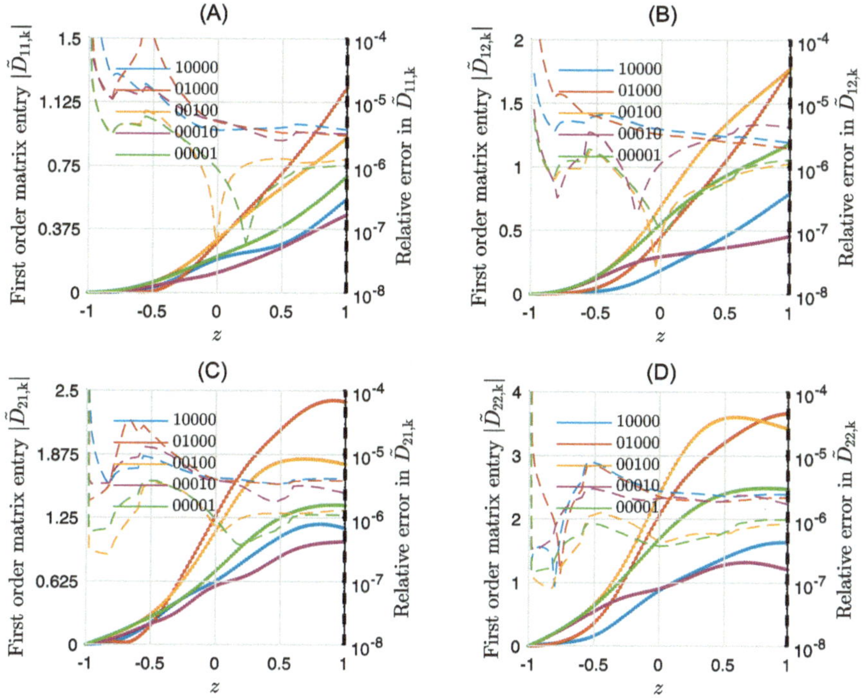

Figure 3 Subplots (A), (B), (C), and (D): left side, a comparison of the analytic (solid line) and the algorithmic (bullet) calculations of the first order matrix coefficients (magnitudes) for the 4 matrix entries, respectively; right side, the relative error between the two calculations (dashed line).

Taylor expanding this expression to 3rd order in h gives

$$|\varepsilon^1 - \varepsilon^0| \approx \frac{1}{6}h^3|\bar{\chi}'''| \equiv (h/\eta)^3 : \eta = \left|\frac{1}{6}\bar{\chi}'''\right|^{-1/3}$$

As described in Appendix C.2, the meshing algorithm chooses h in proportion to the value of η (Eq. (C.5a)) to place an upper bound on the relative error ϵ in the calculation of $\|\bar{\mathrm{D}}\|$ over all segments (Eq. (C.5b)). Hence, as shown in Fig. 3B, the tailored grid points become concentrated in regions where $|\bar{\chi}'''|$ is large. The particular algorithm that is used places the segment lengths h within a factor of 2 of the upper bound $\hat{h}\eta$ with $\hat{h}(\epsilon)$ described by Eq. (C.6a).

6.2 Calculation of the primary nonlinear coefficients

Given that the Helmholtz-Lagrange invariant is an inherent characteristic of the ODE (and of the optics), it stands to reason that placing a bound on

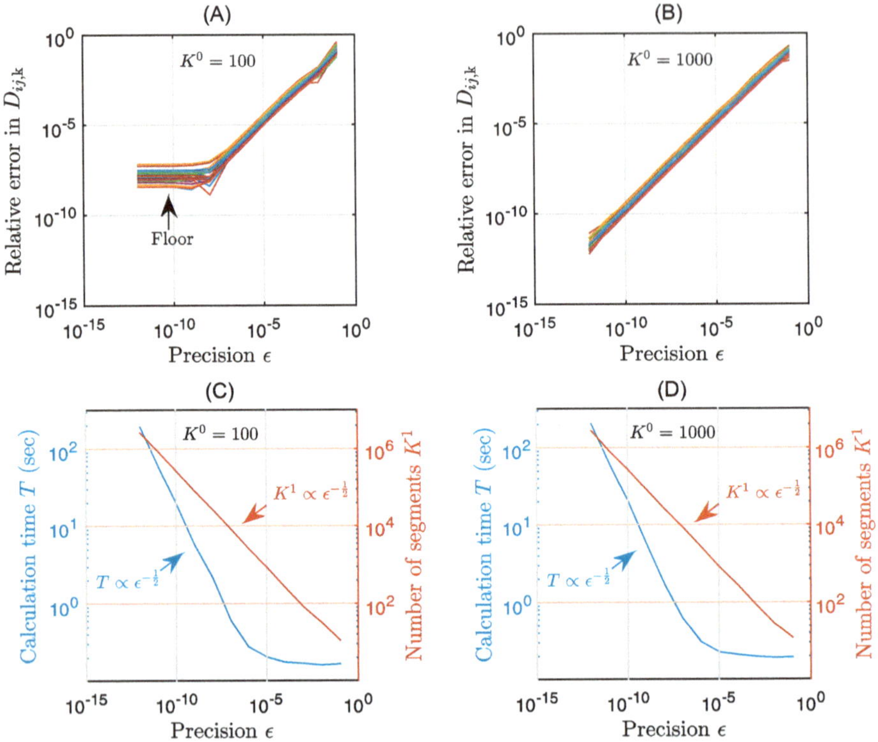

Figure 4 Subplots (A) and (B): the relative error between the analytic and algorithmic calculations of the principal and first order matrix entries evaluated at the image plane ($z = 1$) for a starting grid density K^0 of 100 and 1000, respectively; the data for the axial ray, $D_{12,\mathbb{O}}$, is excluded as its value is 0 at the image plane. Subplots (C) and (D): the dependence of the calculation time T (left side) and the tailored grid density K^1 (right side) on the precision ϵ for the two starting grid densities, respectively.

the error in its calculation will bound the error in the calculation of the series coefficients in the quadratics for the first order matrix. This conjecture is validated in Fig. 3, which compares the analytic calculation of these coefficients, Eq. (E.6), with the algorithmic calculation, Eq. (5.20), for the four matrix entries.

Fig. 4 shows the relative error in the calculation of the coefficients for the principal and first order matrix entries evaluated at the image plane $z = 1$ as a function of the requested precision ϵ. Subplot (A) shows the calculation for a starting grid with $K^0 = 100$ uniform segments with lengths $h^0 = 2/K^0$. As seen, the relative error reduces in commensurate with the requested precision, but then plateaus for $\epsilon < 10^{-8}$. This floor arises from the errors incurred in the SA algorithm by interpolating the starting values of $\bar{\chi}'$ and of the entries in the matrix A onto the tailored grid points. As already stated, the mapping between the two grids is performed by a cubic-spline interpolation, which has an error that scales as the product of the 4th derivative of the function being interpolated with $(h^0)^4$. The impact of the interpolation error vanishes for a starting grid with a higher density, $K^0 = 1000$, as illustrated in subplot (B).

Subplots (C) and (D) show the dependence of the calculation time T and of the tailored segment density K^1 on the requested precision ϵ for the two starting grid densities K^0. As expected from a 2nd order discretization scheme, the tailored grid density scale as $\epsilon^{-1/2}$. For moderate values of ϵ, the computation time remains below 1 second. However, as the demand for precision increases, the computation of Eq. (5.11) eventually dominates the overall calculation to make $T \propto \epsilon^{-1/2}$ as well. A comparison of the subplots (C) and (D) shows that the starting grid density has no material impact on the computation time.

6.3 Calculation of the secondary nonlinear coefficients

Analytic calculations of the higher order nonlinear coefficients are not pursued here. To validate the SA algorithm for the calculation of these coefficients, a comparison is made between the calculation of Eq. (5.22), which is specific to the round lens, and the calculation of Eq. (D.8), which is more general, applicable to a multipole lens. Both array equations describe the series in the coordinates $\boldsymbol{r}^{(2)}$ (Eq. (5.24b)).

The tailored grid for the general array equation differs from that derived for the array equation that is specific to the round lens because the two equations have different forms. The function η associated with the general array equation is derived from the homogeneous equation of Eq. (D.6),

$$\rho'' = -i(\bar{\chi}''\rho + 2\bar{\chi}'\rho')$$

and its complex conjugate. These equations are recast as a matrix differential equation, $C' = AC$, for the four variables in $\boldsymbol{r}^{(2)}$ (Eq. (5.24b)). The resulting 4×4 matrix A is found to have a zero trace, which makes $\varepsilon^0 = 0$ (Eq. (C.1a)). The discretized relation for the evolution of $\|C\|$ gives ε^1 (Eq. (C.3b)) and thereby η (Eq. (C.4b)) as

$$|\varepsilon^1| \approx 2h^3|\bar{\chi}'\bar{\chi}''| \equiv (h/\eta)^3 : \eta = |2\bar{\chi}'\bar{\chi}''|^{-1/3}$$

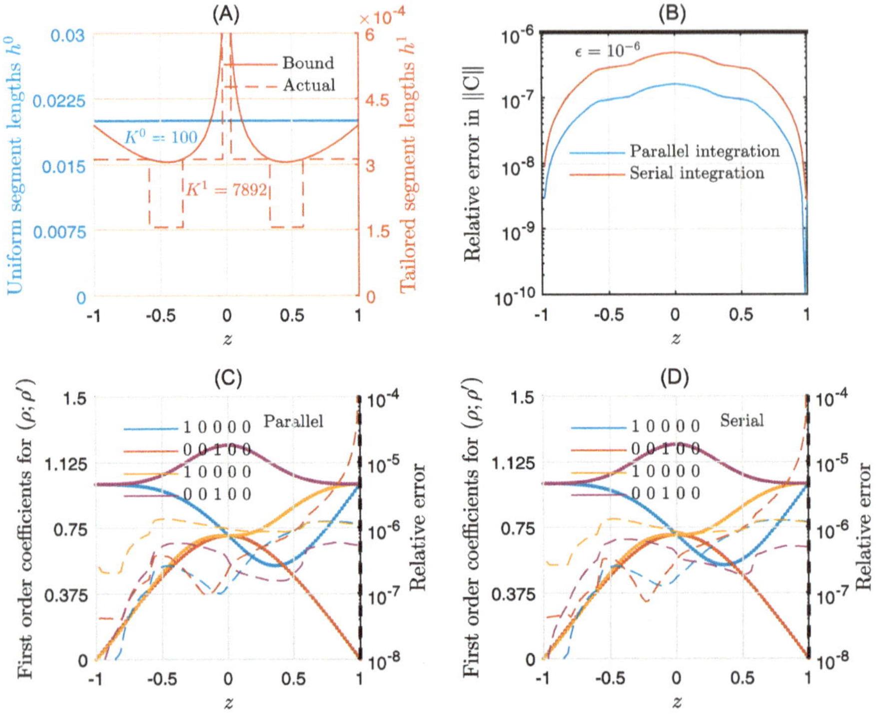

Figure 5 Subplot (A): the segment lengths for the uniform grid (left side) and the tailored grid (right side) for a precision of $\epsilon = 10^{-6}$. Subplot (B): the relative error in the calculation of the matrix determinant between the general (multipole) ray equation and the ray equation specific to the round lens for $\epsilon = 10^{-6}$ with the former derived via parallel and serial integration. Subplots (C) and (D): left side, a comparison of the calculations of the fundamental rays (magnitudes) using the round lens algorithm (solid line) and the multipole lens algorithm (bullet) with the latter derived via parallel and serial integration, respectively; right side, the relative error between the calculations from the round and the multipole lens algorithms for $\epsilon = 10^{-6}$ (dashed line).

The resulting tailored segment lengths are shown in Fig. 5A for a precision of $\epsilon = 10^{-6}$. The grid points are shown to concentrate near $z = \pm 0.5$, where η maximizes.

Eq. (D.8) is solved on the tailored grid using both serial and parallel integration as described in Sections 4.1 and 4.2. As shown in Fig. 5B, the relative error in the calculation of $\|\mathrm{C}\|$ between the general and the round-lens specific algorithms remains below ϵ for both integration methods. Subplots (C) and (D) compare the calculations of the fundamental rays for the position ρ and slope ρ'. The results from the round-lens specific

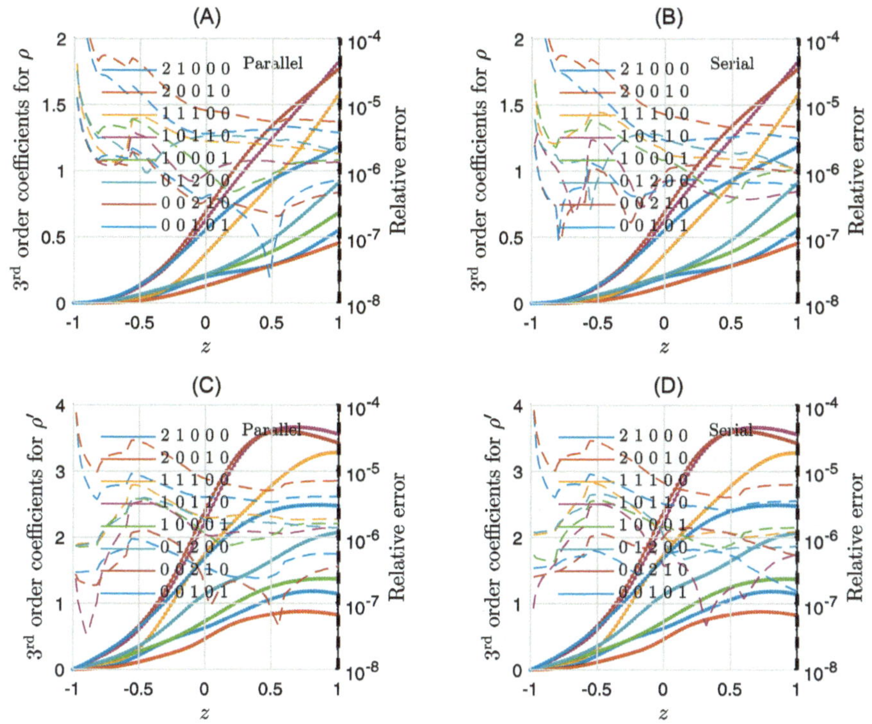

Figure 6 Subplots (A) and (B): left side, a comparison of the calculations of the 3rd order series coefficients (magnitudes) for the position ρ using the round lens algorithm (solid line) and the multipole lens algorithm (bullet) with the latter derived via parallel and serial integration, respectively; right side, the relative error between the calculations from the round and the multipole lens algorithms for $\epsilon = 10^{-6}$ (dashed line). Subplots (C) and (D): a repeat of the analysis in subplots (A) and (B) for the slope ρ'.

and the general algorithms are shown to agree for both integration methods. Figs. 6 and 7 show the agreement between the two algorithms for the calculations of the 3rd and 5th order series coefficients, respectively. Here, the ranking for the chromatic coefficients is derived by giving the coordinate $\tilde{u}/\bar{u}$ a weight of 2 to categorize, for example, the index (10001) as 3rd order and the indices (10002) and (21001) as 5th order.

Figs. 8A and B show the relative error in the calculations of the series coefficients at the image plane between the two algorithms. The relative error in the series coefficients through 5th order is graphed as a function of the precision in the calculation of $\|\mathrm{C}\|$. The graphs show that these errors reduce in commensurate with the demand for precision for both integration methods. The data is limited to $\epsilon \geq 10^{-6}$ because of the demand for

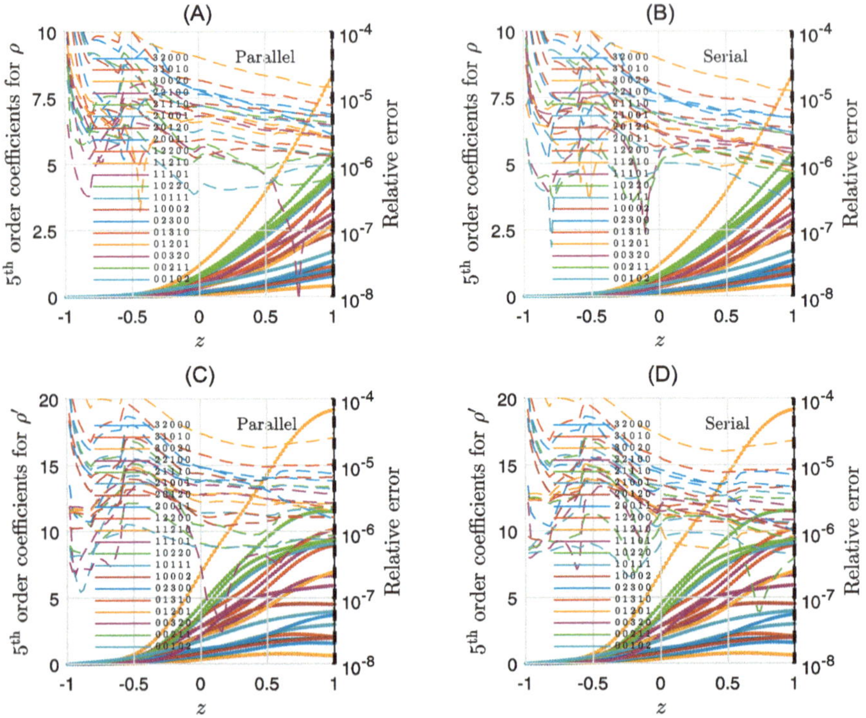

Figure 7 Subplots (A) and (B): left side, a comparison of the calculations of the 5th order series coefficients (magnitudes) for the position ρ using the round lens algorithm (solid line) and the multipole lens algorithm (bullet) with the latter derived via parallel and serial integration, respectively; right side, the relative error between the calculations from the round and the multipole lens algorithms for $\epsilon = 10^{-6}$ (dashed line). Subplots (C) and (D): a repeat of the analysis in subplots (A) and (B) for the slope ρ'.

memory by the calculation of the general algorithm, not that of the algorithm specific to the round lens. Even though both algorithms represent the series in $\boldsymbol{r}^{(2)}$ in the end, the algorithm that is specific to the round lens represents the series more compactly in the quadratic coordinates during the integration step of Eq. (5.11). Furthermore, the integration is performed with fewer tailored segments.

The grid density and the calculation time for the general algorithm are shown in subplots (C) and (D). The grid density is the same for both integration methods. The subplots illustrate the $\epsilon^{-1/2}$ dependencies of both the tailored grid density and the calculation time. A comparison of the two subplots shows that the calculation time of the general algorithm with serial integration is $\sim 10\times$ greater than the calculation time with parallel

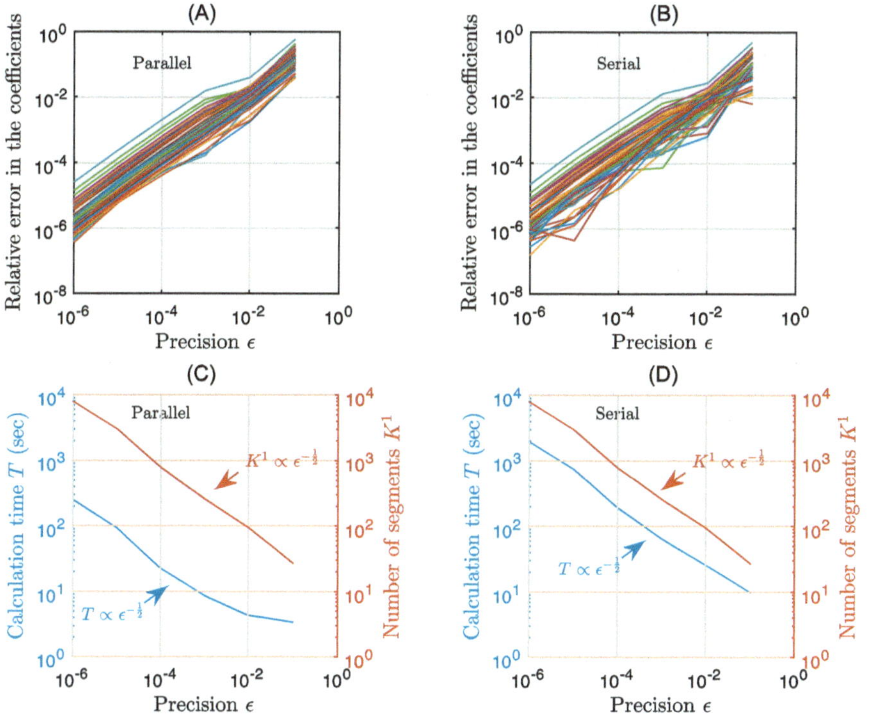

Figure 8 Subplots (A) and (B): the relationship between the precision ϵ and the relative error in the calculations of the 1st, 3rd, and 5th order coefficients for the position and slope at the image plane ($z = 1$) between the round and the multipole lens algorithms with the latter derived via parallel and serial integration, respectively; the data for the axial ray, (00100), is excluded as its value is 0 at the image plane. Subplots (C) and (D): the dependence of the calculation time T (left side) and the tailored grid density K^1 (right side) of the multipole lens algorithm on the precision ϵ for parallel and serial integration, respectively.

integration despite the common, tailored grid. The longer simulation time for serial integration arises from the repeated application of the $\otimes$ operator. The calculation time of the general algorithm with parallel integration is, itself, much longer than the calculation time for the algorithm that is specific to the round lens. The calculation time for the latter (which is not shown in the figure) is about 1 second for all $\epsilon \geq 10^{-6}$. This calculation is relatively fast because, as already mentioned, the truncated series for a given order is described with fewer coefficients in the quadratic coordinates while performing the integration in Eq. (5.11) and because, for a given precision, the tailored grid is not as dense.

The computational efficiency may be improved by employing a higher order difference scheme, such as the 4th order method provided by Runge-Kutta. The resulting scaling, $T \propto \epsilon^{-1/4}$, should reduce the computation time by an additional factor of $(K^1)^{1/4}$.

7. Conclusion

The proper design of charged particle instruments requires accurate simulations of the particle trajectories within the applied electromagnetic fields. The direct computation of the many trajectories comprising a beam of particles can be time consuming. Furthermore, what is of interest in many cases are the optical properties of the instrument, which are described by small deviations in the particle trajectories from the optical axis. Consequently, methods have been developed to characterize the optical properties themselves, circumventing the need to simulate the individual particle trajectories. Traditional methods use perturbation theory to manually derive expressions for properties like the primary aberration coefficients of a lens using, as input, the behavior of the electromagnetic fields about the optical axis. However, the manual procedure is difficult to extend for the computation of the higher order aberrations as the algebra becomes intractable.

Here, an algorithm has been presented that, like the DA approach, calculates these coefficients automatically. The algorithm makes use of the binomial theorem to codify the calculation of the aberration coefficients to any order to produce an algebraic solution for the particle coordinates in the neighborhood of the optical axis. This procedure is not limited to a specific instrument or geometry; in other words, it can be applied to many types of charged particle microscopes, prisms, spectrometers, energy filters, and accelerators with straight or curvilinear axes. Furthermore, it is not limited to the relativistic Newton's equation. It can also be applied to higher order nonlinear ordinary differential equations for the evolution of the variables in the neighborhood of some chosen path in the coordinate system. Hence, the algorithm should also be of interest to researchers in disciplines outside of charged particle optics.

Appendix A. Generalization of the algorithm

The solution to an N'th order nonlinear ODE can be derived analogously to what has been presented for $N = 2$. The analysis begins with a single variable p,

$$p^{(N)} \equiv \frac{d^N p}{dz^N} = f(\boldsymbol{q})$$

with $\boldsymbol{q}$ representing the set of N lower derivatives of p,

$$\boldsymbol{q} = (p^{(0)}; p^{(1)}; \ldots p^{(N-1)})$$

The solution via serial integration is virtually identical to what has been presented in Section 4.1 with

$$\boldsymbol{q}' = (p^{(1)}; p^{(2)}; \ldots f) : f = \boldsymbol{q} \cdot \underline{f}$$

In parallel integration, the function f is decomposed as

$$f = f_{\mathbb{0}} + \sum_n q_n f_{\mathbb{1}(n)} + \boldsymbol{q} \cdot \underline{f_{\mathbb{2}}}$$

where $f_{\mathbb{0}}$ is the zeroth order coefficient, $\mathbb{1}$ represents the first order indices, and $\underline{f_{\mathbb{2}}}$ houses the higher order coefficients. The linear contributions enter into the homogeneous equation

$$p^{(N)} = \sum_n q_n f_{\mathbb{1}(n)}$$

whose solution is expressed as a linear combination of the initial conditions, $\boldsymbol{q} = \mathrm{C}\boldsymbol{q}_0$ with C an $N \times N$ matrix housing the N first order coefficients and their $N-1$ derivatives,

$$C_{mn} = c_{\mathbb{1}(n)}^{(m-1)} : c_{\mathbb{1}}^{(m)} \equiv \frac{d^m c_{\mathbb{1}}}{dz^m}$$

and satisfying $\mathrm{C}' = \mathrm{AC}$ where

$$\begin{aligned} \mathrm{A}_{mn} &= \delta_{m+1,n} : m \in 1 \ldots N-1 \\ \mathrm{A}_{Nn} &= f_{\mathbb{1}(n)} \end{aligned} \tag{A.1}$$

The general solution is again derived using the method of variation of parameters, in which the coordinates $\boldsymbol{q}$ are expressed as linear combinations of the variable coefficients $\boldsymbol{\theta}$, $\boldsymbol{q} = \mathrm{C}\boldsymbol{\theta}$. Taking the derivative of this equation and noting the relation

$$\boldsymbol{q}' = \mathrm{A}\boldsymbol{q} + \boldsymbol{g} : \boldsymbol{g} \equiv (0; 0; \ldots 0; f_{\mathbb{0}} + \boldsymbol{q} \cdot \underline{f_{\mathbb{2}}})$$

where $\boldsymbol{g}$ is a column vector containing $N-1$ leading zeros, the first order differential equation for $\boldsymbol{\theta}$, Eq. (4.7b), and the resulting solution for $\boldsymbol{q}$, Eqs. (4.9c) and (4.9d), are recovered.

The solution is readily extended to a set of I variables $\boldsymbol{p}$. The ODE for I such variables is given by

$$\boldsymbol{p}^{(N)} = \boldsymbol{f}(\boldsymbol{q}) : \boldsymbol{f} \equiv (f_1; f_2; \ldots f_I)$$

with the coordinate $\boldsymbol{q}$ extended to contain the I variables $\boldsymbol{p}$,

$$\boldsymbol{q} = (\boldsymbol{p}^{(0)}; \boldsymbol{p}^{(1)}; \ldots \boldsymbol{p}^{(N-1)})$$

The resulting extension of the ODE for serial integration is trivial. For parallel integration, the solution is expressed with the matrix C extended from size N to size $J = NI$ for the J coordinates and first order indices. Concurrently, the 0 and 1 entries in the matrix A are extended to the zero and identity matrices of size I, respectively, and the last row of A is extended to an $I \times J$ matrix with the entries $f_{i,\mathbb{1}(j)}$. Last, the column vector $\boldsymbol{g}$ is generalized to a column vector of size J with $(N-1)I$ leading 0 entries with the last I entries provided by $f_{i,\mathbb{0}} + \boldsymbol{q} \cdot \underline{f_{i,\mathbb{2}}}$.

Appendix B. Compound series expansions

Assume that there exists a power series S in the set of I coordinates

$$\boldsymbol{r} = (r_1; r_2; \ldots r_I)$$

Defining the array of coefficients in this series as $\underline{s}$ makes

$$S = \boldsymbol{r} \cdot \underline{s} = \sum_{\mathrm{k}} \boldsymbol{r}^{\mathrm{k}} s_{\mathrm{k}} = \sum_{\mathrm{k}} \frac{1}{\mathrm{k}!} \boldsymbol{r}^{\mathrm{k}} \{s\}_{\mathrm{k}} \tag{B.1a}$$

where k describes the I indices or powers

$$\mathrm{k} \equiv (k_1 k_2 \ldots k_I)$$

$\boldsymbol{r}^{\mathrm{k}}$ is shorthand for

$$\boldsymbol{r}^{\mathrm{k}} \equiv \prod_i r_i^{k_i} \tag{B.1b}$$

and the second expression for the sum in Eq. (B.1a) is the Taylor series representation, using the Taylor coefficients

$$\{s\}_{\mathrm{k}} \equiv \mathrm{k}!s_{\mathrm{k}} \tag{B.1c}$$

where k! is shorthand for the product of the factorials of the indices,

$$\mathrm{k}! \equiv \prod_i k_i! \tag{B.1d}$$

Now assume that the coordinates r_i are themselves described by a power series in the set of J coordinates

$$\boldsymbol{q} = (q_1; q_2; \ldots q_J)$$

Of interest is the power series representation of the sum S in these coordinates. The transformation of the series in $\boldsymbol{r}$ into one in $\boldsymbol{q}$ is derived in two steps.

First, the coordinates r_i are decomposed as

$$r_i = \bar{r}_i + \tilde{r}_i(\boldsymbol{q})$$

with $\bar{r}_i$ defining the contribution that is independent of $\boldsymbol{q}$. This decomposition is inserted into Eq. (B.1a) to give the binomial expansion

$$S = \tilde{\boldsymbol{r}} \cdot (\bar{\boldsymbol{r}} \odot \underline{s}) \tag{B.2a}$$

where the old series has been regrouped as a series in the coordinates $\tilde{\boldsymbol{r}}$ with the new coefficients, themselves, a series in the coordinates $\bar{\boldsymbol{r}}$ as defined by the circled-dot operator $\odot$. The Taylor coefficients of the new series are expressed as

$$\{\bar{\boldsymbol{r}} \odot \underline{s}\}_{\mathrm{k}} \equiv \sum_{\mathrm{j}} \frac{1}{\mathrm{j}!} \bar{\boldsymbol{r}}^{\mathrm{j}} \{s\}_{\mathrm{j+k}} \tag{B.2b}$$

with j and k each representing I indices. For $\mathrm{k} = \mathbb{0}$, where

$$\mathbb{0} \equiv (0 \ldots 0)$$

Eq. (B.2b) reduces to the dot product

$$(\bar{\boldsymbol{r}} \odot \underline{s})_{\mathbb{0}} = \bar{\boldsymbol{r}} \cdot \underline{s} \tag{B.3a}$$

If $\bar{\boldsymbol{r}} = \mathbf{0}$, then only the $\mathrm{j} = \mathbb{0}$ term survives in Eq. (B.2b) to give the identity relation

$$\mathbf{0} \odot \underline{s} = \underline{s} \tag{B.3b}$$

Second, the coordinates $\tilde{r}_i$ are expanded as a power series in the J coordinates q_j,

$$\tilde{r}_i = \boldsymbol{q} \cdot \underline{c_i} = \sum_{\mathrm{l}} \boldsymbol{q}^{\mathrm{l}} c_{i,\mathrm{l}}$$

where $\underline{c_i}$ describes the coefficients in this series and l describes the J indices or powers

$$\mathrm{l} \equiv (l_1 l_2 \ldots l_J)$$

By design, $\tilde{r}_i$ has no contribution from $\mathrm{l} = \mathbb{0}$; that is,

$$c_{i,\mathbb{0}} = 0 \ \forall i$$

This property is required for the next step where the series for $\tilde{r}_i$ is inserted into the expression for S.

Defining the set of arrays

$$\underline{\boldsymbol{c}} \equiv (\underline{c_1}; \underline{c_2}; \ldots \underline{c_I})$$

and the array

$$\underline{d} \equiv \bar{\boldsymbol{r}} \odot \underline{s}$$

Eq. (B.2a) is transformed as

$$S = (\boldsymbol{q} \cdot \underline{\boldsymbol{c}}) \cdot \underline{d} = \boldsymbol{q} \cdot (\underline{\boldsymbol{c}} \otimes \underline{d}) \tag{B.4a}$$

with the new coefficients derived from a convolution of the old as defined by the circled-times operator $\otimes$. The convolution amounts to the multinomial expansion

$$(\underline{\boldsymbol{c}} \otimes \underline{d})_{\mathrm{m}} \equiv \sum_{\underline{\mathrm{k}}} \left[\prod_{i,\mathrm{l}} \frac{1}{k_{i,\mathrm{l}}!} (c_{i,\mathrm{l}})^{k_{i,\mathrm{l}}} \right] \{d\}_{\mathrm{k}} \tag{B.4b}$$

This formula describes a constrained sum over the set of indices

$$\underline{\mathrm{k}} = (\underline{k_1}; \underline{k_2}; \ldots \underline{k_I}) : \underline{k_i} \equiv [\ldots k_{i,\mathrm{l}} \ldots]$$

with k defining the collection of I indices k_i comprising the sums of $k_{i,1}$ over the indices l,

$$\mathrm{k} \equiv (k_1 k_2 \ldots k_I) : k_i = \sum_{\mathrm{l}} k_{i,\mathrm{l}} \tag{B.4c}$$

The values of $k_{i,\mathrm{l}}$ are constrained by the relation

$$\mathrm{m} = \sum_{i,\mathrm{l}} \mathrm{l} k_{i,\mathrm{l}} : \mathrm{m} \equiv (m_1 m_2 \ldots m_J) \tag{B.4d}$$

Whereas the values for the indices in l are bounded by the corresponding values in m, the values for the indices in k are bounded by the sum over all k_i. For $\mathrm{l} \neq \mathbb{0}$, this bound is equal to the sum of the indices in m as shown by the inequality

$$\sum_j m_j = \sum_{i,\mathrm{l}} \Big(\sum_j l_j\Big) k_{i,\mathrm{l}} \geq \sum_{i,\mathrm{l}} k_{i,\mathrm{l}} = \sum_i k_i$$

The constraint on the sum for m does not limit the values of $k_{i,\mathbb{0}}$; however, because of the condition placed on $c_{i,\mathbb{0}}$, the quantity $(\underline{\boldsymbol{c}} \otimes \underline{d})_{\mathrm{m}} = 0$ if any $k_{i,\mathbb{0}} \neq 0$. The contribution from $k_{i,\mathbb{0}} = 0\,\forall i$ only impacts the index $\mathrm{m} = \mathbb{0}$, for which

$$(\underline{\boldsymbol{c}} \otimes \underline{d})_{\mathbb{0}} = d_{\mathbb{0}} \tag{B.5a}$$

If $\underline{\boldsymbol{c}} = \underline{\boldsymbol{0}}$, then the only coefficient that survives is $d_{\mathbb{0}}$ to make

$$(\underline{\boldsymbol{0}} \otimes \underline{d})_{\mathrm{m}} = d_{\mathbb{0}} \delta_{\mathrm{m}\mathbb{0}} \tag{B.5b}$$

The special case of $\underline{\boldsymbol{c}} = \underline{\boldsymbol{I}}$ (the identity coefficients) with $I = J$ simplifies Eq. (B.4a) as

$$S = \boldsymbol{q} \cdot \underline{d} = \boldsymbol{q} \cdot (\underline{\boldsymbol{I}} \otimes \underline{d})$$

to give the identity relation

$$\underline{\boldsymbol{I}} \otimes \underline{d} = \underline{d} \tag{B.5c}$$

The circled-dot and circled-times operations, like the dot operation, are linear in the second array of coefficients as well as applicable to more than one such array. Hence, the coefficients in the multinomial expansion for a set of K sums

$$\boldsymbol{S} = (S_1; S_2; \ldots S_K)$$

may be expressed with $\underline{s}$ and $\underline{d}$ replaced by

$$\underline{\boldsymbol{s}} = (\underline{s_1};\ \underline{s_2};\ \ldots \underline{s_K})$$

and

$$\underline{\boldsymbol{d}} \equiv (\bar{\boldsymbol{r}} \odot \underline{\boldsymbol{s}})$$

Here, the special case of $\underline{\boldsymbol{d}} = \underline{\boldsymbol{I}}$ with $K = I$ makes

$$\boldsymbol{S} = \boldsymbol{q} \cdot \underline{\boldsymbol{c}} = \boldsymbol{q} \cdot (\underline{\boldsymbol{c}} \otimes \underline{\boldsymbol{I}})$$

to give a second identity relation,

$$\underline{\boldsymbol{c}} \otimes \underline{\boldsymbol{I}} = \underline{\boldsymbol{c}} \tag{B.5d}$$

B.1 Computation of the coefficients

The computation of the coefficients in the multinomial expansion is not a trivial task owing to the complex relation between the index m and the set of indices $k_{i,1}$. A description of methods for deriving all combinations of indices $k_{i,1}$ that contribute to a given index m will not be given here. Instead, a derivation is presented that determines the coefficients via a recursive application of the binomial expansion.

The derivation starts with Eq. (B.2a):

$$S = \tilde{\boldsymbol{r}} \cdot \underline{d} : \underline{d} = (\bar{\boldsymbol{r}} \odot \underline{s})$$

with the components of $\tilde{\boldsymbol{r}}$ represented as

$$\tilde{r}_i = \boldsymbol{q}^{\mathrm{l}_1} c_{i,\mathrm{l}_1} + \boldsymbol{q}^{\mathrm{l}_2} c_{i,\mathrm{l}_2} + \ldots \boldsymbol{q}^{\mathrm{l}_M} c_{i,\mathrm{l}_M}$$

where l_m can be any one of the M indices (excluding $\mathbb{0}$) that contributes to the truncated power series. Next, a partial sum of $\tilde{r}_i$ is defined as

$$\tilde{r}_i^{(m)} \equiv \boldsymbol{q}^{\mathrm{l}_{m+1}} c_{i,\mathrm{l}_{m+1}} + \boldsymbol{q}^{\mathrm{l}_{m+2}} c_{i,\mathrm{l}_{m+2}} + \ldots \boldsymbol{q}^{\mathrm{l}_M} c_{i,\mathrm{l}_M}$$

which makes $\tilde{r}_i^{(0)} = \tilde{r}_i$. Using the definition of the partial sum, S can be expressed as

$$S = (\boldsymbol{q}^{\mathrm{l}_1} \mathrm{c}_{\mathrm{l}_1} + \tilde{\boldsymbol{r}}^{(1)}) \cdot \underline{d}$$

which, by applying the binomial expansion, becomes

$$S \to \tilde{\boldsymbol{r}}^{(1)} \cdot \underline{d}^{(1)} : \underline{d}^{(1)} = (\boldsymbol{q}^{\mathrm{l}_1} \mathrm{c}_{\mathrm{l}_1}) \odot \underline{d}$$

with the Taylor coefficients for $\underline{d}^{(1)}$ given by

$$\{d^{(1)}\}_{\mathrm{k}} = \sum_{\mathrm{j}} \frac{1}{\mathrm{j}!} \boldsymbol{q}^{\mathrm{l}_1 j} (\boldsymbol{c}_{\mathrm{l}_1})^{\mathrm{j}} \{d\}_{\mathrm{j}+\mathrm{k}} : j \equiv \sum_i j_i$$

These coefficients are, themselves, a power series in $\boldsymbol{q}$ with the series containing a single coefficient: that is,

$$\{d^{(1)}\}_{\mathrm{k}} = \sum_{\mathrm{l}} \boldsymbol{q}^{\mathrm{l}} \left(\{d^{(1)}\}_{\mathrm{k}}\right)_{\mathrm{l}}$$

with

$$\left(\{d^{(1)}\}_{\mathrm{k}}\right)_{\mathrm{l}} = \sum_{\mathrm{j}} \frac{1}{\mathrm{j}!} \delta_{\mathrm{l},\mathrm{l}_1 j} (\boldsymbol{c}_{\mathrm{l}_1})^{\mathrm{j}} \{d\}_{\mathrm{j}+\mathrm{k}}$$

The above process may be repeated by replacing $\tilde{\boldsymbol{r}}^{(1)}$ with $\boldsymbol{q}^{\mathrm{l}_2}\boldsymbol{c}_{\mathrm{l}_2} + \tilde{\boldsymbol{r}}^{(2)}$ to make $S \to \tilde{\boldsymbol{r}}^{(2)} \cdot \underline{d}^{(2)}$. In general, S can be expressed as

$$S \to \tilde{\boldsymbol{r}}^{(m)} \cdot \underline{d}^{(m)}$$

with the Taylor coefficients for $\underline{d}^{(m)}$ derived from the following recursion relation:

$$\left(\{d^{(m)}\}_{\mathrm{k}}\right)_{\mathrm{l}} = \sum_{\mathrm{j}} \frac{1}{\mathrm{j}!} (\boldsymbol{c}_{\mathrm{l}_m})^{\mathrm{j}} \left(\{d^{(m-1)}\}_{\mathrm{j}+\mathrm{k}}\right)_{\mathrm{l}-\mathrm{l}_m j}$$

with the starting value given by

$$\left(\{d^{(0)}\}_{\mathrm{k}}\right)_{\mathrm{l}} = \delta_{\mathrm{l},\mathbb{0}} \{d\}_{\mathrm{k}}$$

Taking the above sequence to its limit gives

$$S \to \tilde{\boldsymbol{r}}^{(M)} \cdot \underline{d}^{(M)} = \mathbf{0} \cdot \underline{d}^{(M)}$$

The dot product with $\mathbf{0}$ isolates the contribution from $\mathrm{k} = \mathbb{0}$ to make

$$S = \boldsymbol{q} \cdot (\underline{c} \otimes \underline{d}) : (\underline{c} \otimes \underline{d})_{\mathrm{l}} = \left(\{d^{(M)}\}_{\mathbb{0}}\right)_{\mathrm{l}}$$

B.2 Inversion

Given the power series relation

$$\boldsymbol{q}_1 = \boldsymbol{q}_0 \cdot \underline{c} : \boldsymbol{c}_{\mathbb{0}} = \mathbf{0} \tag{B.6a}$$

it may be of interest to know the inverse relation

$$q_0 = q_1 \cdot \underline{d} : \underline{d} \equiv \underline{c}^{-1} \tag{B.6b}$$

Such is the case for a system containing two optical elements that are mirror symmetric, where the conjugate planes of the second element are the inverse of those of the first. The relation for the second optical element need not be found by simulating the optics of the entire element. Instead, it can be derived from the existing relation, Eq. (B.6a). The derivation is akin to the derivation of the inverse map in DA (Berz et al., 2015, chapter 5).

First, Eq. (B.6b) is inserted into Eq. (B.6a) and Eq. (3.3b) is applied to give

$$\underline{d} \otimes \underline{c} = \underline{I} \tag{B.7}$$

Next, the array $\underline{c}$ is decomposed into its linear and nonlinear contributions:

$$\underline{c} = \mathrm{C}\underline{I} + \underline{c_2} \tag{B.8}$$

Inserting Eq. (B.8) into Eq. (B.7) and using Eq. (B.5d) yields

$$\underline{d} = \mathrm{C}^{-1}(\underline{I} - \underline{d} \otimes \underline{c_2}) \tag{B.9}$$

This equation is solved iteratively starting with $\underline{d} \equiv \underline{0}$. The solution is found after K iterations, where K is the order of the truncated power series.

Eq. (B.6b) gives the relation between the coordinates at the conjugate planes of the second optical element. In general, what is of interest is a relation between q_1 and the coordinates throughout the second optical element. Starting with the existing relation for the coordinates in the first optical element,

$$q(z) = q_0 \cdot \underline{c}(z)$$

and introducing the inverse relation

$$q_0 = q_1 \cdot \underline{c}^{-1}(z_1)$$

the relation for the coordinates in the second optical element is found as

$$q(z) = q_1 \cdot \left(\underline{c}^{-1}(z_1) \otimes \underline{c}(z)\right)$$

Appendix C. Numerical integration

The Euler difference scheme will produce an overall error in $\boldsymbol{q}$ that scales linearly with the segment lengths, h. This linear dependence places a stringent requirement on the minimum number of segments and thereby on the minimum number of computations for a desired precision. The number of segments may be reduced by increasing the order of the precision of the difference scheme. As the demand for precision increases, the reduction in the computational cost due to the fewer segments should outweigh the computational cost per segment due to the higher order difference scheme.

In what is to follow, a second order difference scheme is presented. The description of the discretization scheme is followed by a prescription for choosing the segment lengths in commensurate with the gradients in the forcing function of the ODE. The segmentation levels the discretization errors over all segments to further reduce their number for a desired precision.

C.1 Discretization

For a second order difference scheme, the change in $\boldsymbol{q}$ across a segment is approximated by the trapezoid rule; that is, by averaging the derivative of $\boldsymbol{q}$ over the two ends to make

$$\Delta \boldsymbol{q} = h\langle \boldsymbol{q}' \rangle : \boldsymbol{q} \equiv (q_1; q_2; \ldots q_J)$$

with

$$\langle \boldsymbol{u} \rangle \equiv \frac{1}{2}\big(\boldsymbol{u}(z, \boldsymbol{q}) + \boldsymbol{u}(z + h, \boldsymbol{q} + \Delta \boldsymbol{q})\big)$$

The second term is Taylor expanded in $\boldsymbol{q}$, but not in z as this functional dependence is predefined over all segments. The expansion of the second term in $\boldsymbol{q}$ need only be performed to first order in h to give a result that is precise to order h^2. Hence,

$$\boldsymbol{u}(z + h, \boldsymbol{q} + \Delta \boldsymbol{q}) \approx \Big(1 + \sum_j \Delta q_j \partial_j\Big)\boldsymbol{u}(z + h, \boldsymbol{q}) : \partial_j \equiv \frac{\partial}{\partial q_j}$$

where the difference in each coordinate q_j is approximated using the first order Euler scheme: $\Delta q_j \approx h q_j'$. Expressing $\langle \boldsymbol{q}' \rangle$ as a power series in $\boldsymbol{q}$ gives

$$\Delta \boldsymbol{q} = \boldsymbol{q} \cdot \underline{\Delta \boldsymbol{q}} : \underline{\Delta \boldsymbol{q}} \approx h\underline{\langle \boldsymbol{q}' \rangle}$$

where

$$\underline{\langle \boldsymbol{u} \rangle} \equiv \frac{1}{2}\left(\underline{\boldsymbol{u}}(z) + \underline{\boldsymbol{u}}(z+h)\right) + \frac{1}{2} h \sum_j \underline{q_j'}(z) \times \underline{\partial_j \boldsymbol{u}}(z+h)$$

with the convolution applied individually to each array of coefficients in the set $\underline{\partial_j \boldsymbol{u}}$. The Taylor coefficients for the partial derivatives of $\boldsymbol{u}$ are found as

$$\{\partial_j \boldsymbol{u}\}_{\mathbb{1}} = \{\boldsymbol{u}\}_{\mathbb{1}+\mathbb{1}(j)}$$

with $\mathbb{1}$ representing the collection of first order indices.

C.2 Segmentation

Here, a prescription is given for choosing the length of each segment h to level the discretization errors in the calculation of the determinant of the matrix C for the linear coefficients ($\boldsymbol{q} = \mathrm{C}\boldsymbol{q}_0$). For this invertible matrix, Jacobi's formula gives the following relation for the derivative of $\|\mathrm{C}\|$:

$$\frac{d \ln \|\mathrm{C}\|}{dz} = \mathrm{tr}(\mathrm{C}^{-1}\mathrm{C}')$$

where tr signifies the trace operation. Inserting the relation

$$\mathrm{C}' = \mathrm{AC}$$

into the trace and noting the trace identity

$$\mathrm{tr}[\mathrm{A}, \mathrm{B}] = 0 : [\mathrm{A}, \mathrm{B}] \equiv \mathrm{AB} - \mathrm{BA}$$

makes

$$\frac{d \ln \|\mathrm{C}\|}{dz} = \mathrm{tr}(\mathrm{C}^{-1}\mathrm{AC}) = \mathrm{tr}\,\mathrm{A}$$

Hence, the evolution of $\|\mathrm{C}\|$ can be expressed exactly as

$$\|\mathrm{C}(z+h)\| = \exp(\varepsilon^0)\|\mathrm{C}(z)\|$$
$$\varepsilon^0 \equiv \int_z^{z+h} dz\,\mathrm{tr}\,\mathrm{A} \tag{C.1a}$$

Eq. (C.1a) is compared to the expression found from the discretization of C across the segment,

$$\Delta \mathrm{C} = \int_z^{z+h} dz\,\mathrm{C}' \approx h\langle \mathrm{A}\rangle \mathrm{C}$$

where, consistent with the Trapezoid rule used to obtain $\Delta \boldsymbol{q}$, the value of the integrand is averaged at the two ends of the segment to make

$$\langle \mathrm{A} \rangle = \frac{1}{2}\big(\mathrm{A}(z) + \mathrm{A}(z+h)\big) + \frac{1}{2} h \mathrm{A}(z+h)\mathrm{A}(z)$$

The discretized relation for the evolution of $\|\mathrm{C}\|$ across a segment becomes

$$\begin{aligned} \|\mathrm{C}(z+h)\| &\approx \exp(\varepsilon^1)\|\mathrm{C}(z)\| \\ \varepsilon^1 &\equiv \ln \|\mathrm{I} + h\langle \mathrm{A} \rangle\| \end{aligned} \tag{C.1b}$$

with I representing the identity matrix.

The evolution of $\|\mathrm{C}\|$ as determined by the exact (Eq. (C.1a)) and discretized (Eq. (C.1b)) relations will agree to 2nd order in h. Hence, $h\langle \mathrm{A} \rangle$ is Taylor expanded to 3rd order in h to determine the error:

$$\langle \mathrm{A} \rangle \approx \mathrm{A} + \frac{1}{2} h(\mathrm{A}' + \mathrm{A}^2) + \frac{1}{4} h^2 (\mathrm{A}'' + 2\mathrm{A}'\mathrm{A}) \tag{C.2}$$

To derive ε^1 to 3rd order in h, it is first realized that for a matrix B, $\|e^{\mathrm{B}}\| = e^{\mathrm{tr\,B}}$: a relation that follows from the fact that the determinant of a matrix is given by the product of its eigenvalues. Setting $\mathrm{B} = \ln(\mathrm{I} + h\langle \mathrm{A} \rangle)$ and using the Mercator series expansion for the natural-log yields

$$\varepsilon^1 = -\sum_{n \geq 1} \frac{(-h)^n}{n} \operatorname{tr} \langle \mathrm{A} \rangle^n \tag{C.3a}$$

This series converges in practice, where the eigenvalues of $h\langle \mathrm{A} \rangle$ fall within the unit circle of the complex domain. Introducing Eq. (C.2) into Eq. (C.3a) and expanding ε^1 to 3rd order in h gives

$$\varepsilon^1 \approx h \operatorname{tr} \mathrm{A} + \frac{1}{2} h^2 \operatorname{tr} \mathrm{A}' + h^3 \Big(\frac{1}{4} \operatorname{tr} \mathrm{A}'' - \frac{1}{6} \operatorname{tr} \mathrm{A}^3 \Big) \tag{C.3b}$$

Expanding ε^0 to 3rd order in h,

$$\varepsilon^0 \approx h \operatorname{tr} \mathrm{A} + \frac{1}{2} h^2 \operatorname{tr} \mathrm{A}' + \frac{1}{6} h^3 \operatorname{tr} \mathrm{A}''$$

and subtracting it from the expansion for ε^1 gives the relative error between the exact and approximate relations for $\|\mathrm{C}(z+h)\|$ as

$$\big|\exp(\varepsilon^1 - \varepsilon^0) - 1\big| \approx \big|\varepsilon^1 - \varepsilon^0\big| \equiv (h/\eta)^3 \tag{C.4a}$$

with

$$\eta = \left| \frac{1}{12} \operatorname{tr} A'' - \frac{1}{6} \operatorname{tr} A^3 \right|^{-1/3} \tag{C.4b}$$

Inserting the general expression for A, Eq. (A.1), into Eq. (C.4b) and noting that the trace isolates the diagonal elements yields

$$\begin{aligned} \eta &= \left| \frac{1}{6} \operatorname{tr} g \right|^{-1/3} \\ g &= \frac{1}{2} f''_{\mathbb{1}(N)} - f^3_{\mathbb{1}(N)} - 3 f_{\mathbb{1}(N)} f_{\mathbb{1}(N-1)} - 3 f_{\mathbb{1}(N-2)} \end{aligned} \tag{C.4c}$$

for a single variable p. The trace operation is kept in the above expression for the case of I variables, where $f_{\mathbb{1}(M)}$ is replaced by a square matrix of size I with elements

$$F_{ij} = f_{i,\mathbb{1}((M-1)I+j)}$$

The length of each segment is restricted to

$$\alpha \hat{h} \leq \frac{h}{\eta} \leq \hat{h} \tag{C.5a}$$

with α chosen to limit the total number of segments in the lens, K, and $\hat{h}$ chosen to limit the overall error ϵ in the calculation of $\|C\|$. Using the upper bound in Eq. (C.5a), Eq. (C.4a) is accumulated over all segments to give

$$\epsilon = \left| e^{\sum_{k=1}^{K} \varepsilon_k^1 - \varepsilon_k^0} - 1 \right| \leq \left| e^{K\hat{h}^3} - 1 \right| \approx K\hat{h}^3 \tag{C.5b}$$

Using the lower bound in Eq. (C.5a), the span of the lens H is evaluated as the sum of the segment lengths to give the inequality

$$\alpha \langle \eta \rangle K \hat{h} \leq H \tag{C.5c}$$

where $\langle \eta \rangle$ denotes the average value of η over all segments.

Combining Eqs. (C.5b) and (C.5c) and ascribing to ϵ the acceptable precision in the calculation of $\|C\|$, an upper bound is found for the choice of $\hat{h}$:

$$\hat{h} \leq \sqrt{\frac{\alpha \langle \eta \rangle}{H} \epsilon} \tag{C.6a}$$

The maximum value of $\hat{h}$ is chosen to limit the number of segments. Inserting this value into Eq. (C.5c) gives

$$K \leq \left(\frac{H}{\alpha\langle\eta\rangle}\right)^{3/2} \frac{1}{\sqrt{\epsilon}} \tag{C.6b}$$

Eq. (C.6b) reflects the fact that for a 2nd order discretization scheme, the number of segments required is inversely proportional to the square-root of the desired precision.

If Eq. (C.5a) were ignored and instead all segment lengths were chosen the same, $h = H/K$, then the constant of proportionality in Eq. (C.6b) would be modified as $(\langle\eta\rangle)^{-3/2} \rightarrow (\langle\eta^{-3}\rangle)^{1/2}$. The resulting sensitivity to the small values of η would increase the number of segments K for a desired precision ϵ.

Appendix D. Equation of motion for a round lens

The relativistic equation of motion is best described in the proper time and velocity, here reduced by the speed of light c:

$$d\tau \equiv cdt/\gamma$$
$$\dot{\vec{r}} \equiv d\vec{r}/d\tau : \vec{r} \equiv (x, y, z)$$

where γ defines the relativistic mass increase,

$$\gamma = \sqrt{1 + |\dot{\vec{r}}|^2} \tag{D.1}$$

In concert, the initial kinetic energy U and the electric and magnetic potential energies, Φ^e and Φ^b, are reduced by the mass energy, mc^2, as

$$(u, \phi^e, \phi^b) \equiv \frac{1}{mc^2}(U, \Phi^e, c\Phi^b)$$

For many practical cases, the effects of the beam space charge and current on the electromagnetic potentials may be neglected. In such cases, the potentials become governed by the Laplace equation:

$$\nabla^2\phi = 0$$

Parameterizing the optical axis with z and expressing the transverse position of the particle in complex coordinates,

$$\rho \equiv x + iy = |\rho|e^{i\varphi}$$

$$\nabla_\perp \equiv \partial_x + i\partial_y = 2\partial_{\rho^*} : \partial_q \equiv \frac{\partial}{\partial q}$$
$$\left|\dot{\vec{r}}\right|^2 = |\dot{\rho}|^2 + w^2 : w \equiv \dot{z}$$

the equation of motion is expressed as

$$\begin{aligned}\ddot{\rho} &= f \\ f &\equiv \partial_{\rho^*}\gamma^2 - iw\mathcal{D}_\rho\phi^b\end{aligned} \tag{D.2a}$$

with γ expressed as a function of the kinetic energy, Eq. (5.3), and

$$\mathcal{D}_\rho \equiv 2\partial_{\rho^*} - \rho'\partial_z : \rho' \equiv \frac{d\rho}{dz} \tag{D.2b}$$

Solving Eq. (D.1) for the axial speed gives

$$\begin{aligned}w &= \mathcal{W}_1(\gamma^2 - 1, |\dot{\rho}|) \\ \mathcal{W}_1(\alpha, \beta) &\equiv \sqrt{\alpha - \beta^2}\end{aligned} \tag{D.2c}$$

For a round lens, where the geometry possesses rotational invariance, the evolution of the coordinates becomes partly governed by the constancy of the canonical angular momentum,

$$P = |\rho|^2(\dot{\varphi} + \dot{\chi})$$

where $\dot{\chi}$ can be interpreted as the frequency of gyration in a magnetic field as measured from the perimeter of the orbit. The gyro-frequency is proportional to the magnetic flux per radian ψ as described by Eqs. (5.4a) and (5.4b).

The constancy of P gives $\dot{\varphi}$ as a function of $|\rho|$, thereby reducing the number of dependent variables in the ODE to one. The resulting ODE for $|\rho|$, however, has a forcing function with singular terms, scaling inversely with $|\rho|$, which are not amenable to the standard algebraic algorithm. These singularities can be sidestepped by working in a coordinate system that rotates with the gyro-phase χ:

$$\varrho \equiv \rho e^{i\chi} : \chi \equiv \int d\tau\, \dot{\chi}$$

In the rotating coordinate system, the expression for the canonical angular momentum reduces to

$$P = \Im(\varrho^*\dot{\varrho})$$

the ODE reduces to

$$\begin{aligned} \ddot{\varrho} &= 2\partial_{\varrho^*}\Gamma \\ \Gamma &\equiv \frac{1}{2}(\gamma^2 - 1) + P\dot{\chi} - \frac{1}{2}|\varrho|^2\dot{\chi}^2 \end{aligned} \tag{D.3}$$

and by decomposing $|\dot{\rho}|$,

$$|\dot{\rho}|^2 = |\dot{\varrho} - i\dot{\chi}\varrho|^2 = |\dot{\varrho}|^2 - 2P\dot{\chi} + \dot{\chi}^2|\varrho|^2$$

the axial speed can be expressed as

$$w = \mathcal{W}_1(2\Gamma, |\dot{\varrho}|)$$

D.1 The ray equation

If the particle travel time is not of interest, then the parameter *time* may be eliminated for a direct description of the ray path. While it is possible to derive the ray equation from the principle of least action, a more direct approach is found by transforming the time-derivative in the equation of motion as

$$\begin{aligned} \frac{d}{d\tau} &= w\frac{d}{dz} \\ \frac{d^2}{d\tau^2} &= w^2\frac{d^2}{dz^2} + \frac{1}{2}(w^2)'\frac{d}{dz} \end{aligned} \tag{D.4}$$

where, by inserting the relation $\dot{\rho} = w\rho'$ into Eq. (D.2c), the axial speed is expressed as a function of the axial derivative of ρ:

$$\begin{aligned} w &= \mathcal{W}_2(\gamma^2 - 1, |\rho'|) \\ \mathcal{W}_2(\alpha, \beta) &\equiv \sqrt{\frac{\alpha}{1+\beta^2}} \end{aligned} \tag{D.5}$$

Eq. (D.4) is introduced into Eq. (D.2a) to transform the derivatives in time to those in space. An inspection of Eq. (D.5) reveals that the expression for $(w^2)'$ in Eq. (D.4) has a dependence on ρ''. To unravel this dependence, Eq. (D.2a) is multiplied by $(\rho^*)'$, and the real component is extracted to make

$$(w^2)' = (\gamma^2)' - 2\Re(\rho' f^*)$$

with the complete derivative of γ^2 expressed as

$$(\gamma^2)' = \left(\partial_z + 2\Re(\rho'\partial_\rho)\right)\gamma^2$$

Introducing the expression for $(w^2)'$ into Eq. (D.2a), the general ray equation is found as

$$\rho'' = \frac{1}{2} w^{-2} \mathcal{D}_\rho \gamma^2 - w^{-1}(i\mathcal{D}_\rho - \rho'\mathcal{I})\phi^b \qquad \mathcal{I} \equiv \Im\left((\rho^*)'\mathcal{D}_\rho\right) \tag{D.6}$$

with $w = \mathcal{W}_2(\gamma^2 - 1, |\rho'|)$.

A similar derivation transforms Eq. (D.3) to give the ray equation for a round lens in the rotating frame of reference:

$$\varrho'' = w^{-2}\mathcal{D}_\varrho \Gamma : w = \mathcal{W}_2(2\Gamma, |\varrho'|) \tag{D.7}$$

D.2 The array equation

The array equation for the round lens is described in Section 5. Here, a description is given for deriving a more general array equation, which may also be applied to a multipole optical element. Defining

$$\boldsymbol{p} \equiv (\rho; \rho^*)$$

the ODE of Eq. (D.6) is expressed formally as

$$\boldsymbol{p}'' = \boldsymbol{f} = \boldsymbol{r}^{(2)} \cdot \underline{\boldsymbol{f}} : \underline{\boldsymbol{f}} = (\underline{f_1}; \underline{f_2}) \tag{D.8}$$

with the driving force represented as a series in the coordinates $\boldsymbol{r}^{(2)}$, defined by Eq. (5.24b).

The set of arrays $\underline{\boldsymbol{f}}$ is derived by using the prescriptions of Section 3.2. The products of functions in Eq. (D.6) are translated to convolutions of their series coefficients:

$$\underline{f_1} = \frac{1}{2}\underline{w^{-2}} \times \underline{\mathcal{D}_\rho \gamma^2} - \underline{w^{-1}} \times (i\underline{\mathcal{D}_\rho} - \underline{I_3} \times \underline{\mathcal{I}})\phi^b \tag{D.9a}$$

where $\underline{I_j}$ describes the identity coefficient for the jth coordinate and

$$\underline{\mathcal{I}} \equiv \frac{1}{2i}(\underline{I_4} \times \underline{\mathcal{D}_\rho} - \underline{I_3} \times \underline{\mathcal{D}_{\rho^*}}) \tag{D.9b}$$

The array of coefficients for the differential operator, Eq. (D.2b), and its complex conjugate are expressed as

$$\underline{\mathcal{D}_\rho} = 2\underline{\partial_2} - \underline{I_3} \times \underline{\partial_z} \qquad \underline{\mathcal{D}_{\rho^*}} = 2\underline{\partial_1} - \underline{I_4} \times \underline{\partial_z} \tag{D.9c}$$

where the partial derivatives with respect to the coordinates shift the index of the Taylor coefficients as described by Eq. (3.4a).

The derivation of the arrays involving the axial speed follow from similar derivations in Section 5.5. The notable difference here is that $|\rho'|^2$ is described as the product of the 3rd and 4th coordinates. Hence, using Eq. (D.5), the array for w^{-2} is expressed as

$$\underline{w^{-2}} = (\underline{1} + \underline{I_3} \times \underline{I_4}) \times \underline{(\gamma^2 - 1)^{-1}} \tag{D.9d}$$

with the second array in the convolution calculated from Eq. (3.5) with $g = \gamma^2 - 1$ and $\nu = -1$. The array for w^{-1} is then derived from Eq. (3.5) with $g = w^{-2}$ and $\nu = 1/2$.

The second array in the set $\underline{f}$ is derived similarly from the complex-conjugate of Eq. (D.6) as

$$\underline{f_2} = \frac{1}{2}\underline{w^{-2}} \times \underline{\mathcal{D}_{\rho^*}\gamma^2} + \underline{w^{-1}} \times (i\underline{\mathcal{D}_{\rho^*}} + \underline{I_4} \times \underline{\mathcal{I}})\phi^b \tag{D.9e}$$

D.3 The electromagnetic potentials

In an axisymmetric, cylindrical geometry, the smoothness of the electromagnetic potentials about the axis requires that their profiles be functions of $|\rho|^2 = x^2 + y^2$. Hence, the potential is expanded as a power series in $|\rho|^2$, and the result is inserted into the Laplace equation to derive a recursion relation for the coefficients. The result is a series representation of the potential energies with coefficients that are functions of the axial derivatives of their on-axis (base) profiles:

$$\phi(z, |\rho|^2) = \sum_l \phi_l(z)|\rho|^{2l} \tag{D.10a}$$

$$\phi_l(z) = \frac{(-1)^l}{(2^l l!)^2}\frac{d^{2l}}{dz^{2l}}\bar{\phi}(z) : \bar{\phi}(z) \equiv \phi(z, 0) \tag{D.10b}$$

Appendix E. Primary nonlinear coefficients

Eq. (5.20) is validated to first order in the quadratics $\boldsymbol{R}_0$ against the analytic calculation of the primary nonlinear coefficients for Glaser's electromagnetic configuration. This calculation is derived from the expression of Eq. (5.20) limited to first order in the quadratics. For this expression, it is sufficient to approximate the matrices A, B, C, and D to first order in their respective quadratic coordinates, where now these coordinates have

$\tilde{u}/\bar{u}$ appended for calculating the chromatic coefficients. Using the matrix A as the example, the first order approximation is expressed as

$$\underline{\mathrm{A}} = \underline{1}\bar{\mathrm{A}} + \underline{\tilde{\mathrm{A}}} : \underline{\tilde{\mathrm{A}}} \approx \sum_j \mathrm{A}_{\mathbb{1}(j)} \underline{I_j}$$

with $j \in 1...5$ for the five coordinates in **Q** (Eq. (5.23b)).

The approximation for the matrices is combined with the identity relation (Eq. (B.5d))

$$\underline{\boldsymbol{c}} \otimes \underline{I_j} = \underline{c_j}$$

to simplify the expressions in Sections 5.4 and 5.5 for transforming a series in **Q** to one in $\mathbf{Q}_0$. In particular, the expression for the primary nonlinear coefficients in the rotating frame of reference, Eq. (5.11), is reduced to

$$\bar{\mathrm{C}}^{-1}\underline{\tilde{\mathrm{C}}} \approx \sum_j \int dz \, \underline{c_j} \bar{\mathrm{C}}^{-1} \mathrm{A}_{\mathbb{1}(j)} \bar{\mathrm{C}} \tag{E.1}$$

with $\underline{c_j} : j \in 1, 2, 3$ obtained from Eq. (5.9b), evaluated with $\mathrm{C} = \bar{\mathrm{C}}$,

$$\underline{c_{i+j-1}} = \sum_{kl} \bar{C}_{ik} \bar{C}_{jl} \underline{I_{k+l-1}} : i, j, k, l \in 1, 2$$

and with $\underline{c_j} = \underline{I_j} : j \in 4, 5$ for the two constants of motion $(P, \tilde{u}/\bar{u})$.

The expressions for the nontrivial entries of $\mathrm{A}_{\mathbb{1}}$ are derived from Eq. (5.2b) with Γ given by Eq. (5.2c) and approximated as

$$\bar{\Gamma} = \frac{1}{2}(\bar{\gamma}^2 - 1)$$
$$\tilde{\Gamma} \approx \bar{\gamma}\tilde{\gamma} + P(\dot{\chi}_{\mathbb{0}} + \dot{\chi}_{\mathbb{1}(1)}|\varrho|^2) - \frac{1}{2}|\varrho|^2(\dot{\chi}_{\mathbb{0}}^2 + 2\dot{\chi}_{\mathbb{0}}\dot{\chi}_{\mathbb{1}(1)}|\varrho|^2)$$

The expressions for γ and $\dot{\chi}$ are derived from Eq. (5.3) with $u = \bar{u} + \tilde{u}$ and from Eqs. (5.4a) and (5.4b), respectively. They are approximated as

$$\bar{\gamma} = 1 + \bar{u} - \bar{\phi}^e$$
$$\tilde{\gamma} \approx \tilde{u} - \phi^e_{\mathbb{1}(1)}|\varrho|^2 - \phi^e_{\mathbb{2}(1)}|\varrho|^4 : \mathbb{2}(1) = 20000$$

and

$$\dot{\chi}_{\mathbb{0}} = -\frac{1}{2}(\bar{\phi}^b)'$$

$$\dot{\chi}_{\mathbb{1}(j)} = -\frac{1}{4}|\varrho|^2(\phi^b_{\mathbb{1}(1)})'\delta_{j1}$$

The coefficients in the series for the electromagnetic potentials are related to the axial derivatives of their base values as described by Eq. (D.10b). The second order terms in $\tilde{\Gamma}$, those proportional to $P|\varrho|^2$ and $|\varrho|^4$, are included (and only needed) to evaluate $\partial_{|\varrho|^2}\Gamma$ to first order in **Q**. The expression for Γ^{-1} is approximated to first order using the formula

$$f^\nu \approx \bar{f}^\nu(1 + \nu\bar{f}^{-1}\tilde{f})$$

To express the primary nonlinear coefficients in the fixed frame, Eq. (5.16a) for matrix B is evaluated to first order in $\mathbf{Q}_0$ to make

$$\bar{\mathrm{B}} = e^{-i\bar{\chi}}(\mathrm{I} - i\bar{\chi}'\mathrm{b}) \tag{E.2}$$

$$i\underline{\mathrm{B}} \approx \bar{\mathrm{B}}\int dz\,\underline{\tilde{\chi}'} + e^{-i\bar{\chi}}\underline{\tilde{\chi}'}\mathrm{b} \tag{E.3}$$

with the first order coefficients for the axial derivative of the gyro-phase given by

$$\underline{\tilde{\chi}'} \approx \sum_j \underline{\varsigma_j}\chi'_{\mathbb{1}(j)} : \chi'_{\mathbb{1}(j)} = (w^{-1})_{\mathbb{1}(j)}\dot{\chi}_{\mathbb{0}} + \bar{w}^{-1}\dot{\chi}_{\mathbb{1}(1)}\delta_{j1}$$

Using Eq. (5.12c) and the first order approximation to f^ν, w^{-1} is approximated to first order in **Q** as

$$w^{-1} \approx \frac{1}{\sqrt{2\bar{\Gamma}}}\left(1 + \frac{1}{2}(|\varrho'|^2 - \bar{\Gamma}^{-1}\tilde{\Gamma})\right)$$

The arrays for the matrices B and C are convolved as prescribed by Eq. (5.17) to give the array for matrix D to first order in $\mathbf{Q}_0$:

$$\bar{\mathrm{D}} = \bar{\mathrm{B}}\bar{\mathrm{C}}\bar{\mathrm{B}}_0^* \tag{E.4}$$

$$\underline{\tilde{\mathrm{D}}} \approx \underline{\tilde{\mathrm{B}}}\bar{\mathrm{C}}\bar{\mathrm{B}}_0^* + \bar{\mathrm{B}}\underline{\tilde{\mathrm{C}}}\bar{\mathrm{B}}_0^* + \bar{\mathrm{B}}\bar{\mathrm{C}}\underline{\tilde{\mathrm{B}}}_0^* \tag{E.5}$$

Last, the coordinates $\mathbf{Q}_0$ are expressed as a series in their counterparts in the fixed frame, $\boldsymbol{R}_0$, as described by Eq. (5.19a) with the array of coefficients approximated to first order as

$$\underline{d_j} = \underline{I_j} : j \in 1, 2, 5$$

$$\underline{d_3} \approx \underline{I_3} + (\bar{\chi}')^2\underline{I_1} + 2\bar{\chi}'\underline{I_4}$$

$$\underline{d_4} \approx \bar{w}\underline{I_4} - \dot{\chi}_{\odot}\underline{I_1}$$

and evaluated in the initial plane, z_0, to approximate Eq. (5.20) as

$$\boldsymbol{r} \approx \boldsymbol{R}_0 \cdot \left(\underline{1}\bar{\mathrm{D}} + \sum_j \left(\underline{d_j}\right)_0 \mathrm{D}_{\mathbb{1}(j)}\right)\boldsymbol{r}_0 \tag{E.6}$$

The transformation of $\boldsymbol{R}$ to $\boldsymbol{r}^{(2)}$ is not entertained here. For the purpose of validating the algorithm for the primary nonlinear coefficients, a comparison of the algorithmic calculation of Eq. (5.20) with the analytic calculation of Eq. (E.6) will suffice.

E.1 Glaser's configuration

Glaser's electromagnetic configuration (Grivet, 1965, chapter 9) is one with no electric field and a bell-shaped axial magnetic field profile: that is,

$$\bar{e}_z \equiv -(\bar{\phi}^e)' = 0$$
$$\bar{b}_z \equiv -(\bar{\phi}^b)' = \frac{b_0}{1+z^2}$$

Here, the axial and transverse coordinates are normalized to the characteristic length scale for the magnetic field. Calling this length λ, the homogeneous equation in the rotating frame is expressed as

$$\varrho'' + \alpha^2 \varrho = 0 : \alpha = \frac{\lambda \dot{\chi}_{\odot}}{\bar{w}}$$

with

$$\bar{w} = \sqrt{\bar{\gamma}^2 - 1} : \bar{\gamma} = 1 + \bar{u}$$
$$\dot{\chi}_{\odot} = \frac{1}{2}\bar{b}_z$$

By making the following change of variables

$$z = \cot\theta$$

and looking for a solution of the form $\varrho \propto \csc\theta g(\theta)$, the ODE simplifies to give the homogeneous solution as

$$\varrho \propto \csc\theta e^{\pm i\beta\theta} : \beta^2 \equiv 1 + \left(\frac{b_0\lambda}{2\bar{w}}\right)^2$$

This expression is encapsulated in matrix form to make

$$\bar{\mathrm{C}}(\theta) = \mathrm{c}(\theta)\mathrm{c}^{-1}(\theta_0)$$
$$\mathrm{c}(\theta) \equiv \begin{bmatrix} \csc\theta e^{+i\beta\theta} & \csc\theta e^{-i\beta\theta} \\ (\cos\theta - i\beta\sin\theta)e^{+i\beta\theta} & (\cos\theta + i\beta\sin\theta)e^{-i\beta\theta} \end{bmatrix}$$

The relation between the conjugate planes of this lens are found by setting $\bar{\mathrm{C}}_{12} = 0$:

$$\theta_n = \theta_0 - \frac{n\pi}{\beta} : \frac{\pi}{\beta} < \theta_0 \le \pi$$

where the bound on θ_0 describes the object locations for which the lens forms at least one image ($n \ge 1$).

The principal contribution to the gyro-phase is derived from the expression of its axial derivative,

$$\bar{\chi}' = \alpha = \sqrt{\beta^2 - 1}\sin^2\theta$$

whose integral gives

$$\bar{\chi} = \int dz\,\bar{\chi}' = -\int d\theta\,\csc^2\theta\,\alpha = \sqrt{\beta^2 - 1}(\theta_0 - \theta)$$

Last, $\tilde{\Gamma}$ is evaluated in the Glaser configuration and the resulting expression is introduced into the two non-trivial matrix entries of $\mathrm{A}_{\mathbb{1}}$ and into $\chi'_{\mathbb{1}}$ to give their five, first order coefficients in **Q** (Eq. (5.23b)) as

$$\begin{bmatrix} A_{21} \\ A_{22} \\ \chi' \end{bmatrix}_{\mathbb{1}} = \begin{bmatrix} \frac{1}{2}\alpha\alpha'' - \alpha^4 & 0 & -\alpha^2 & 2\alpha^3 - \frac{1}{4}\alpha'' & \frac{2\bar{\gamma}}{1+\bar{\gamma}}\alpha^2 \\ \alpha\alpha' & 0 & 0 & -\alpha' & 0 \\ \frac{1}{2}\alpha^3 - \frac{1}{8}\alpha'' & 0 & \frac{1}{2}\alpha & -\alpha^2 & -\frac{\bar{\gamma}}{1+\bar{\gamma}}\alpha \end{bmatrix}$$

with the axial derivatives of α (with respect to z) expressed as

$$\alpha' = -\alpha\sin 2\theta$$
$$\alpha'' = 2\alpha\sin\theta\sin 3\theta$$

Even though the integrals for the primary nonlinear coefficients in the rotating frame (Eq. (E.1)) can be solved in closed form, the myriad of terms involved make it more practical to codify the analysis presented in this section and to perform the resulting integrals using numerical quadrature. The precision of this numerical integration method is more than adequate for a comparison of the analytic calculations obtained with Eq. (E.6) with the algorithmic calculations derived from Eq. (5.20).

References

Berz, M. (1989). Differential algebraic description of beam dynamics to very high orders. *Particle Accelerators*, *24*, 109–124.

Berz, M. (1999). *Modern map methods in particle beam physics*. San Diego: Academic Press.

Berz, M., Makino, K., & Wan, W. (2015). *An introduction to beam physics*. Boca Raton: CRC Press.

Grivet, P. (1965). *Electron optics*. New York: Pergamon Press.

Hawkes, P., & Kasper, E. (2018). *Principles of electron optics, Vol. 1* (2nd ed.). San Diego: Academic Press.

Munro, E., Wang, L., Rouse, J., & Liu, H. (2006). Theory and application of the differential algebraic method for computing the aberrations of electron optical systems. In *IMC-16, Sapporo, Vol. 2* (p. 588).

PART TWO

The Nano-Aperture Ion Source

CHAPTER SEVEN

Introduction to focused ion beams, ion sources, and the nano-aperture ion source

Leon van Kouwen
Department of Imaging Physics, Delft University of Technology, Delft, Netherlands
e-mail address: lvankouwen@gmail.com

Contents

1. Why ion beams are fun

1.1 What is a focused ion beam system

A directed stream of ions traveling in a vacuum is what is called an ion beam. Excitement emerges from the precise control we can attain over such a beam by using electric and magnetic fields. It is common practice to focus an ion beam down to within a 10 nm diameter by using a high-end focused ion beam (FIB) system. Fig. 1 shows what such a FIB system usually looks like and what optics are involved in essence.

Sure enough such control of fields and particles is an achievement in itself, but it does have very practical applications indeed. To understand

Advances in Imaging and Electron Physics, Volume 212
ISSN 1076-5670
https://doi.org/10.1016/bs.aiep.2019.09.001

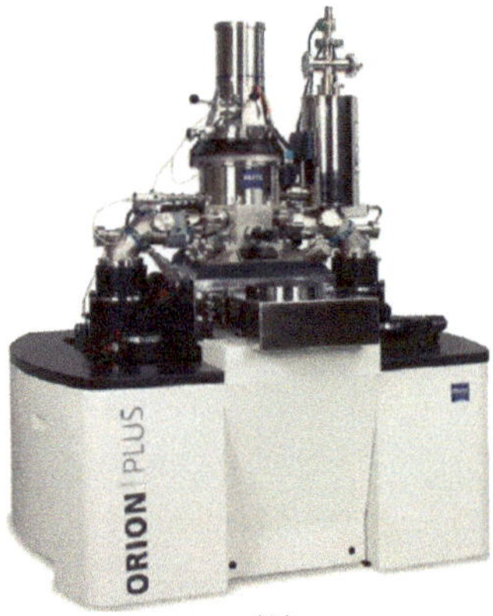

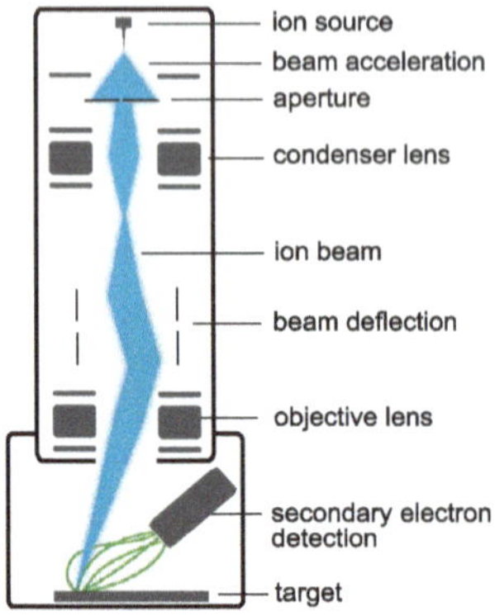

Figure 1 Focused ion beam systems consist of an optical column and the supporting elements such as vacuum pumps and electronics, together claiming one or two cubic meters of space. (A) The Carl Zeiss Orion is known for its unprecedented high resolution and for supplying helium ions. Image courtesy of Carl Zeiss Microscopy GmbH, printed with permission. (B) The essence of any focused ion beam system is the optical system. The beam emerges from an ion source after which the ions are accelerated up to 1 to 40 kV. A series of lenses and apertures accommodate high resolution spot formation. Beam deflection electrodes steer the beam to achieve patterning and imaging by raster scanning while simultaneously collecting secondary electrons as signal input.

what can be achieved we should look at the interactions ions have with a target material.

1.2 Ion-surface interactions

Besides a few exceptions, we like to accelerate the beam up to a high energy between 1 and 40 keV. This means a lot of energy is available to initiate various physical processes. The most important interactions are visualized in Fig. 2. The subject of ion-solid interaction is broad and literature is abundant (see for example Dake, King, Pitts, and Czanderna (2002) or Giannuzzi and Stevie (2005) or Schmidt and Wetzig (2012)). In this section the unfamiliar reader can get up to speed with some basics.

Very comparable to electron beam impact, is the energy transfer to electrons bound in the material. Some of these can escape into the vacuum to become so-called secondary electrons (SE's). This can be relatively efficient, for example the SE yield of Ga^+ 30 kV on silicon is 3.3 electrons per incident ion (Lipp et al., 1996) while for 30 keV electrons it is only 0.138 (Reimer & Tollkamp, 1980). In general the SE yield is in the order of 0.5

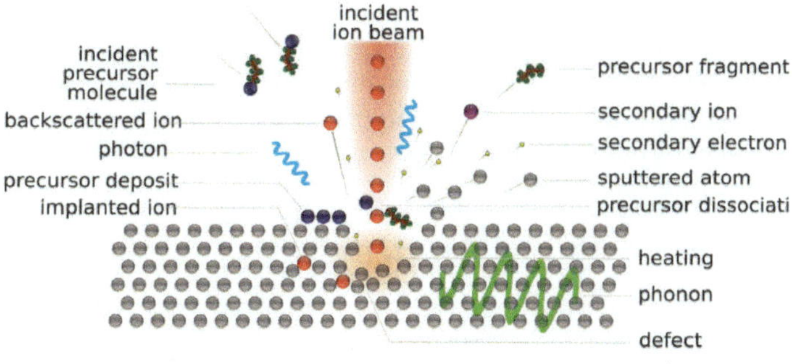

Figure 2 When a highly energetic ion beam hits a solid, a plurality of interactions occurs. The most utilized interactions are the generation of secondary electrons, sputtering of the solid and deposition by dissociation of precursor molecules. The relative occurrence of the interactions depends strongly on the materials involved, the ionic species, and the beam characteristics.

Table 1 The sputter yield (number of removed atoms per incident ion) for a 30 kV ion beam. Calculated using the empirical model by Yamamura and Tawara (1996). The material is assumed to be amorphous, the incidence of the beam is normal to the surface, and no redeposition effects are taken into account.

	C	Al	Si	Au
He^+	0.016	0.040	0.020	0.128
Ar^+	1.147	3.221	1.650	11.139
Ga^+	1.796	4.757	2.412	17.701
Au^+	2.529	6.747	3.379	28.225

to 15 depending on the ion species and the solid, and for a great deal on the angle of incidence (Castaldo, Withagen, Hagen, Kruit, & van Veldhoven, 2011).

Quite different from electron impact is the removal of material. Just like a cannonball removing bricks from a castle wall, an ion hitting a sample will knock out atoms. This is commonly referred to as sputtering and when performed in a controlled fashion, ion milling. Tables 1 and 2 can be used to get acquainted with the typical rate and efficiency with which this phenomenon occurs. What we can see directly from the table is that efficiency of sputtering is greatly dependent on the impact material and the ionic species involved.

The energy required to remove an atom from the top surface (surface binding energy) is in the order of some eV's, for example 4.66 eV for silicon

Table 2 The sputtering rate (volume removal rate) in um^3/s for a 30 kV ion beam with a current of 1 nA. Calculated using the empirical model by Yamamura and Tawara (1996). The material is assumed to be amorphous, the incidence of the beam is normal to the surface, and no redeposition effects are taken into account.

	C	Al	Si	Au
He^+	0.0006	0.0042	0.0025	0.0136
Ar^+	0.0407	0.3345	0.2058	1.1810
Ga^+	0.0637	0.4941	0.3008	1.8767
Au^+	0.0896	0.7007	0.4215	2.9926

(Kudriavtsev, Villegas, Godines, & Asomoza, 2005). Comparing this energy to the beam energy of keV's, while considering the values for the sputtering yields from Table 1, we conclude that only a fraction of the beam energy is transferred into sputtering.

Sputtering is not a simple process as many aspects such as re-deposition, angle of incidence dependent sputtering yield, and probe current-distributions with wide beam tails make matters complex (Lehrer, Frey, Petersen, & Ryssel, 2001; Frey, Lehrer, & Ryssel, 2003; Lindsey & Hobler, 2012). Furthermore, the presence of certain gases near the beam impact site can influence the sputtering rate and other aspects of sputtering to great extent (Utke, Hoffmann, & Melngailis, 2008).

The introduction of special gases called precursors gases allows a very useful type of ion-surface interaction. Namely, ion-beam-induced deposition (IBID). The gas molecules consist of a volatile part and a part that may attach to the surface. The bonds between the two parts can be broken using the ion beam and thereby a pattern can be deposited locally. Two examples of precursor gases are tungsten hexacarbonyl $W(CO)_6$ for depositing tungsten and trimethyl platinum $C_9H_{16}Pt$ for platinum deposits (Schmidt & Wetzig, 2012, p. 272).

The interactions involved in IBID are rather complicated (Chen, Alkemade, & Salemink, 2008; Alkemade & Miro, 2014). Primary ions, secondary electrons, secondary atoms are all involved in IBID and it is not entirely clear what the most dominant mechanism is (Chen, Salemink, & Alkemade, 2009). Inevitably during deposition, sputtering occurs, counteracting the growth. The supply of precursor molecules is governed by molecules adsorbed to the surface and surface diffusion into the beam, which balances the depletion of the precursor molecules by the ion beam.

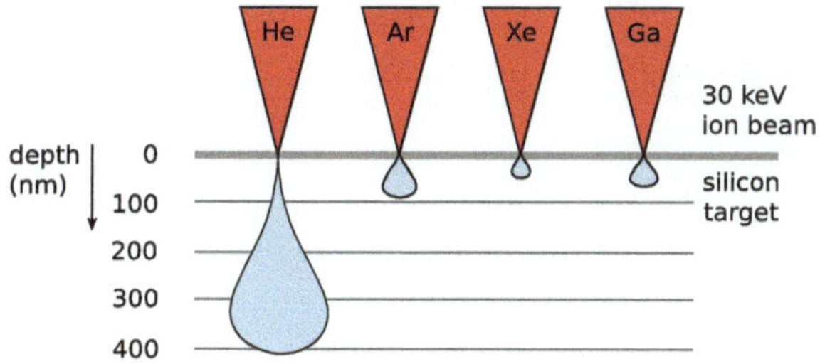

Figure 3 Sketch of the region in which ions penetrate and atoms get dislocated in a flat silicon substrate after bombardement by 30 keV ions. The tendency is that heavy mass particles spread out more rapidly and penetrate less deeply compared to lighter elements. Despite the typical penetration depth being in the order of hundreds of nanometers, the lateral resolution is not limited by this measure. Only from the first few nanometers below the surface atoms and electrons can escape. This figure is based on results by Tan, Livengood, Shima, Notte, and McVey (2010).

Most certainly we have not covered all surface interactions as we can induce plasmons (Baragiola, Ritzau, Monreal, Dukes, & Riccardi, 1999), phonons, amorphization (Henkel, Heera, Kögler, Skorupa, & Seibt, 1997; Huh, Hong, & Shin, 2013), we may implant ions (Joshi-Imre & Bauerdick, 2014), have ions scatter back into the vacuum (Lindsey & Hobler, 2012), create new ions (secondary ions) (Stevie, 2005, p. 269-280), and certainly there are more interactions possible, depending on the target material and the ion species. Eventually most energy is transferred into heat, although this rarely leads to a temperature rise of more than a few degrees Kelvin (Giannuzzi & Stevie, 2005, p. 47).

High energy ion impact on a solid is characterized by the many interactions a single ion experiences during the process of deceleration. Therefore, ion impact has often been described by a cascade model (Sigmund, 1969), meaning that the complete interaction is modeled as a series of binary events. This method has been used to study sputtering for example (Hofsass, Zhang, & Mutzke, 2014). Fig. 3 shows a sketch of one aspect of what is learned from such studies, namely the typical region in which the ions penetrate and dislocate atoms. A comparison is made between different mass ions hitting silicon at 30 keV and we see the obvious trend that lighter ions penetrate more deeply. Such trajectories are also dependent on the primary ion beam energy, but in any case we can state that the penetration is far shallower than we see with electrons, which is in the order of microns.

(A) (B) (C)

Figure 4 Focused ion beams are indispensable as shown by these examples. (A) A gold nano-antenna is made by FIB milling. Reprinted with permission from (Kollmann et al., 2014), with a redrawn scale bar. Copyright 2014 American Chemical Society. (B) For a quick development cycle, integrated circuits are often adjusted using focused ion beam milling and deposition. This procedure is known as circuit edit. In this image, re-routing of some wires is shown. Image courtesy of Lloyd Peto of NanoScope Services Ltd - www.nanoscopeservices.co.uk. Reprinted with permission. (C) Very thin membranes can be imaged relatively well with a focused ion beam due to the high surface sensitivity. (Beyer et al., 2015), the scale bar is redrawn.

1.3 Applications of focused ion beams

Although the ion-surface interactions are quite complicated, they do open up a variety of possibilities. Fig. 4 gives an impression of some applications.

A very basic, but important use of the FIB tool is making cross sections. By removing material one can see structures that are otherwise hidden. Opening up micro-machined devices to see if the structure is according to plan is common practice. It is also possible to investigate, for example, the internal structure of a cell-nucleus by milling away tissue at a very precisely defined location, something that can never be achieved by using macroscopic tools. This unprecedented cross sectioning capability has led to the development of the dual-beam machine, a device which comprises both a focused ion and a focused electron beam. The ion beam is used for milling while the electron beam provides well-known imaging possibilities. One can not imagine transmission electron microscopy without the aid of dual-beams. Properly preparing thin slices of the substrate, the so-called TEM-lamella, is essential for achieving good quality TEM images (Mayer, Giannuzzi, Kamino, & Michael, 2007).

The removal of material can also be patterned in a more complicated fashion to achieve functional devices. In addition, we can deposit structures by using dissociation of precursor gases to grow material: ion beam induced deposition (IBID). This deposition technique is very comparable to the electron optical equivalent in terms of procedure and results, see

Rius (2014) for a comparison. The combination of milling and deposition makes the FIB tool a cut-and-paste device for the nano-scale. One example of combining milling and deposition is the direct fabrication of a hard mask for reactive-ion-etching (Waid, Wanzenboeck, Muehlberger, Gavagnin, & Bertagnolli, 2014). While these techniques are mostly used for manipulating surfaces, sometimes the fabrication of three-dimensional structures is possible (Li et al., 2013). Free standing deposits can be grown if the beam patterning speed is matched carefully with the deposition growth rate (Matsui, 2012, p. 872).

The most important industrial branch that uses this cut-and-paste possibility is circuit edit. When micro-chips are designed and a prototype is developed, in practice there is always the need for some tweaking to get the right functionality. It is impractical to develop a million dollar lithography mask for every tweak and this is where FIB nano-surgery can save the day.

Making functional devices with IBID can be challenging, as achieving high purity is not obvious. Most precursors result in carbon deposition as well, which causes poor resistivity or other unwanted properties. In fact, typically one finds only 30 to 40% metallic content in a deposit (Fernández-Pacheco, De Teresa, Córdoba, & Ibarra, 2009). This is still better than what is commonly achieved with electron-beam-induced deposition, though. The choice of precursor, the conditions of the sample, writing strategy and the beam characteristics all influence the purity. In industrial circuit edit systems, 1 to 2×10^{-6} Ωm is routinely achieved (Remes, Vahakangas, & Uusimaki, 2009) from the precursor $W(CO)_6$, which is about 25 times higher than pure tungsten. One example of an excellent result is the demonstration of 10 nm cobalt lines with 0.5×10^{-6} Ωm resistivity (Wu et al., 2014). If the vacuum conditions and the cleanliness of the sample are poor though, or the precursor flux and the beam current are not carefully chosen, one easily ends up with many orders of magnitude poorer resistivity. Although the resistivity is commonly believed to be determined by the ratio of intended deposit versus the carbon fraction, some research suggest that it is the residual gallium content that determines the resistivity (Bandaru, Faraby, & DiBattista, 2015). For very small platinum-based nanowires, the deposit can transition from insulator-like to metal-like depending on the size of the structure (Fernández-Pacheco et al., 2009).

Besides purity control, precise spatial control may be difficult as well due to the wide-tailed beam profile and proximity effects. Unwanted deposition up to 2 um away from the irradiated area may occur (Lan, Chang,

Chang, Chang, & Chen, 2015). Despite the difficulties, ion beam induced deposition is often successfully employed, for example to fabricate superconducting nano-wires (Dhakal et al., 2010) or a nano-scale hall sensor (Gabureac, Bernau, Utke, & Boero, 2010).

Although focused ion beams are mainly fabrication tools, they are also used for imaging. Just like the well-known scanning electron microscope (SEM), we can raster scan the ion beam and measure at every position the response of the sample; most commonly one measures the electrons that are freed from the material. The two most noteworthy differences between scanning electron and ion microscopy is the difference in penetration depth, and the inevitable removal of material associated with ion beam impact. In practice often one prefers a SEM over a SIM for all-round imaging work, however, sometimes the specific advantages of ions are employed. Some examples of imaging techniques with helium ions are ionoluminescence (Boden, Franklin, Scipioni, Bagnall, & Rutt, 2012), scanning reflection ion microscopy (Petrov & Vyvenko, 2015), employing channeling contrast (Hlawacek et al., 2012; Veligura, Hlawacek, van Gastel, Zandvliet, & Poelsema, 2012), and imaging ultra thin layers (Hlawacek et al., 2012; Beyer et al., 2015).

One noteworthy aspect of scanning ion microscopy, and this is underappreciated, is the relation between sputtering and resolution. More often than not the resolution is directly coupled to the spot size, but this neglects the act of sputtering while gathering enough statistics for an image. Simply put, a small structure may already be gone before enough signal is acquired to distinguish it properly from random signal fluctuations. One manifestation of this problem is the difficulty of imaging graphene with helium ions (Fox et al., 2013). The interested reader is referred to Castaldo (2011) for a full analysis of this matter. Ultimately, ions can never beat the resolution obtained in for example high energy scanning transmission electron microscopy.

An analytical technique that does not have an electron based analogue is secondary ion mass spectroscopy (Stevie, 2005). One acquires secondary ions that are generated by the impact of primary ions and determines their masses, by for instance time-of-flight spectroscopy. Often cesium or oxygen ions are preferred because of the high secondary ion yield; typically a factor 30 higher than gallium or argon. If the spatial resolution does not need to be very high, SIMS is unparalleled for its sensitivity, reaching parts per million even down to parts per billion as a detection limit.

2. A closer look at FIB systems

Bright and monochromatic charged particle sources have enabled high resolution scanning-probe microscopy and nano-fabrication. Poor sources have orders of magnitude less current in the final probe, leading to unworkable acquisition times at high resolution, hence bright and monochromatic sources are crucial. In this section it is explained how the source determines the final probe quality, what sources are available, and what we would like from a new source.

2.1 How to get high quality ion beams

Ultimately, the assessment for focused ion beam systems is the quality of the probe when it hits the sample, for that is what matters in applications. The final probe quality is influenced by many aspects of which the ion source and the optical column are the most obvious ones. Some more practical aspects are sample stage stability, magnetic shielding, electronic stability, vibration suppression, etc. Fortunately, these practical aspects can be addressed with mature technology, so it is usually a matter of money and adequate engineering to ensure the optics are living up to their potential.

From astronomy to electron optics, the most infamous limit in optical systems is diffraction. Next in line are the aberrations; the inability of lenses to form perfect images. We can estimate the spot blurring diameter due to diffraction and aberration in terms of the final beam divergence angle as follows:

$$d_{\mathrm{a}} = 0.54\frac{\lambda}{\alpha}, \qquad d_{\mathrm{s}} = 0.18 C_{\mathrm{s}} \alpha^3, \qquad d_{\mathrm{c}} = 0.6 C_{\mathrm{c}} \frac{\Delta \Phi}{\Phi} \alpha, \qquad d_{\mathrm{g}} = M d_{\mathrm{v}}. \quad (1)$$

We use d for the diameter of a particular contribution, the subscript a, s and c indicates diffraction, spherical aberration and chromatic aberration respectively. The subscript g indicates the source image or geometrical contribution and v the virtual source. α represents the beam angle (or half opening angle), M is the total magnification of the system, λ the de Broglie wavelength, C_{s} the spherical aberration coefficient, and C_{c} the chromatic aberration coefficient. If we use the full width 50 characteristic for the probe contributions, they can conveniently be added using the Barth and Kruit (1996) power-sum law.

$$d^2 = \left(d_{\mathrm{g}}^{1.3} + \left(d_{\mathrm{A}}^4 + d_{\mathrm{s}}^4 \right)^{\frac{1.3}{4}} \right)^{\frac{2}{1.3}} + d_{\mathrm{c}}^2. \quad (2)$$

We assess diffraction with an example. Suppose we have a 1 keV hydrogen ion beam with a small beam angle of 1 mrad. The diffraction contribution to the spot size is about 1 nm. This example is for the lightest atom possible, a relatively low energy and a small aperture angle so that usually the diffraction contribution is much smaller. We can understand from this numerical example why focused ion beam opticians are usually not bothered too much with diffraction.

Far more relevant are the aberrations of the optical system. But since we can simply minimize them by using only a fraction of the beam and obtain a very small beam angle, it is no longer a question of what resolution we can achieve, but rather what resolution we can get at a certain beam current. The amount of current in a beam is conveniently calculated using the reduced brightness, a measure of intensity that is, under reasonable circumstances, constant throughout the optical system. This quantity will be discussed in more detail in the following chapters, but a quick definition for the reduced brightness B_r is: the total current in a probe, divided by the solid angle of the beam cone, divided by the area of illumination, divided by the beam voltage. Using the brightness, the current I of the final probe is given by

$$I = \tfrac{1}{4}\pi^2 B_r \Phi d^2 \alpha^2, \tag{3}$$

in which Φ is the beam energy.

Eqs. (1), (2) and (3) are the essential ingredients needed for optimizing a FIB system, but they do not directly reveal the optimized situation. It is very convenient to use the probe-current relation of an optimized two-lens optical column. This means that the beam limiting aperture and the system's magnification are finely balanced to get the smallest spot size for a particular required beam current. The general optimization is not so easily expressed in one explicit equation nor would it be very enlightening. By looking at four limiting cases we can get a good grasp of the situation, though. The optimal current-probe size relations of a two-lens system as derived by Hagen, Fokkema, and Kruit (2008) are:

$$I = 1.71 \frac{d^4 \Phi^3 B_r}{\Delta\Phi^2 C_c^2} \quad \text{objective lens chromatic} \tag{4}$$

$$I = 2.47 \frac{d^{\frac{8}{3}} \Phi B_r}{C_s^{\frac{2}{3}}} \quad \text{objective lens spherical} \tag{5}$$

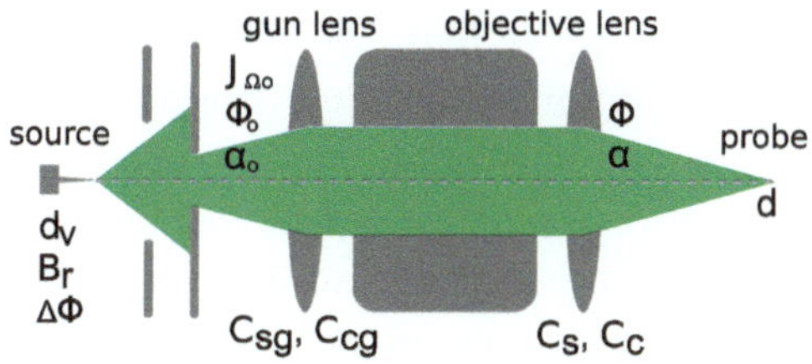

Figure 5 The basic parameters of a two-lens probe forming optical system are indicated. In an optimized system, the beam final beam angle α and the total magnification M of the system are tuned to get the most current in the smallest final spot.

$$I = 2.18\frac{d^2 J_{\Omega o}\Phi_o^{\frac{1}{2}}\Phi^{\frac{3}{2}}}{\Delta\Phi^2 C_{cg} C_c} \qquad \text{gun lens + objective lens chromatic} \qquad (6)$$

$$I = 6.86\frac{d^{\frac{2}{3}} J_{\Omega o}\Phi^{\frac{1}{4}}}{C_{sg}^{\frac{1}{2}} C_s^{\frac{1}{6}} \Phi_o^{\frac{1}{4}}} \qquad \text{gun lens + objective lens spherical} \qquad (7)$$

In these equations we have I the current in the final beam, d the diameter of the final spot, Φ the landing electrostatic energy, $\Delta\Phi$ the electrostatic energy spread, B_r the reduced brightness, C_s the spherical aberration of the objective lens, C_c the chromatic aberration of the objective lens, C_{sg} the spherical aberration of the gun lens, C_{cg} the chromatic aberration of the gun lens, Φ_o the electrostatic energy at the object side, and $J_{\Omega o}$ the angular current density at the object side. See Fig. 5 for the system schematic. The pre-factors relate to the probe size diameter and the energy spread both measured in the full width 50 characteristic.

These equations give excellent insight in how to optimize a focused ion beam system. In all equations we see that the objective lens aberrations should be as low as possible. Unfortunately there is even with the best lens design a theoretical limit to minimizing the aberrations of a lens (Scherzer, 1936). Only aberration correctors can offer solace, but these are still very inconvenient and expensive. Secondly, we see that simply increasing the landing energy is always beneficial, although this consideration should go hand in hand with the ion-surface interactions one desires, and in addition, very high voltages become technically more challenging.

Besides an indication of what a good optical column can do, one can see from these equations that a good ion source is a very effective way of achieving high performance from a FIB-system. Eqs. (4) and (5) describe the low current regime (alternatively: the high resolution regime, or source de-magnifying regime), where we may be limited by spherical or chromatic aberration of the objective lens. We see from the equations that the current in the beam is proportional to the brightness. The brightness of ion sources can vary orders of magnitude so the source can make a huge impact on the final performance. The same accounts for the energy spread in the chromatic limit. For the highest resolution, the chromatic aberration dominates and high brightness low energy spread sources can outperform lesser sources greatly as the current in a given probe size scales with $\frac{B_r}{\Delta\Phi^2}$.

In the higher current regime described by Eqs. (6) and (7), loosely speaking when the source needs to be magnified, the gun lens aberrations become important. This can be understood by inevitably magnifying the gun lens aberration along with magnifying the source image. Again the energy spread squared comes into play and a low energy spread can be highly advantageous. We see that not the brightness, but the angular current density becomes the figure of merit for source intensity. This is a somewhat tricky situation, for it seems that increasing the brightness is futile. However, when increasing the brightness without changing the virtual source size, effectively the angular current density is increased and thereby this does lead to enhanced performance. So, interpretation of these equations requires some care, but in any case it should be evident that the better source can make the final beam perform better.

It is interesting to look also at the power of d in the spherically limited case for the higher current regimes. We can see that we need to make the probe diameter quite a bit bigger for some increase in current, when compared to the other regimes. This regime is sometimes avoided all together by using a different source for high current applications (Kellogg et al., 2010).

One aspect that is not apparent from these equations is the importance of the virtual source size. It is always possible, in principle, to design an optical system with the magnification needed to arrive at the optimum so the virtual source size may seem relatively unimportant. However, if the virtual source size is very large, say microns, one needs huge demagnification to arrive at nanometer probe size. This is no easy task for optical system designers. On the other hand, if the virtual source size is very small, one will very quickly arrive at the source magnifying regime, making such a source only useful for the very low current regime.

It should be clear that the source quality to great extent determines the quality of the final probe in terms of current and size. This explains the abundance of research on better sources and it motivates research on the NAIS as well.

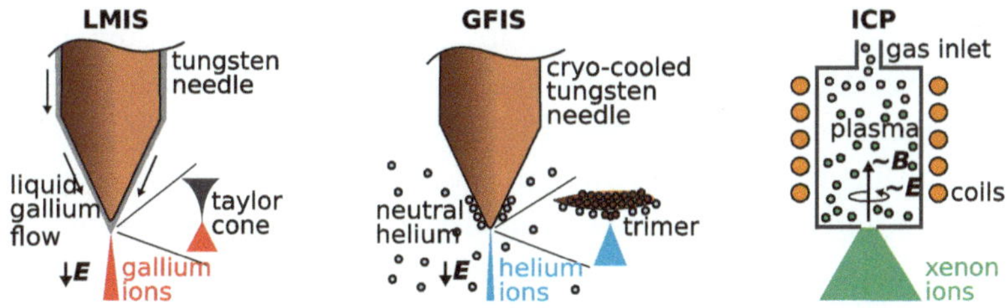

Figure 6 Cartoons of three industrially available methods for generating ions indicating their main concept and differences. LMIS and GFIS technology are similar as they both use a tungsten needle to get initial field enhancement. Furthermore they both employ additional sharpening of the final tip to get even more field enhancement. For the LMIS, the additional field enhancement is automatically achieved by the formation of a Taylor cone in the liquid metal. The supply of liquid metal runs along the tungsten needle. For the GFIS, the additional field enhancement is achieved by creating a 3-atom supertip (trimer) at the end of the tungsten needle. The supply of helium is governed by supplying the needle-surrounding vacuum by a trace amount of helium gas. The neutral atoms are polarized by the strong electric field and will be drawn towards the needle. The ICP does not comprise a tip, but relies on a plasma volume maintained by an alternating magnetic and consequentially an electric field.

2.2 State-of-the-art FIB systems

For many years, the gallium liquid metal ion source (LMIS) has been the industrial standard. It is bright, reliable, cheap, and provides a relatively high current; a sensible all-round choice. See Fig. 6 for a cartoon of the principle mechanism. The source employs electrospray to generate ions. A sharp tungsten needle is provided with a flow of liquid metal towards its tip and a strong electric field is applied in the direction of the needle. The sharpness of the tip causes field enhancement. In addition, the liquid metal is deformed by opposing forces of the surface tension and the electric field into an even sharper shape called a Taylor cone (Taylor, 1964). The field at the tip becomes about 1×10^{10} V m^{-1} and the radius as small as 1.5 nm (Forbes & Mair, 2008). One would expect an amazing brightness from such a small emitter, however, the virtual source size is not determined by this small radius, but by the ion-ion coulomb forces the ions endure right after emission (Hagen et al., 2008), yielding an effective source diameter of 50 nm or so. One obtains a brightness up to 1×10^6 A/m^2 sr V, which is still fairly good.

Using gallium is not a strict necessity for a LMIS to operate and other metals, mostly alloys, can be used as well. Gallium is usually preferred for its low melting point and low vapor pressure. Furthermore, alloy sources are considered more difficult to operate and require a mass filter (Bischoff, 2005). Nevertheless, liquid metal ion sources of other species have been developed and found applications (Bischoff, 2005, 2008). Some examples of species available are B, Be, Si, Sn, Au, Fe, Ni, Cr; all positively charged and sometimes multi-ionized. When comparing brightness, energy spread and total current, as well as lifetime and stability, the gallium LMIS is superior. Lighter species such as beryllium can show performance that comes close to gallium though (Schwind & Swanson, 2007).

One intrinsic aspect of liquid metal ion sources is that they yield metal ions. This may or may not be desirable, but the restriction is certainly a demerit. For example, when repairing UV nano-imprint masks using gallium ions, the gallium contamination can decrease the transparency of the mask down to unacceptable levels (Waid et al., 2013). For mesoscopic thin-films, even the slightest gallium dose can destroy the functionality (Barzola-Quiquia et al., 2010). Gallium ions being the industrial standard is mostly the result of excellent source performance, not because gallium is always the preferred ion species.

The helium ion microscope (Ward, Nottea, & Economou, 2006) received a warm welcome when it got introduced to the focused ion beam community. Many scientific endeavors are associated with the tool, which can be attributed to the wide variety of applications it enabled (Hlawacek, Veligura, van Gastel, & Poelsema, 2014). In concept it is comparable to the liquid metal ion source in that it uses a sharp tip and field ionization. However, the neutral helium atoms awaiting promotion to ions do not flow across the needle but are inserted into the vacuum around the needle. When a helium atom reaches the tip, it may become ionized due to the strong field at the tip surface. The key trick to make the performance exceptional is the manipulation of the end form of the tip until it assumes the shape of a three sided pyramid. This concentrates the electric field at the apex of the pyramid, hence the field ionization takes place predominantly at only a few atoms. A very small virtual source size is the result. Employing the high brightness and low energy spread with good optical column design gives a small probe size of 0.4 nm at a beam current of 0.25 pA and 35 keV landing energy (Hill & Rahman, 2011).

The GFIS providing helium ions may, in some cases, prevent the defunctionalization that is seen with gallium due to the inert nature of helium. However, the low sputtering yield (see Table 1) and the low total current makes the technology somewhat limited in application. Note that the low sputtering yield for a great deal counter-acts the very high brightness. It should also be noted that sub-surface bubble formation is an issue one has to deal with (Siegele, Weatherly, Haugen, Lockwood, & Howe, 1995; Fox, Chen, Faulkner, & Zhang, 2012).

On the other end of the spectrum in terms of brightness and total current, are the plasma based ion sources. Different techniques exist to acquire and maintain the plasma, but they all use an (alternating) electric field in an enclosed volume to generate charged species. Typically these sources have a low brightness making them not the obvious choice for high resolution applications. They can deliver staggering mA's of current though and are often versatile in the delivered species.

One well-known plasma source is the duoplasmatron, which is used for a variety of applications, notably particle colliders and spacecraft propulsion. This source may be operated with a brightness of 1×10^3 A/m^2 sr V (Smith, Tesch, Martin, & Kinion, 2008), but this is an absolute maximum for argon. Commercially available sources usually operate at 250 to 500 A/m^2 sr V and for O^- it can be as low as 25 to 50 A/m^2 sr V. Energy spreads are typically 5 to 15 eV.

Another example is the multi-cusp plasma source, which can deliver, with some effort, 2×10^3 A/m^2 sr V (Ji et al., 2002). In pulsed mode one may reach a peak brightness of 1×10^5 A/m^2 sr V (Guharay, Douglass, & Orloff, 2004), but with a duty time of 1 ms at 12 Hz repetition rate the brightness is on average almost two orders of magnitude lower.

More recently, effort has been put in the development of a bright inductively-coupled plasma source (ICP) in order to enable reasonably high resolution while maintaining the merit of high current (Smith et al., 2006). It delivers xenon ions, which are heavy and inert. The achieved brightness is about 1×10^4 A/m^2 sr V. At around 20 to 50 nA it becomes favorable to use such an ICP source rather than a gallium LMIS in terms of current versus probe size (Kellogg et al., 2010). Decent performance has also been shown for argon ions (Nabhiraj, Menon, Rao, Mohan, & Bhandari, 2010) and other gaseous ionic species are in principle possible too.

Table 3 makes a comparison of three ion sources for high resolution applications currently available on the market.

2.3 Future FIB sources

Developing new ion sources is an ongoing quest. Making ion sources even brighter and more mono-chromatic, is an obvious motivation for source development. Acquiring more total current with a decent resolution is also

Table 3 Comparison of established ion sources by listing nominal specifications. The maximum beam current may differ from the total emission current because the outermost part of the beam can be difficult to focus into a small spot. **LMIS**: Liquid metal ion source, data from Forbest (1997) and Hagen et al. (2008). The source properties are in general worse for other species than gallium. **GFIS**: gas field ionization source, data from Ward et al. (2006). The denoted measure of the energy spread is claimed as an upper bound. Values of 0.25 to 0.5 eV were measured by Ernst, Bozdech, Schmidt, Schmidt, and Larkins (1993) in a field ionization setup, but not directly in a helium ion microscope in its present form. **ICP**: inductively-coupled plasma source, data from Smith et al. (2006). The source properties are in general worse for other species than xenon.

	LMIS	**GFIS**	**ICP**	
Brightness	6×10^5	1.0×10^9	9.1×10^3	A/m^2 sr V
Energy spread fwhm	5	≤ 1	6.7	eV
Virtual source size	38	≤ 0.25	7.2×10^3	nm
Max. beam current	30	0.1	250	nA
Main ion species	Ga	He	Xe	
Compatible species	B, Be, Si, Sn, Au,...	Ne	H, He, Ne, Ar, O_2,...	

desirable for certain applications. Note that considerations of lifetime, stability, reliability and costs can not be excluded when trying to improve upon commercial devices. Perhaps even most driving, is the desire for different ionic species. For example, for efficient milling we like to use heavy ions, while for imaging, preventing sputtering as much as possible is beneficial. Preventing or employing chemical reactions determine the desired ion species as well.

By using Table 3 we can make the goals for new source development more concrete. The GFIS offers incredible brightness and low energy spread but delivers only light elements and offers a low total current. While the ICP can deliver heavy noble gas ions, it reaches a brightness that is two orders of magnitude lower than the LMIS. We begin to see that one may like to have all three sources at the same time, or at least a source that scores reasonable on all fronts. A source offering a brightness of, say, 1×10^6 A/m^2 sr V, an energy spread of 1 eV or smaller, at least 1 nA of current, and of course a variety of available species, is likely to become very popular.

A natural follow-up of the GFIS technology is the extension to heavier species than helium, as is often desired (Tan et al., 2011). In fact, a system delivering neon ions next to helium is now on the market. However, the GFIS technology is not easily extended to higher mass noble gases (Rahman et al., 2012). Furthermore, this technology seems not to be extendable to beam currents above 100 pA, thus GFIS will likely continue to find its main purpose in the highest resolution applications. Note furthermore that using a relatively high current of 100 pA from such a small emitter usually leads to the problem of large gun lens aberrations, as explained earlier in this section.

Making bright, mono energetic ion beams using laser cooling was proposed by Freinkman, Eletskii, and Zaitsev (2004). The ultra-cold ion source (UCIS) (van der Geer et al., 2007) and the magneto-optical trap ion source (MOTIS) (Hanssen, McClelland, Dakin, & Jacka, 2006; Knuffman, Steele, Orloff, & McClelland, 2011) are two manifestations of using laser cooling in combination with three dimensional magneto-optical trapping and laser ionization. The anticipated high brightness was never achieved due to the slow diffusion rate in the trap, limiting the supply of atoms to be ionized. One way to overcome this problem seems to be using an atomic beam rather than relying on diffusion. Calculations of this approach look promising (Wouters et al., 2014; ten Haaf, Wouters, van der Geer, Vredenbregt, & Mutsaers, 2014; Knuffman, Steele, & McClelland, 2013).

More ideas have been proposed, for instance an ion source based on ionic liquids (Zorzos & Lozano, 2008). However, none of the novel ideas for making high performance ion beams has been rigorously demonstrated experimentally. Even if high performance would be shown, no single source will be able to offer the complete set of desired ion species. We are therefore driven to develop a high-end ion source that may very well open a variety of applications for a wide audience.

3. The nano-aperture ion source

We propose an ion source based on electron impact gas ionization inside a sub-micron sized gas chamber (Tondare, 2006; Jun, Kutchoukov, & Kruit, 2011; Jun, 2014; van Kouwen & Kruit, 2018). The gas chamber consists of two very thin membranes of about 100 nm thick, which are separated by a small distance of 100 nm to 1000 nm, as shown in Fig. 7. A small aperture of 100 to 500 nm in the membranes allows the ions to escape and a focused electron beam to enter the ionization region while maintaining a high pressure inside the structure. An electric field between the membranes directs the ions towards the exit aperture to ensure fast emission. Outside the membrane structure an electric field further accelerates the beam up to the desired high energy. The combination of a high

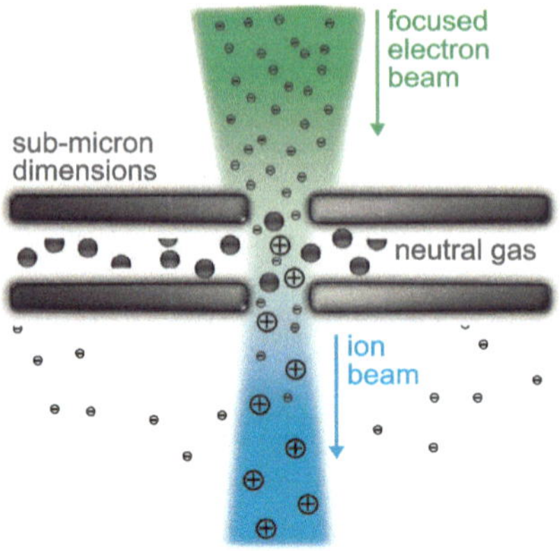

Figure 7 The nano-aperture ion source: forming high brightness beams of various species by ionizing a gas confined in a sub-micron volume using electron beam impact ionization. An electric field across the membranes extract the ions out of the double-membrane structure.

current density electron beam and a highly confined ionization region is believed to give excellent optical performance.

Fig. 8 shows the concept with more technical details than Fig. 7. A key aspect of the concept is using a complete electron column on top of an ion column. This enables a well focused and well positioned electron beam into the electron-entrance aperture. The required voltages on the optical elements of the complete system can become somewhat tricky as indicated by the example in the figure for creating 30 keV ions at the sample from a 1 keV electron impact energy. The double membrane structure is so small, that fabrication of the device relies on micro-machining methods. This nano-fluidic and electrical device is referred to as the chip. The use of a differential vacuum is needed to maintain a good enough vacuum for the Schottky electron source, while leaking out gas from the chip. The chip needs to mediate between the sub-micron scale of the source and the macroscopic scale of wires and tubes.

Many aspects of the NAIS are motivated by obtaining a high brightness, so we begin by elaborating the estimation of the brightness. Langmuir (1937) derived an equation for the emission of electrons from a flat cathode, from which we can deduce the following expression for the reduced brightness:

$$B_r = \frac{eJ}{\pi kT} \tag{8}$$

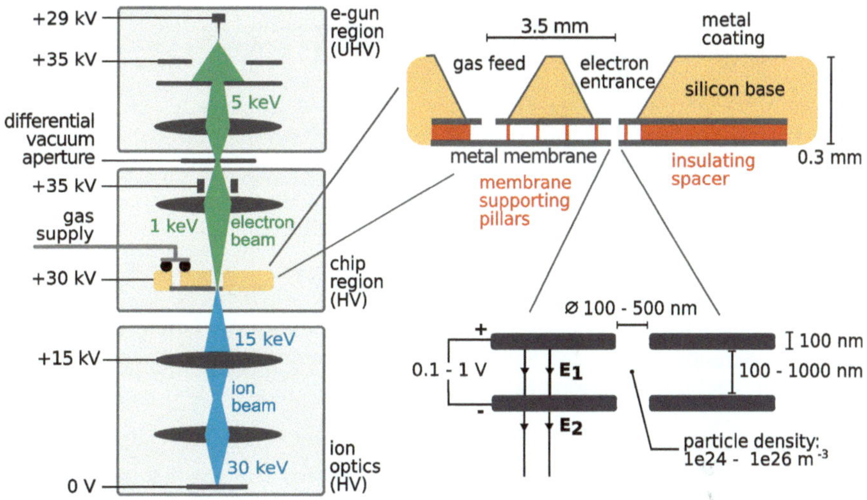

Figure 8 A more detailed implementation of the concept shown in Fig. 7. A fully capable SEM is used to focus and steer a high intensity electron beam into the double membrane structure. Below the chip an ion optical column is needed to control the ion beam for practical use. As an example, the applied high voltages for creating 30 keV ions from a gas that is hit by 1 keV electrons are indicated. It illustrates that one needs to operate the electron gun, including beam scanning, at high voltage, even though the final electron beam energy is only 1 keV. We intend to apply a bias voltage across the chip membranes so the membrane spacer needs to be of insulating material.

We may try to use this equation for our configuration, as is seen throughout the literature for various configurations. We consider very small membrane spacing such that we can assume emission from a plane rather than from a volume. We note that a crucial aspect of the equation is the assumption that the initial transverse momenta are thermally distributed. This may be clear when the particles are still neutral and in thermal equilibrium, but when ionized and extracted, this is not necessarily the case. What we need to do in order to get a meaningful estimate using this equation is determine the current density we can reasonably expect under the condition that the transverse momenta are still thermally distributed. The next two sections are working towards that goal.

3.1 Electron impact ionization and ion-neutral scattering

Essential for the concept is the efficiency at which we can generate ions from the impinging electrons on the neutral gas. Electron impact ionization is thoroughly studied and cross sections describing the ionization probabil-

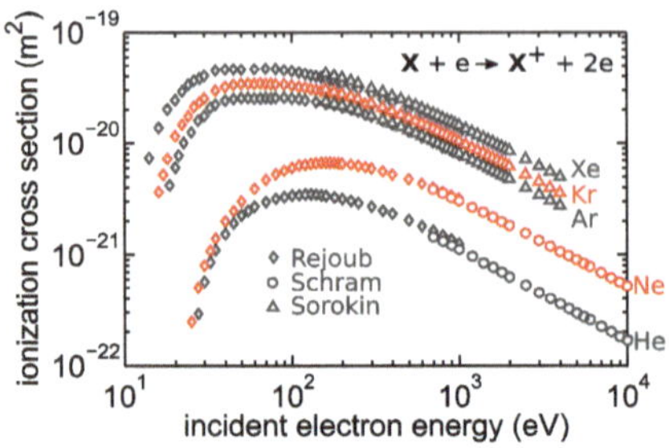

Figure 9 Single charged ionization cross sections for electrons hitting a noble gas. Data from Rejoub, Lindsay, and Stebbings (2002), Schram, de Heer, van der Wiel, and Kistemaker (1965), and Sorokin et al. (2000).

ity (σ_+) are widely available. As first candidates for ion species in the NAIS we consider noble gases as they show little chemical activity and are available in a nicely spaced variety of masses. In Fig. 9 a set of noble gas single ionization cross sections are plotted. We observe that heavier gases are easier to ionize and secondly we see that the maxima for the single ionization are in the range 50 to 100 eV. The double ionization cross sections are one to two orders of magnitude lower.

A second important cross section is associated with ion-neutral scattering ($\sigma_{\text{i-n}}$). Once an ion is created somewhere in the gaseous volume, it is accelerated towards the exit aperture and in its path may encounter one or several neutrals. Previously, a molecular diameter has been used to describe this (Jun et al., 2011), however this does not take into account the polarization of the neutral and thereby not the energy dependence of the cross section, nor does it consider the possibility of exchanging charge. We will discuss better alternatives later, but for introductory purposes use the molecular diameter anyhow.

Ion neutral scattering events will greatly affect the transverse momentum distribution as energy acquired in the extraction field can be transferred to transverse energy. The acquired kinetic energy from the field in the downstream direction is in the order of 0.5 eV, a typical voltage that is needed for good extraction. The transverse energy spread is in the order of kT. Comparing the energy scales of $kT = 0.025$ eV at room temperature to the 0.5 eV of the membrane voltage, we conclude that scattering events will indeed greatly affect the transverse thermal momenta distributions. Hence, the brightness acquired will be much lower than predicted by Eq. (8) if a large fraction of the generation ions have experienced scattering.

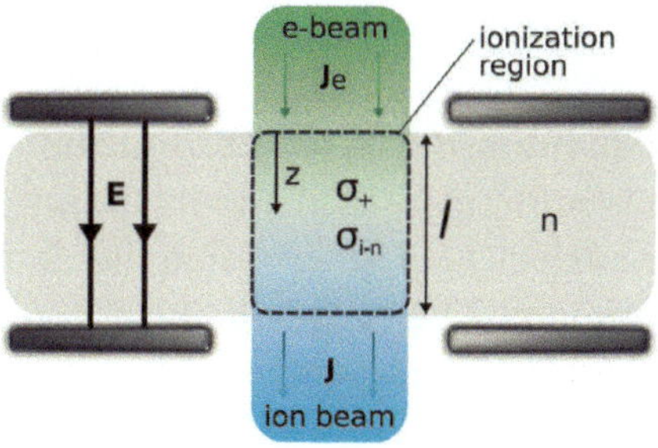

Figure 10 A simple model for ionization in the NAIS assumes a cylindrical ionization region at uniform particle density. Inside this region all ionization and ion-neutral scattering occurs.

We may try to use a sufficiently rarefied configuration such that the ion-neutral scattering effects are not dominating. We can estimate the conditions for such a situation by a simple balance equation of ion generation. We assume a top-hat distribution of the gas of length l in the z-direction while stretching to infinity in x and y. The electron irradiation follows a radial top-hat distribution and is directed parallel along the z-axis. See Fig. 10 for a sketch of this model. We assume that once an ion scatters on a neutral it is no longer contributing to the brightness, so we consider it lost. Under these conditions we can write

$$\mathrm{d}J = dJ_{\text{gen}} - dJ_{\text{scat}} = \left(J_e\sigma_+ - J\sigma_{\text{i-n}}\right) n\mathrm{d}z. \tag{9}$$

We use J for the unscattered ion current density, J_{gen} and J_{scat} for the generated and scattered ion current density, n for the particle density, J_e the electron current density. There is some ambiguity regarding the electron current density as a function of z. If an electron engages in an ionization process, it returns two electrons, although their energies and directions are different. We will ignore this for now and assume the electron current density to be constant in z. The solution to Eq. (9) is then $J(z) = J_{\text{e}}\frac{\sigma_+}{\sigma_{\text{i-n}}}\left(1 - e^{-n\sigma_{\text{i-n}}z}\right)$ and the total generated current is simply $J_{\text{gen}}(z) = J_{\text{e}}n\sigma_+ z$.

Intuitively, we may tune the particle density of the gaseous region such that the mean free path of the ions equals the length of the gas region, after all, we were aiming for unscattered ions. So, we set $z = \lambda = (n\sigma_{\text{i-n}})^{-1}$ to give for the total generated current $J_{\text{gen}}(z) = J_e\frac{\sigma_+}{\sigma_{\text{i-n}}}$. Because $\sigma_+ \ll \sigma_{\text{i-n}}$, we conclude that most of the electrons will never engage in ionization at all and the assumption of constant electron current density is quite reasonable; most electrons fly through the gaseous region unaffected. The total unscattered ion current is slightly lower due to some scattering and is given by $J = J_e \frac{\sigma_+}{\sigma_{\text{i-n}}}(1 - \frac{1}{e})$.

We may ask the question if it is useful to further increase the particle density. If we find the optimum mathematically from $\frac{dJ}{dz} = 0$, we find for the total unscattered ion current $J = J_e \frac{\sigma_+}{\sigma_{\text{i-n}}}$. A value that is only slightly different from the intuitive optimum. The particle density increase required to reach this value goes to infinity. Surely, now the assumption that the current density is constant is violated.

We can try a slightly different version of the model. If we assume that the electrons that have ionized an atom can no longer contribute to any other ionization process, the electron current density follows a simple exponential decay: $J_e = J_{e,0} e^{-\sigma_+ n z}$. Physically this means that after ionization the two electrons scatter away from the region from which ionized particles can be extracted or don't have enough energy for ionization. This is rather simplified as well, but prevents the unphysical situation of the optimum occurring for an infinite particle density. Note that this is quite a conservative approach because both electrons that ionize and ions that gas-scatter are removed in the model, while in reality they may still contribute a little. Under this assumption the solution to Eq. (9) is $J(z) = J_{e,0} \frac{\sigma_+}{\sigma_{\text{i-n}} - \sigma_+}\left(e^{-n\sigma_+ z} - e^{-n\sigma_{\text{i-n}} z}\right)$. If we compute the optimum from $\frac{dJ}{dz} = 0$ we obtain an optimum when $nl = (\sigma_{\text{i-n}} - \sigma_+)^{-1} \ln \frac{\sigma_{\text{i-n}}}{\sigma_+}$. This optimum gives for the unscattered ion current density $J = J_{e,0} \frac{\sigma_+}{\sigma_{\text{i-n}}}^{\frac{\sigma_{\text{i-n}}}{\sigma_{\text{i-n}} - \sigma_+}}$. Again, we use $\sigma_{\text{i-n}} \gg \sigma_+$, to simplify and we get $J \approx J_{e,0} \frac{\sigma_+}{\sigma_{\text{i-n}}}$.

From the analysis presented here we conclude that a reasonable estimate for the best attainable unscattered ion current density is simply $J_e \frac{\sigma_+}{\sigma_{\text{i-n}}}$. Even though the models used do not consider the details of the system, they capture the essence of balancing ionization and ion-neutral scattering. That is to say, the product of pressure and gaseous length can be increased to yield more ionization, but if the ion mean free path becomes in the order of the gaseous length or smaller, a further increase is hardly beneficial for extracting more unscattered ions. Note that the total generated current will in fact continue to rise because the scattered ions can still be extracted; they are simply not expected to contribute much to a high brightness.

3.2 An electron beam for a NAIS system

There may be little remedy for ion-neutral scattering, however, we can optimize the ion current density by optimizing the electron beam. Since the ion current density is proportional to the electron current density, we seek an electron optical configuration that yields the highest current density. This is different from electron microscopes, where usually you would like a certain resolution and then try to get the highest current in that particular probe size. In this case, a small probe size is not directly what we are after. The ion current density is also proportional to the ionization cross section, which is dependent on the impact energy. Therefore, the most ions are generated when the product of the electron current density and the ionization cross section is highest.

We can use Eqs. (4)–(7) to investigate what this means for an optical system if we substitute for the current density $J = \frac{4I}{\pi d^2}$. In the objective lens chromatically and spherically limited cases the current scales with d^4 and $d^{8/3}$ respectively, so in both cases the current density is increased by increasing the probe size. This means that this is never an optimized regime as increasing the probe is beneficial. The current in the regimes where the chromatic aberration of the gun lens becomes important scales with d^2, so in that case the current density is constant. The gun lens spherical aberration limited situation scales with $d^{2/3}$ so increasing the probe size starts to decrease the current density. Considering these scaling effects there must be some optimum where the high and low current regimes intersect.

We can see from Eqs. (4)–(7) that in order to achieve the best performance, we need low aberrations for the objective lens as well as for the gun lens. Furthermore, high brightness, low energy spread and high angular current density all improve the current density. A natural choice for an electron source becomes the Schottky electron gun. Besides its excellent optical properties, it is also highly stable and durable. We also consider a LaB_6 source and a cold field emitter.

In order to get an estimate for the current density we may expect, we will optimize a hypothetical column based on realistic optical characteristics. Eqs. (4)–(7) give certain limiting conditions but in order to do a full optimization we use Eqs. (1), (2), (3) and balance the beam angle and magnification for achieving the smallest spot size for a particular current. This method is well described in the literature (Kruit et al., 2006). The chosen values for the characterizing the column are listed in Table 4. The properties of the considered electron sources are shown in Table 5.

Since we haven't decided on the electron beam impact energy, we need to bring the energy dependence of the objective lens aberrations to attention. Suppose we scale down a particular lens configuration, effectively we scale down all trajectories proportionally and so the aberrations scale down accordingly too. It is well-known that for obtaining minimal aberrations,

Table 4 The characteristics of a hypothetical electron optical column as used in the optimization of highest current density. These values are considered realistic for high-end lenses and are comparable to what is used elsewhere (Kruit, Bezuijen, & Barth, 2006).

C_{sg}	10	mm
C_{cg}	10	mm
$C^*_{s,e}$	2	mm
$C^*_{c,e}$	2	mm
$C^*_{s,m}$	1	mm
$C^*_{c,m}$	1	mm
V^*	1	kV

Table 5 The most important characteristics of three electron sources that may be considered in a current density optimized electron column. The values are taken from Bogner, Jouneau, Thollet, Basset, and Gauthier (2007) and Swanson and Schwind (2008).

Source	B_r	$\Delta\epsilon$	d_v
	$A/m^2\,sr\,V$	eV	nm
Schottky	7.5×10^7	0.6	30
LaB_6	5.0×10^5	1.0	10^4
Cold field emitter	5.0×10^8	0.3	5

the smallest lenses are best. However, for electrostatic lenses, a maximum field that can be maintained before break-down occurs should be respected, often 10 kV/mm is used as a rule of thumb. This means that there is a limit to how far we can downscale. We can only downscale beyond this limit if we scale the beam energy and the lens potential accordingly. If we consider only good lenses that are operated at the field strength limit, we can expect the aberration coefficients to scale proportionally to the landing energy. For magnetic lenses there is a similar limit, namely that the metal can maintain a certain magnetic field before saturation effects occur. This leads to the aberration coefficients scaling proportionally to the square root of the landing energy, as explained by Riecke (1982, p. 178). So, we have for the scaling of the aberrations:

$$C_{s,m} = C^*_{s,m}\sqrt{\frac{V}{V^*}}, \qquad C_{c,m} = C^*_{c,m}\sqrt{\frac{V}{V^*}} \qquad \text{(magnetic lenses)} \tag{10}$$

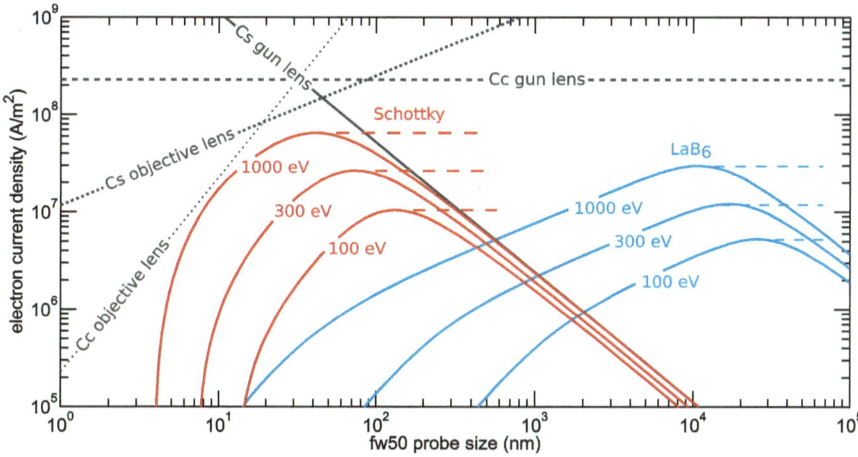

Figure 11 Optimization curves of a two-lens electron column with a magnetic objective lens. The electron current density is shown, for that is the figure of merit of a NAIS devoted electron column. For 1 keV electrons from a Schottky source the curves of limiting regimes as calculated by Eqs. (1) are shown. The dashed red and blue lines indicate that the current density in the center of the probe can always be made at least as high as the maximum for probe sizes larger than where the maximum occurs even though the average current density of the probe will decrease.

$$C_{s,e} = C^*_{s,e}\frac{V}{V^*}, \qquad C_{c,e} = C^*_{c,e}\frac{V}{V^*} \qquad \text{(electrostatic lenses)} \qquad (11)$$

in which the starred aberration coefficients are known reference values of a lens that is operated at a beam voltage V^*.

The results of the electron column optimization assuming a magnetic objective lens are shown in Fig. 11. Clearly a maximum at the point where both gun and objective lens aberrations become important is visible, as was explained earlier. For the Schottky gun we find the maximum somewhere between 40 and 100 nm full width 50 probe size. This sets the typical dimension of the inlet aperture in the double membrane structure. For 1 keV electrons, the maximum current density is 6.5×10^7 A m^{-2}, the probe size is 40 nm, and the total current of the beam is 167 nA. Interestingly, if we would use a SEM under normal operation conditions and focus down to a few nanometer, we easily loose an order of magnitude or more in current density compared to this optimum. In addition, this would also mean a much lower total current. A NAIS devoted e-beam is quite different from a high resolution SEM.

Arguably the best choice is a Schottky electron gun as the e-gun for the NAIS, even though a LaB_6 cathode gives comparable current density. The problem with a LaB_6 cathode is that the optimum occurs at a very high probe size, which leads to several complications. For high resolution applications we would need a huge demagnification. In contrast, the spot size of the Schottky optimum is very close to the virtual source size of LMIS emitters, enabling the use of already existing optical columns. The big spot size of the LaB_6 also requires a very large aperture, which gives a big amount of gas leaking into the vacuum. Furthermore, a 10 μm ion beam virtual source size is inconvenient for high resolution applications.

A cold field emitter can give a very high current density of 8.7×10^7 $\mathrm{A\,m^{-2}}$, but with only a 10 nA of total current. These specs are interesting for certain applications, but a cold field emitter is rather sensitive and unreliable so it is disregarded anyhow. See Tondare (2006) for a detailed analysis of using a cold field emitter for the NAIS.

We should make the remark that in the calculations so far we have looked at the average current density in the probe. If we, however, would look at the current density in the center of the probe, the curves would look different. Suppose from some particular optical configuration we would increase the beam angle, then, the current density in the center of the probe can only stay the same or increase. It is impossible to lower a current density by adding more current. However, the total probe size full width 50 increases more rapidly than the current, so the average current density does decrease. In practical terms this means we could use a very large opening angle and use a small electron-entrance aperture in the top membrane to select the high current density region of the focused probe. In Fig. 11 this notion is indicated by dashed lines continuing from the optima. Using such a configuration would mean a very high current hitting the membrane continuously, which may cause deformation by heating, or give rise to charging effects. While it may be possible, this seems a little risky.

From this optimization we could conclude that we want to use the highest possible energy. This, however, would neglect the energy dependence of the ionization cross section. By looking at the product of the ionization cross section and the electron current density we get the best figure of merit. Fig. 12 shows the results of multiplying the optima as obtained in Fig. 11 for the Schottky electron gun with the ionization cross section.

In Fig. 12 we can see a rather shallow optimum, meaning that a very specific energy is not required. The position of the optima for the two different gases is quite comparable, so a particular e-gun optimization is

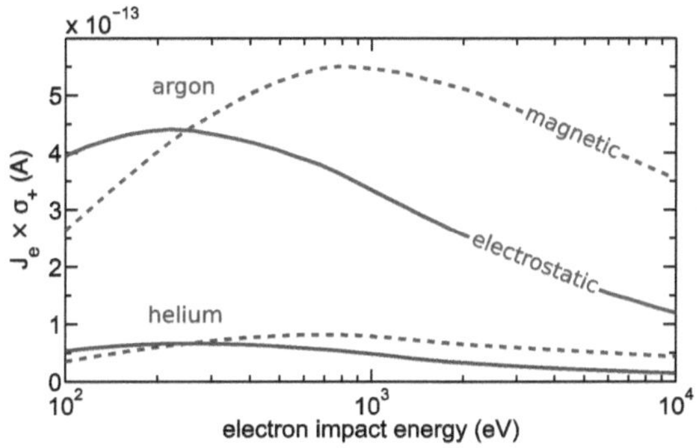

Figure 12 Optimizing the electron beam for the NAIS by finding the highest combination electron current density and ionization cross section. A magnetic or electrostatic objective lens performs differently because the aberrations scale differently with the impact energy. We show two example gases to indicate that the difference in ionization cross section makes the performance species dependent. The ionization cross sections used for this calculation are shown in Fig. 9.

good for different gases. The optimal energy does depend a little on the choice of a magnetic or electrostatic objective lens, but roughly speaking we find the optimal energy at a few hundred eV. We are, however, inclined to choose 1 keV. While 1 keV seems sufficiently low to obtain a good ionization efficiency, it is high enough to comfortably work with. Low energy beams are preferably avoided as they are more prone to magnetic disturbances, are easier affected by parasitic charging, and are more troubled by statistical coulomb interactions.

Speaking of coulomb interactions, the values of this optimization should make every charged particle optician a little worried. We are suggesting to operate an electron beam at 1 keV with about 100 nA of current; that is a very high current at a relatively low energy. It becomes necessary to at least estimate the coulomb interactions. We use the slice method (Kruit & Jansen, 2008; Jiang, 1996) to estimate the probe deterioration due to coulomb interactions.

Fig. 13 shows two manifestations one may encounter in a typical column design; a single cross over and a parallel beam. We denote the pre-final lens as the column lens, this may be the gun lens or an intermediate lens. In the most simple configuration the beam energy in the region after the column lens, through the objective lens, up to the sample, is equal to the landing energy.

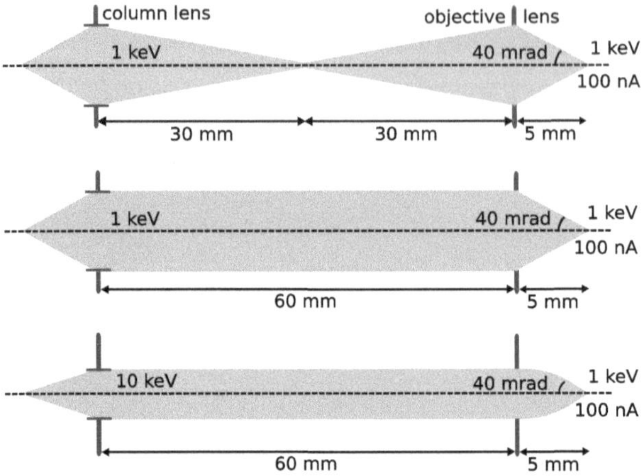

Figure 13 Three different beam segments are considered for a statistical coulomb interaction calculation of a NAIS electron gun. All configurations have the same final probe characteristics: 40 mrad opening angle, 1 keV landing energy, 100 nA of total current, and 40 nm fw50 probe size, if no coulomb interactions would be present. Two configurations are maintained at 1 keV, while the bottom configuration runs the electrons at 10 keV until the beam is decelerated in the last 5 mm.

We consider a relatively short, but feasible length of 60 mm between the column lens and the objective lens. This is usually the region where we find a deflection system, a stigmator, the objective lens coils, a valve, detectors, etc. A 5 mm distance from the objective lens center to the sample is quite typical. We assume that any coulomb interactions prior to the column lens are negligible. In a conventional SEM, a so-called coulomb tube achieves this; a region at which the electrons are maintained at very high voltage. Alternatively one simply designs a very short electron column where the gun is placed right in front of the column lens. In the gun region we usually do have coulomb interactions but these are already accounted for in source characterizing parameters of brightness, energy spread, and virtual source size.

For the beam properties we use values close to what was found in the optimization demonstrated in Fig. 11, namely I = 100 nA, V = 1 kV, $\alpha = 40$ mrad, $d_{\text{probe}} = 40$ nm. We calculate the contribution of the trajectory displacement at the probe forming location. We find in the single cross over configuration a trajectory displacement of 199 nm and a Boersch effect contribution to the energy spread of 144 meV. In the parallel beam configuration we find 168 nm and 18 meV. The Boersch effect contribution is quite acceptable, but the trajectory displacement becomes the probe size limiting factor when comparing to our 40 nm target value. Interestingly it doesn't matter a great deal if one prevents a cross over or not for the trajectory displacement (Jansen, 1988; Jiang & Kruit, 1996). This indicates that we can not solve the issue by simply reconfiguring the optics a bit.

When running a parallel beam at 10 keV with deceleration at the last moment we obtain a total trajectory displacement fw50 disk of 29.6 nm. This is comprised by 7.2 nm in the deceleration region and 22.4 nm in the beam line. This is quite alright, but getting it below 20 nm would make it insignificant. Further increasing the intermediate energy and shortening the beam line are means that help to this end.

One may argue that these few calculations are not representative for all possible optical configurations, and this is true. The point is mainly to create awareness of coulomb interactions of NAIS optimized electron columns. More specifically, we see that beam segments of tens of millimeters at 1 keV are just not going to cut it.

3.3 Expected performance

From the previous section we found that we may get an electron beam with 100 nA of current in a 50 nm fw50 spot at 1 keV. This still requires skillful optical design, but is certainly feasible. We have assumed that the coulomb interactions are sufficiently reduced to not be dominating. We also learned that a decent guess for the ionization efficiency of brightness contributing ions is $\frac{\sigma_+}{\sigma_{\text{i-n}}}$. This gives typically around 1 nA of total current. Using Eq. (8) we get for the brightness

$$B_r = \frac{\sigma_+}{\sigma_{\text{i-n}}} \frac{eJ_e}{\pi kT}. \tag{12}$$

We find for all noble gases a brightness higher than 1×10^6 A/m^2 sr V. In Table 6 the estimated brightnesses are listed more concisely. As a first approximation, we can assume that the virtual ion source is an image of the e-beam spot profile, giving 50 nm for the fw50 virtual source size.

We note that on one hand, the brightness given here is quite conservative. We have taken aberration coefficients as often encountered in a SEM. However, a dedicated electron beam column could have lower aberrations, as we could decelerate at the last moment, or put the chip much closer to the objective lens than a sample in a SEM. We may also improve the gun lens aberrations. Furthermore we have neglected the brightness contribution of scattered particles, which may contribute anyway.

Table 6 Estimated ionization efficiency and brightness of noble gas ions by using a 1 keV, 50 nm, 100 nA electron beam. The ion-neutral cross sections are based on van der Waals diameters (Bondi, 1964). Ionization cross sections are taken from Rejoub et al. (2002).

	$\sigma_{\text{i-n}}$ m^2	σ_+ m^2	η_I^+ %	Estimated B_r $\text{A/m}^2\,\text{sr V}$
Helium	2.46×10^{-19}	1.24×10^{-21}	0.50	1.9×10^6
Neon	2.98×10^{-19}	3.21×10^{-21}	1.08	3.2×10^6
Argon	4.44×10^{-19}	7.83×10^{-21}	1.76	5.5×10^6
Krypton	5.13×10^{-19}	9.84×10^{-21}	1.92	6.2×10^6
Xenon	5.86×10^{-19}	1.21×10^{-20}	2.06	6.7×10^6

On the other hand, we haven't looked in detail at ion neutral interactions, the curved electric fields inside the chip, and the coulomb interactions of the ions. It is perhaps good to start of a little conservative.

The energy spread is directly related to the voltage across the membranes. Because the ions are generated at different locations and thus at different equipotential planes, an energy spread is created. The purpose of the bias voltage is two-fold. Firstly, it helps extracting the ions by steering them towards the exit aperture. Secondly, the membrane bias is extracting the ions quickly before they can repel each other. If we equate the average chip residential time to the average time between the generation of two ions and if we assume Poissonian statistics for ion generation, we can make the following statement: 58% of the time an ion is present in the chip, it is the only one. We consider this to be a situation where we expect limited influence of coulomb interactions in the chip. An estimate for the bias voltage required to achieve $\langle t_{\text{res}} \rangle = \langle t_{\text{gen}} \rangle$ is $V_b = \frac{8}{9}\frac{m}{q^3} m l^2 I^2$. If $l = 200$ nm and $I = 1$ nA we find that we need 0.58 V for argon. The effective fw50 energy spread will in general be lower than the bias voltage, so the prospect of getting a low energy spread is quite good.

Considering the expected performance we can state that the NAIS can very well be a welcome addition to the available ion sources. The other chapters in this publication discuss the theoretical aspects introduced in this chapter in more detail. Experimental results can be found elsewhere, notably progress on chip fabrication (Jun, Kutchoukov, Heerkens, & Kruit, 2012; Jun, 2014), measurements of the energy spectrum (Jun, 2014), and brightness measurements (van Kouwen & Kruit, 2018).

List of symbols

d	probe size	m
d_g	geometrical probe contribution	m
d_a	diffraction probe contr.	m
d_s	spherical aberr. probe contr.	m
d_c	chromatic aberr probe contr.	m
d_v	virtual source size	m
λ	wavelength	m
α	half beam angle	rad
$\Delta\Phi$	energy spread	V
$\Delta\epsilon$	energy spread	eV
Φ	ion landing energy	V
M	magnification	
I	ion beam current	A
B_r	reduced brightness	$A/m^2\,sr\,V$
$J_{\Omega o}$	angular current dens. gun lens	$A\,sr^{-1}$
Φ_o	beam voltage gun lens	V
C_s	spherical aberration coeff.	m
C_c	chromatic aberration coeff.	m
C_{sg}	gun lens spher. aberr. coeff.	m
C_{cg}	gun lens chrom. aberr. coeff.	m
$C_{s,m}$	magnetic lens spher. aberr. coeff.	m
$C_{s,e}$	electrost. lens spher. aberr. coeff.	m
$C_{c,m}$	magnetic lens chrom. aberr. coeff.	m
$C_{c,e}$	electrost. lens chrom. aberr. coeff.	m
n	neutral particle density	m^{-3}
J	ion current density	$A\,m^{-2}$
J_{scat}	scattered ion current density	$A\,m^{-2}$
J_{gen}	generated ion current density	$A\,m^{-2}$
J_e	electron current density	$A\,m^{-2}$
σ_+	ionization cross section	m^2
$\sigma_{i\text{-}n}$	ion-neutral cross section	m^2
η_I^+	unscattered ionization efficiency	%
V	lens voltage	V
V^*	lens voltage reference	V

Acknowledgment

I would like to thank Pieter Kruit for his guidance and contributions. This work is part of the Industrial Partnership Program (IPP) of the Stichting voor Fundamenteel Onderzoek der Materie (FOM), which is financially supported by the Nederlandse Organisatie voor Wetenschappelijk Onderzoek (NWO). The IPP program is co-financed by FEI Company. This work is based on the dissertation titled *The Nano-Aperture Ion Source*, by Leon van Kouwen.

References

Alkemade, P. F. A., & Miro, H. (2014). Focused helium-ion-beam-induced deposition. *Appl. Phys. A*, *117*(4), 1727–1747.

Bandaru, P., Faraby, H., & DiBattista, M. (2015). Electrical contacts to nanomaterials. *J. Nanosci. Nanotechnol.*, *15*(12), 9315–9329.

Baragiola, R., Ritzau, S., Monreal, R., Dukes, C., & Riccardi, P. (1999). Mechanisms for ion-induced plasmon excitation in metals. *Nucl. Instrum. Methods Phys. Res. B*, *157*(1–4), 110–115.

Barth, J., & Kruit, P. (1996). Addition of different contributions to the charged particle probe size. *Optik*, *101*(3), 101.

Barzola-Quiquia, J., Dusari, S., Bridoux, G., Bern, F., Molle, A., & Esquinazi, P. (2010). The influence of Ga+ irradiation on the transport properties of mesoscopic conducting thin films. *Nanotechnology*, *21*(14), 145306.

Beyer, A., Vieker, H., Klett, R., zu Theenhausen, H. M., Angelova, P., & Gölzhäuser, A. (2015). Imaging of carbon nanomembranes with helium ion microscopy. *Beilstein J. Nanotechnol.*, *6*, 1712–1720.

Bischoff, L. (2005). Alloy liquid metal ion sources and their application in mass separated focused ion beams. *Ultramicroscopy*, *103*(1), 59–66.

Bischoff, L. (2008). Application of mass-separated focused ion beams in nano-technology. *Nucl. Instrum. Methods B*, *266*(8), 1846–1851.

Boden, S. A., Franklin, T. M., Scipioni, L., Bagnall, D. M., & Rutt, H. N. (2012). Ionoluminescence in the helium ion microscope. *Microsc. Microanal.*, *18*(6), 1253–1262.

Bogner, A., Jouneau, P.-H., Thollet, G., Basset, D., & Gauthier, C. (2007). A history of scanning electron microscopy developments: Towards wet-STEM imaging. *Micron*, *38*(4), 390–401.

Bondi, A. (1964). Van der Waals volumes and radii. *J. Phys. Chem.*, *68*(3), 441–451.

Castaldo, V. (2011). *High resolution scanning ion microscopy* (Doctoral dissertation). Delft University of Technology. Retrieved from repository.tudelft.nl.

Castaldo, V., Withagen, J., Hagen, C., Kruit, P., & van Veldhoven, E. (2011). Angular dependence of the ion-induced secondary electron emission for He+ and Ga+ beams. *Microsc. Microanal.*, 17(4), 624–636.

Chen, P., Alkemade, P. F. A., & Salemink, H. W. M. (2008). The complex mechanisms of ion-beam-induced deposition. *Jpn. J. Appl. Phys.*, *47*(6S), 5123.

Chen, P., Salemink, H. W. M., & Alkemade, P. F. A. (2009). Roles of secondary electrons and sputtered atoms in ion-beam-induced deposition. *J. Vac. Sci. Technol. B*, *27*(6), 2718–2721.

Dake, L., King, D., Pitts, J., & Czanderna, A. (2002). Ion beam bombardment effects on solid surfaces at energies used for sputter depth profiling. In A. W. Czanderna, T. E. Madey, & C. J. Powell (Eds.), *Beam effects, surface topography, and depth profiling in surface analysis*. New York, Boston, Dordrecht, London, Moscow: Kluwer Academic Publishers.

Dhakal, P., McMahon, G., Shepard, S., Kirkpatrick, T., Oh, J. I., & Naughton, M. J. (2010). Direct-write, focused ion beam-deposited, 7 K superconducting C-GA-O nanowire. *Appl. Phys. Lett.*, *96*(26), 262511.

Ernst, N., Bozdech, G., Schmidt, H., Schmidt, W., & Larkins, G. L. (1993). On the full-width-at-half-maximum of field ion energy distributions. *Appl. Surf. Sci.*, *67*(1–4), 111–117.

Fernández-Pacheco, A., De Teresa, J. M., Córdoba, R., & Ibarra, M. R. (2009). Metal-insulator transition in PT-C nanowires grown by focused-ion-beam-induced deposition. *Phys. Rev. B, 79*(17), 174204.

Forbes, R., & Mair, G. (2008). Liquid metal ions sources. In J. Orloff (Ed.), *Handbook of charged particle optics* (second edition). CRC Press.

Forbest, R. (1997). Understanding how the liquid-metal ion source works. *Vacuum, 48*(1), 85–97.

Fox, D., Chen, Y., Faulkner, C. C., & Zhang, H. (2012). Nano-structuring, surface and bulk modification with a focused helium ion beam. *Beilstein J. Nanotechnol., 3*, 579–585.

Fox, D., Zhou, Y. B., O'Neill, A., Kumar, S., Wang, J. J., Coleman, J. N., ... Zhang, H. Z. (2013). Helium ion microscopy of graphene: Beam damage, image quality and edge contrast. *Nanotechnology, 24*(33), 335702.

Freinkman, B., Eletskii, A., & Zaitsev, S. (2004). A proposed laser source of ions for nanotechnology. *Microelectron. Eng., 73–74*, 139–143.

Frey, L., Lehrer, C., & Ryssel, H. (2003). Nanoscale effects in focused ion beam processing. *Appl. Phys. A, 76*(7), 1017–1023.

Gabureac, M., Bernau, L., Utke, I., & Boero, G. (2010). Granular CO-C nano-hall sensors by focused-beam-induced deposition. *Nanotechnology, 21*(11), 115503.

Giannuzzi, L., & Stevie, F. (2005). *Introduction to focused ion beams: Instrumentation, theory, techniques, and practice*. Springer Science+Business Media, Inc.

Guharay, S., Douglass, S., & Orloff, J. (2004). High-resolution primary ion beam probe for SIMS. *Appl. Surf. Sci., 231–232*, 926–929.

Hagen, C. W., Fokkema, E., & Kruit, P. (2008). Brightness measurements of a gallium liquid metal ion source. *J. Vac. Sci. Technol. B, 26*(6), 2091.

Hanssen, J. L., McClelland, J. J., Dakin, E. A., & Jacka, M. (2006). Laser-cooled atoms as a focused ion-beam source. *Phys. Rev. A, 74*(6), 063416.

Henkel, T., Heera, V., Kögler, R., Skorupa, W., & Seibt, M. (1997). Kinetics of ion-beam-induced interfacial amorphization in silicon. *J. Appl. Phys., 82*(11), 5360–5373.

Hill, R., & Rahman, F. F. (2011). Advances in helium ion microscopy. *Nucl. Instrum. Methods A, 645*(1), 96–101.

Hlawacek, G., Veligura, V., Lorbek, S., Mocking, T. F., George, A., van Gastel, R., ... Poelsema, B. (2012). Imaging ultra thin layers with helium ion microscopy: Utilizing the channeling contrast mechanism. *Beilstein J. Nanotechnol., 3*, 507–512.

Hlawacek, G., Veligura, V., van Gastel, R., & Poelsema, B. (2014). Helium ion microscopy. *J. Vac. Sci. Technol. B, 32*(2), 020801.

Hofsass, H., Zhang, K., & Mutzke, A. (2014). Simulation of ion beam sputtering with SDTrimSP, TRIDYN and SRIM. *Appl. Surf. Sci., 310*, 134–141.

Huh, Y., Hong, K. J., & Shin, K. S. (2013). Amorphization induced by focused ion beam milling in metallic and electronic materials. *Microsc. Microanal., 19*(Suppl. S5), 33–37.

Jansen, G. (1988). *Coulomb interactions in particle beams* (Doctoral dissertation). Delft University of Technology. Retrieved from repository.tudelft.nl.

Ji, Q., Jiang, X., King, T.-J., Leung, K.-N., Standiford, K., & Wilde, S. B. (2002). Improvement in brightness of multicusp-plasma ion source. *J. Vac. Sci. Technol. B, 20*(6), 2717–2720.

Jiang, X. (1996). *Coulomb interactions in charged particle optical columns* (Doctoral dissertation). Delft University of Technology. Retrieved from repository.tudelft.nl.

Jiang, X., & Kruit, P. (1996). Comparison between different imaging modes in focussed ion beam instruments. *Microelectron. Eng., 30*(1–4), 249–252.

Joshi-Imre, A., & Bauerdick, S. (2014). Direct-write ion beam lithography. *J. Nanotechnol.*, *2014*, 170415.
Jun, D. (2014). *Development of the nano-aperture ion source (NAIS)* (Doctoral dissertation). Delft University of Technology. Retrieved from repository.tudelft.nl.
Jun, D., Kutchoukov, V., Heerkens, C., & Kruit, P. (2012). Design and fabrication of a miniaturized gas ionization chamber for production of high quality ion beams. *Microelectron. Eng.*, *97*, 134–137.
Jun, D., Kutchoukov, V. G., & Kruit, P. (2011). Ion beams in SEM: An experiment towards a high brightness low energy spread electron impact gas ion source. *J. Vac. Sci. Technol. B*, *29*(6), 06F603.
Kellogg, S., Schampers, R., Zhang, S., Graupera, A., Miller, T., Laur, W., & Dirriwachter, A. (2010). High throughput sample preparation and analysis using an inductively coupled plasma (ICP) focused ion beam source. *Microsc. Microanal.*, *16*(Suppl. S2), 222–223.
Knuffman, B., Steele, A. V., & McClelland, J. J. (2013). Cold atomic beam ion source for focused ion beam applications. *J. Appl. Phys.*, *114*(4), 044303.
Knuffman, B., Steele, A. V., Orloff, J., & McClelland, J. J. (2011). Nanoscale focused ion beam from laser-cooled lithium atoms. *New J. Phys.*, *13*(10), 103035.
Kollmann, H., Piao, X., Esmann, M., Becker, S. F., Hou, D., Huynh, C., ... Lienau, C. (2014). Toward plasmonics with nanometer precision: Nonlinear optics of helium-ion milled gold nanoantennas. *Nano Lett.*, *14*(8), 4778–4784.
Kruit, P., Bezuijen, M., & Barth, J. E. (2006). Source brightness and useful beam current of carbon nanotubes and other very small emitters. *J. Appl. Phys.*, *99*(2), 024315.
Kruit, P., & Jansen, G. (2008). Space charge and statistical coulomb effects. In J. Orloff (Ed.), *Handbook of charged particle optics* (second edition). CRC Press.
Kudriavtsev, Y., Villegas, A., Godines, A., & Asomoza, R. (2005). Calculation of the surface binding energy for ion sputtered particles. *Appl. Surf. Sci.*, *239*(3–4), 273–278.
Lan, Y., Chang, W., Chang, Y., Chang, C., & Chen, C. (2015). Effect of focused ion beam deposition induced contamination on the transport properties of nano devices. *Nanotechnology*, *26*(5), 055705.
Langmuir, D. (1937). Theoretical limitations of cathode-ray tubes. *Proc. Inst. Radio Eng.*, *2*(8), 977–991.
Lehrer, C., Frey, L., Petersen, S., & Ryssel, H. (2001). Limitations of focused ion beam nanomachining. *J. Vac. Sci. Technol. B*, *19*(6), 2533–2538.
Li, W., Gu, C., Cui, A., Fenton, J., Jiang, Q., Warburton, P., & Shen, T. (2013). Three-dimensional nanostructures by focused ion beam techniques: Fabrication and characterization. *J. Mater. Res.*, *28*(22), 3063–3078.
Lindsey, S., & Hobler, G. (2012). The significance of redeposition and backscattering in nanostructure formation by focused ion beams. *Nucl. Instrum. Methods B*, *282*, 12–16.
Lipp, S., Frey, L., Lehrer, C., Demm, E., Pauthner, S., & Ryssel, H. (1996). A comparison of focused ion beam and electron beam induced deposition processes. *Microelectron. Reliab.*, *36*(11–12), 1779–1782.
Matsui, S. (2012). Focused ion beam-induced deposition (FIB-ID). In B. Bhushan (Ed.), *Encyclopedia of nanotechnology*. Netherlands: Springer.
Mayer, J., Giannuzzi, L. A., Kamino, T., & Michael, J. (2007). TEM sample preparation and FIB-induced damage. *Mater. Res. Soc. Bull.*, *32*, 400–407.
Nabhiraj, P., Menon, R., Rao, G. M., Mohan, S., & Bhandari, R. (2010). Characterization of compact ICP ion source for focused ion beam applications. *Nucl. Instrum. Methods A*, *621*(1–3), 57–61.

Petrov, Y. V., & Vyvenko, O. F. (2015). Scanning reflection ion microscopy in a helium ion microscope. *Beilstein J. Nanotechnol.*, *6*, 1125–1137.

Rahman, F. H. M., McVey, S., Farkas, L., Notte, J. A., Tan, S., & Livengood, R. H. (2012). The prospects of a subnanometer focused neon ion beam. *Scanning*, *34*(2), 129–134.

Reimer, L., & Tollkamp, C. (1980). Measuring the backscattering coefficient and secondary electron yield inside a scanning electron microscope. *Scanning*, *3*(1), 35–39.

Rejoub, R., Lindsay, B. G., & Stebbings, R. F. (2002). Determination of the absolute partial and total cross sections for electron-impact ionization of the rare gases. *Phys. Rev. A*, *65*(4), 042713.

Remes, J., Vahakangas, J., & Uusimaki, A. (2009). Improving the yield and turn-around time of focused ion beam microsurgery of integrated circuits by LCVD method. *Adv. Packaging IEEE Trans.*, *32*(2), 497.

Riecke, W. D. (1982). Practical lens design. In P. Hawkes (Ed.), *Magnetic electron lenses*. Berlin, Heidelberg: Springer-Verlag.

Rius, G. (2014). Technology basis and perspectives on focused electron beam induced deposition and focused ion beam induced deposition. *Nucl. Instrum. Methods Phys. Res., Sect. B*, *341*, 37–43.

Scherzer, O. (1936). Über einige Fehler von Elektronenlinsen. *Z. Phys.*, *101*(9), 593–603.

Schmidt, B., & Wetzig, K. (2012). *Ion beams in materials processing and analysis*. Vienna: Springer.

Schram, B., de Heer, F., van der Wiel, M., & Kistemaker, J. (1965). Ionization cross sections for electrons (0.6–20 keV) in noble and diatomic gases. *Physica*, *31*, 94–112.

Schwind, G. A., & Swanson, L. W. (2007). Emission characteristics of $Au_{60}Be_{40}$ and $Au_{62}Si_{23}Be_{15}$ liquid metal ion sources. *J. Vac. Sci. Technol. B*, *25*(6), 2586.

Siegele, R., Weatherly, G. C., Haugen, H. K., Lockwood, D. J., & Howe, L. M. (1995). Helium bubbles in silicon: Structure and optical properties. *Appl. Phys. Lett.*, *66*(11), 1319–1321.

Sigmund, P. (1969). Theory of sputtering. I. Sputtering yield of amorphous and polycrystalline targets. *Phys. Rev.*, *187*, 768.

Smith, N., Tesch, P., Martin, N., & Kinion, D. (2008). A high brightness source for nanoprobe secondary ion mass spectrometry. *Appl. Surf. Sci.*, *255*(4), 1606–1609.

Smith, N. S., Skoczylas, W. P., Kellogg, S. M., Kinion, D. E., Tesch, P. P., Sutherland, O., ... Boswell, R. W. (2006). High brightness inductively coupled plasma source for high current focused ion beam applications. *J. Vac. Sci. Technol. B*, *24*(6), 2902–2906.

Sorokin, A. A., Shmaenok, L. A., Bobashev, S. V., Mobus, B., Richter, M., & Ulm, G. (2000). Measurements of electron-impact ionization cross sections of argon, krypton, and xenon by comparison with photoionization. *Phys. Rev. A*, *61*(1), 022723.

Stevie, F. A. (2005). Focused ion beam secondary ion mass spectrometry (FIB-SIMS). In L. Giannuzzi, & F. Stevie (Eds.), *Introduction to focused ion beams: Instrumentation, theory, techniques, and practice*. Springer Science+Business Media, Inc.

Swanson, L. W., & Schwind, G. A. (2008). Review of ZrO/W Schottky cathode. In J. Orloff (Ed.), *Handbook of charged particle optics* (second edition). CRC Press.

Tan, S., Livengood, R., Hack, P., Hallstein, R., Shima, D., Notte, J., & McVey, S. (2011). Nanomachining with a focused neon beam: A preliminary investigation for semiconductor circuit editing and failure analysis. *J. Vac. Sci. Technol. B*, *29*(6), 06F604.

Tan, S., Livengood, R., Shima, D., Notte, J., & McVey, S. (2010). Gas field ion source and liquid metal ion source charged particle material interaction study for semiconductor nanomachining applications. *J. Vac. Sci. Technol. B*, *28*(6), C6F15.

Taylor, G. (1964). Disintegration of water drops in an electric field. *Proc. R. Soc. A*, *280*(1382), 383–397.

ten Haaf, G., Wouters, S. H. W., van der Geer, S. B., Vredenbregt, E. J. D., & Mutsaers, P. H. A. (2014). Performance predictions of a focused ion beam from a laser cooled and compressed atomic beam. *J. Appl. Phys.*, *116*(24), 244301.

Tondare, V. (2006). *Towards a high brightness, monochromatic electron impact gas ion source* (Doctoral dissertation). Delft University of Technology. Retrieved from repository.tudelft.nl.

Utke, I., Hoffmann, P., & Melngailis, J. (2008). Gas-assisted focused electron beam and ion beam processing and fabrication. *J. Vac. Sci. Technol. B*, *26*(4), 1197–1276.

van der Geer, S. B., Reijnders, M. P., de Loos, M. J., Vredenbregt, E. J. D., Mutsaers, P. H. A., & Luiten, O. J. (2007). Simulated performance of an ultracold ion source. *J. Appl. Phys.*, *102*(11), 094312.

van Kouwen, L., & Kruit, P. (2018). Brightness measurements of the nano-aperture ion source. *J. Vac. Sci. Technol. B*, *36*(6), 06J901.

Veligura, V., Hlawacek, G., van Gastel, R., Zandvliet, H. J. W., & Poelsema, B. (2012). Channeling in helium ion microscopy: Mapping of crystal orientation. *Beilstein J. Nanotechnol.*, *3*, 501–506.

Waid, S., Wanzenboeck, H. D., Gavagnin, M., Langegger, R., Muehlberger, M., & Bertagnolli, E. (2013). Focused ion beam induced ga-contamination—An obstacle for UV-nanoimprint stamp repair? *J. Vac. Sci. Technol. B*, *31*(4), 041602.

Waid, S., Wanzenboeck, H. D., Muehlberger, M., Gavagnin, M., & Bertagnolli, E. (2014). Focused ion beam direct patterning of hardmask layers. *J. Vac. Sci. Technol. B*, *32*(4), 041602.

Ward, B. W., Nottea, J. A., & Economou, N. P. (2006). Helium ion microscope: A new tool for nanoscale microscopy. *J. Vac. Sci. Technol. B*, *24*(6), 2871.

Wouters, S. H. W., ten Haaf, G., Notermans, R. P. M. J. W., Debernardi, N., Mutsaers, P. H. A., Luiten, O. J., & Vredenbregt, E. J. D. (2014). Performance predictions for a laser-intensified thermal beam for use in high-resolution focused-ion-beam instruments. *Phys. Rev. A*, *90*(6), 063817.

Wu, H., Stern, L. A., Xia, D., Ferranti, D., Thompson, B., Klein, K. L., ... Rack, P. D. (2014). Focused helium ion beam deposited low resistivity cobalt metal lines with 10 nm resolution: Implication for advanced circuit editing. *J. Mater. Sci., Mater. Electron.*, *25*(2), 587–595.

Yamamura, Y., & Tawara, H. (1996). Energy dependence of ion-induced sputtering yields from monatomic solids at normal incidence. *At. Data Nucl. Data Tables*, *62*(2), 149–253.

Zorzos, A. N., & Lozano, P. C. (2008). The use of ionic liquid ion sources in focused ion beam applications. *J. Vac. Sci. Technol. B*, *26*(6), 2097–2102.

CHAPTER EIGHT

Nano-fluidic flow in the nano-aperture ion source

Leon van Kouwen
Department of Imaging Physics, Delft University of Technology, Delft, Netherlands
e-mail address: lvankouwen@gmail.com

Contents

1. Introduction

Essential to the nano-aperture ion source concept is the delivery of gas to the electron beam irradiated area. The local particle density at the ion-producing region is critical for the performance of the NAIS. At the same time, this is not a directly controllable nor easily measurable quantity. Therefore, adequate analysis of the nano-fluidic flow properties is needed.

With the present ideas on chip design, it is required to have a relatively long supply channel (mm's) to connect between the macroscopic tubing and the nano-scale aperture membrane construction. This has practical, mostly geometrical reasons, such as an o-ring not blocking the ebeam and tubing that shouldn't interfere with the extraction field. Even though alternatives such as gluing instead of an O-ring are possible, a relatively long supply channel seems a part of many solutions. Fig. 1 shows a general overview of the system. Despite the difficulty of controlling the local particle density near the aperture directly, we can accurately set the inlet pressure of this supply channel with ease. The main question addressed in

Advances in Imaging and Electron Physics, Volume 212
ISSN 1076-5670
https://doi.org/10.1016/bs.aiep.2019.09.002

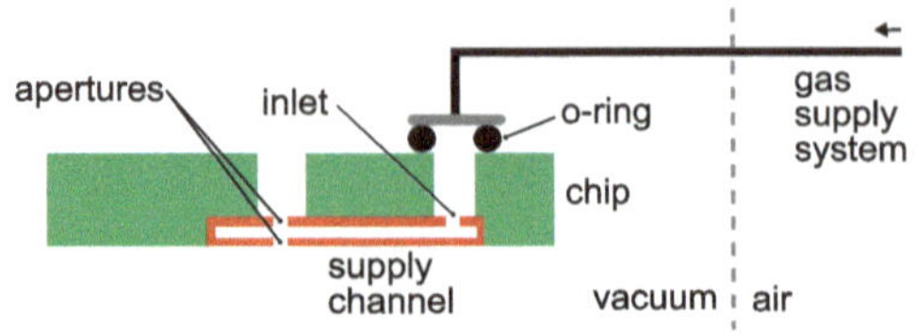

Figure 1 A possible implementation of the fluidic supply chain for the nano-aperture ion source using an o-ring. The required transition from macroscopic tubing to the sub-micron double aperture configuration makes a nano-fluidic channel seem inevitable.

this chapter is the following: when a certain pressure at the channel inlet is applied, how does that translate to the local pressure?

Some secondary objectives relating to flow properties can be identified as well. The flow rate into the vacuum is important for designing the vacuum system. Pump-down speed indicates how quickly one can exchange gas and what purity can be reached. Furthermore, the required inlet pressure and required mass flow rate enables proper design of the macroscopic part of the gas supply system.

2. Flow model

Three aspects of flow can be identified in our system. The first is the outflow through the double aperture structure. It is characterized by a rapid decrease in density, typically from atmospheric pressures into vacuum across a length of only 100 to 500 nm. Secondly, we identify the region near the apertures inside the channel, where we have rotationally symmetric flow. Thirdly, the flow through the supply channel is identified, characterized by a relatively weak density gradient in one direction over a long distance. By mass conservation, the flow through the channel is constant and determined by the outflow through the apertures. Fig. 2 shows the geometry of the studied flow system.

From an optical perspective, the membranes need only to be narrowly spaced close to the double apertures, but for ease of fabrication it is most obvious to maintain this dimension throughout the channel. Therefore, a constant channel height is assumed in the presented model.

This section develops an analytical model in which only the peak pressure near the apertures and the mass flow rate are ultimately considered. We do not investigate the particle density distribution of the complete system in detail. A more comprehensive approach could be based on direct sim-

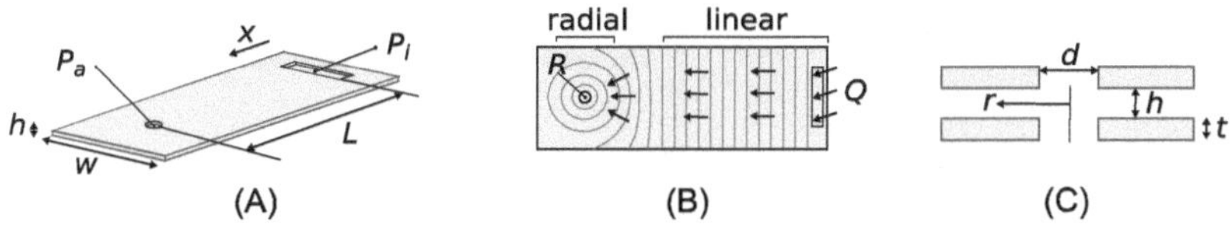

Figure 2 The nano-fluidic channel for the NAIS in its most obvious form is a narrow and long double membrane structure with a double aperture. (A) 3D view of the channel. (B) Top view of the channel with an indication of isobaric lines. A radially symmetric and a linear regime are identified. (C) Cross sectional view of the double aperture structure.

ulation Monte Carlo (DSMC). For long channels (Varoutis, Naris, Hauer, Day, & Valougeorgis, 2009) and apertures (Varoutis, Valougeorgis, Sazhin, & Sharipov, 2008; Varoutis, Valougeorgis, & Sharipov, 2009) such simulations can be found in the literature, however the combination in one simulation might be difficult due to the large difference in scale. In addition, DSMC simulations are notoriously time consuming, so a fast and wide parameter exploration is not easy.

2.1 Dimensionless numbers and flow conductance

Before discussing the transport equations for the configuration in question, we introduce some essential concepts in fluid dynamics and use them to determine the appropriate flow regime.

Dimensionless numbers are often used to characterize a fluidic system. The three most relevant numbers for the NAIS are the Knudsen, Reynolds and Mach number:

$$\mathrm{Kn} = \frac{\lambda}{l}, \qquad \mathrm{Re} = \frac{\rho v l}{\mu}, \qquad \mathrm{Ma} = \frac{v}{v_s}. \tag{1}$$

The Knudsen number indicates the ratio of particle-wall to particle-particle interaction, the Reynolds is a measure for the amount of turbulence and the Mach number is an average flow velocity normalization. Fig. 3 indicates the flow type corresponding to the values of these numbers.

A slight problem with the Knudsen number is that the mean free path is not a directly measurable quantity. We follow Sharipov and Seleznev (1998) in writing the mean free path in terms of the viscosity, molecular mass, pressure, and temperature. Furthermore we express the speed of sound in terms of the molecular properties (Sharipov, 2004) and obtain:

$$\lambda = \frac{\sqrt{\pi}\mu}{2P}\sqrt{\frac{2kT}{m}}, \qquad v_s = \sqrt{\gamma\frac{kT}{m}}. \tag{2}$$

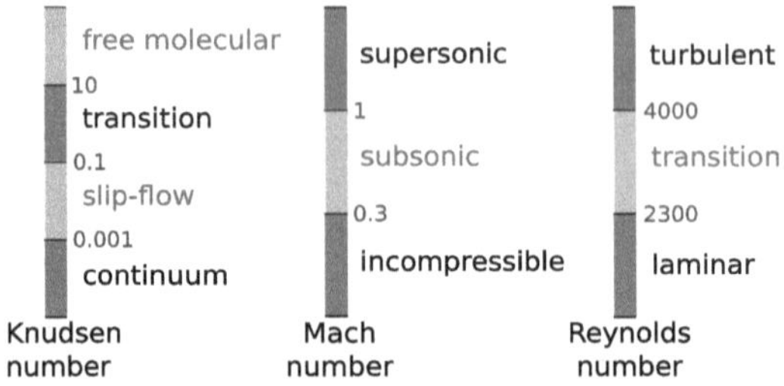

Figure 3 Three dimensionless numbers that are relevant for indicating the type of flow. For the Knudsen number (Colin, 2005) and the Mach number the scales are rather universal, for the Reynolds number the scale is more geometry dependent and this scale relates to flow through a pipe (Holman, 2002). Incompressible refers to very small gradients along the flow direction rather than static compressibility.

For single atom gasses, the specific heat ratio $\gamma = \frac{5}{3}$. A convenient relation between Knudsen, Mach and Reynolds (Colin, 2005) is

$$\mathrm{Kn} = \sqrt{\frac{\pi\gamma}{2}}\frac{\mathrm{Ma}}{\mathrm{Re}}. \tag{3}$$

We intend to operate the NAIS such that the mean free path of the ions is close to or a little shorter than the membrane spacing, as explained in the introductory chapter. The mean free path of ions is slightly shorter than for neutrals, so we may roughly say that the operation regime of the NAIS is characterized by $\mathrm{Kn} \approx 1$. This is only a rough indication and we would like to be able to tune the pressure anyway, so we state that the relevant Knudsen range for the NAIS is 0.1 to 10. In Fig. 3 we can see that this range is called the transition regime. This means that molecular and continuum effects are both relevant and flow analysis is not trivial.

The Mach number is best first examined at the outflow through the apertures. Consider the aperture to be an orifice, i.e. a zero thickness membrane. Continuum outflow into vacuum is said to be choked (Livesey, 2004; Gronych et al., 2010), meaning that a maximum flow speed is attained and (small) changes in the downstream pressure are irrelevant. In fact, for constant cross section tubes the maximum Mach number is simply one (Livesey, 2004). In the case of free molecular flow into vacuum, the Mach number is close to or smaller than one. Since the channel cross section will always be orders of magnitude larger than the aperture cross section, and the Mach number through the aperture is in any case close to one, we can state that the Mach number in the channel is close to zero and incompressible flow conditions apply.

We can use Eq. (3) to estimate the maximum Reynolds number occurring in our system. If we set $\mathrm{Ma} = 1$ and $\mathrm{Kn} = 0.1$ for a worst case estimate, we obtain $\mathrm{Re} \leq 17$. This is in fact an overestimate because at the position were $\mathrm{Ma} = 1$, the local Knudsen number is higher due to a density decrease. We conclude that turbulence effects do not play a role in our system. Surely this was to be expected for such small dimensions. In the linear region of the channel $\mathrm{Re} \ll 1$.

In the high Knudsen regime ($\mathrm{Kn} > 10$), it is customary to work with the flow conductance C and a gas flow rate Q, or a mass flow rate $\dot{M}$. The transport equations are

$$\frac{\mathrm{d}(PV)}{\mathrm{d}t} = Q = -C\Delta P, \qquad \dot{M} = -\frac{m}{kT} C\Delta P. \tag{4}$$

These expressions can easily be derived from Fick's law if the gas is assumed to be ideal. An expression for the conductance of an arbitrarily shaped open area A in an infinitely thin membrane is (Weston, 1985)

$$C_0 = \tfrac{1}{4} A \langle v \rangle, \qquad \langle v \rangle = \sqrt{\frac{8kT}{\pi m}}. \tag{5}$$

To account for the non-zero length of a channel, the concept of a transmission probability α is introduced (van Essen & Heerens, 1976; Livesey, 1998). The flow conductance of the entrance (E. (5)) is multiplied by the probability of a molecule that enters the channel making it out to the other side of the channel:

$$C = \alpha C_0 \tag{6}$$

It can be convenient to relate a transmission probability or a flow conductance to a diffusion coefficient as follows:

$$D = C\frac{L}{A} = \tfrac{1}{4}\alpha L \langle v \rangle. \tag{7}$$

Note that this simple expression is only valid in the case of constant cross section channels because the diffusion coefficient is a local property while the flow conductance is an average quantity for the full channel.

2.2 Flow conductance of apertures

The simplest and most familiar aspect of the flow system is the outflow through the apertures. In the case of short ducts, Dushman (1922) proposed a simple expression that is often found in vacuum technology handbooks (Weston, 1985). It gives a maximum error of 11% (Henning, 1978). A slightly more advanced expression suggested by Henning (1978) gives errors less than 0.6% on the domain $t/d = (0.01, 19)$. We choose Henning's expression as its applicable range and accuracy is excellent, while being fairly simple. Further discussion on the various alternatives can be found elsewhere (van Essen & Heerens, 1976; Santeler, 1986). We get for the free molecular flow conductance C_{fm} of a short duct

$$C_{\mathrm{fm}} = \alpha C_0, \qquad \alpha = \frac{1}{0.996 + 0.94\left(\frac{t}{d}\right)^{0.94}}. \tag{8}$$

An often encountered expression for continuum flow through an orifice derived from the Euler equation is (Sharipov, 2004; Gronych et al., 2010)

$$C_{\mathrm{is}} = \sqrt{\gamma}\left(\frac{2}{\gamma+1}\right)^{\frac{\gamma+1}{2(\gamma-1)}} \sqrt{2\pi}\, C_0. \tag{9}$$

Since for single atom gasses $\gamma = \frac{5}{3}$, and we can simplify to $C_{\mathrm{v}} = 1.82 C_0$. Interestingly, it differs from free molecular flow only by a numerical factor. This expression is based on isentropic flow, requiring $\mathrm{Re} \gg 1$ (Sharipov, 2004), which is not quite met in our case and the results may be doubtful. Furthermore, it does not incorporate the non-zero membrane thickness.

Eqs. (8) and (9) give values for the limiting cases of free molecular flow and continuum flow. However, we would like to have a more generally valid expression. An alternative suiting our needs very well was proposed by Fujimoto and Usami (1984). They devised an interpolation equation that works over a wide range of Knudsen values (see also the work of Sharipov (1986, 2004); Sharipov and Kozak (2009)). In addition, it interpolates over duct aspect ratio. If $\mathrm{Kn} > 0.00025$ and $\mathrm{Re} < 2800$ their empirical formula for the flow conductance of short ducts C_{F} ('F' denotes Fujimoto) is

$$\frac{C_{\mathrm{F}}}{C_0} = \left(\alpha + \frac{0.4733 + 0.907 z^{0.5}}{1 + 10.4 z + 16.1 z^2}\right), \tag{10}$$

Table 1 The flow conductance for a short tube as predicted by different models for a few typical geometries. C_0 denotes the free molecular conductance of zero-thickness orifices (Eq. (5)), C_{fm} is the free molecular conductance of a short tube (Eq. (8)) and C_{is} is the isentropic flow conductance (Eq. (9)). The empirical formula by Fujimoto and Usami's C_F (Eq. (10)) has negligible error compared to their experimental values for Kn > 0.1 (Fujimoto & Usami, 1984). For Kn < 0.1 the error may be up to 5% (Sharipov, 2004) compared to experimental data on mono-atomic gases by Jitschin, Ronzheimer, and Khodabakhshi (1999). All calculations assume a single aperture in combination with a large vessel.

d	t	C_0	C_{fm}/C_0	C_{is}/C_0	C_F/C_0		
nm	nm	$m^3\,s^{-1}$			Kn = 0.1	Kn = 1	Kn = 10
100	0	2.45×10^{-12}	1.00	1.82	1.35	1.06	1.01
100	100	2.45×10^{-12}	0.52	1.82	0.73	0.53	0.52
300	100	2.20×10^{-11}	0.75	1.82	1.04	0.78	0.75
1000	100	2.45×10^{-10}	0.91	1.82	1.24	0.95	0.91

with

$$z = \text{Kn} \cdot \left(\alpha + 0.125 \exp\left[-0.5\,(12\alpha - 11.2)^2\right] + 0.18 \exp\left[-14.7\alpha\right] - 0.08\right)^{-1}. \tag{11}$$

Although it may not be very elegant, it is highly practical. The tested domain of tube length over tube diameter is 0.00025 up to 12.7, quite broad indeed. The precision of this interpolation is more than required and surely a more simple interpolation scheme could be used. We may even get a false sense of accuracy because this model assumes flow out of a large vessel, while we start from a restricted channel and double aperture configuration. Next to that, it will be very difficult to precisely know the critical dimensions. With this in mind, it is the closest model that is readily available; not correcting for tube length and viscous flow effects would give an even worse inaccuracy.

Some examples of calculated flow conductances are given in Table 1. Interestingly, the values acquired from the empirical formula of Fujimoto are never more than a factor two different from C_0. This means that despite operating in the transition regime, decent estimates can be made by just using the free molecular flow conductance. This strokes with findings by Sharipov (1986), who concluded that for orifices, the flow conductance is at most 1.5 times the molecular flow conductance.

We are working towards a model for the full flow characteristics of the NAIS, however we can already analyze the aspect of vacuum load due to aperture outflow. If for example $d = 100$ nm, $t = 100$ nm, and $\mathrm{Kn} = 1$, Eq. (10) gives: $C = 1.3 \times 10^{-12}\ \mathrm{m^3\,s^{-1}}$ yielding a total vacuum load of $Q = 2.5 \times 10^{-7}\ \mathrm{Pa\,m^3\,s^{-1}}$. Suppose we would like to acquire a Schottky-emitter-friendly vacuum of 1×10^{-7} Pa, then we need a pumping speed of $2.5\ \mathrm{m^3\,s^{-1}}$ to cope with the outflow load. This is a near-impossible requirement. A typically used small ion-getter pump delivers a mere $0.025\ \mathrm{m^3\,s^{-1}}$, not even considering the low efficiency for noble gases. Even a high end turbo-molecular-pump can only reach $1.0\ \mathrm{m^3\,s^{-1}}$. Additionally, this example considers just one, quite small, aperture. We conclude that in the chip region we should settle for a more modest pressure of say 1×10^{-4} Pa, hence a Schottky electron gun can never reside in the same vacuum.

2.3 Flow through shallow channels

The flow through a long rectangular channel is considered in this section. One may suspect that a relatively wide channel of say, 100 um in width, will have negligible flow restriction compared to a 100 nm sized aperture. If we assume free molecular flow for the moment, we can use the transmission probability of a long narrow slit to estimate this. Weissler and Carlson (1979) give $\alpha = \left(1 + \frac{3}{8}\frac{(w+h)L}{wh}\right)^{-1}$. With the requirement $L \gg w \gg h$, it is easy to show that $\alpha \approx \frac{8}{3}\frac{h}{L}$. We find for a typical channel of $L \times w \times h = 3000 \times 100 \times 0.1$ μm from Eq. (5) that $C = 8.2 \times 10^{-13}\ \mathrm{m^3\,s^{-1}}$. We compare this to the flow conductance $C = 2.5 \times 10^{-12}\ \mathrm{m^3\,s^{-1}}$ of a 100 nm aperture using Eq. (10). The channel flow conductance is more than 10 times lower, which directly leads to pressure drops of more than a factor 10.

It may even be worse. At $\mathrm{Kn} \approx 1$, molecular collisions lower the flow conductance compared to free molecular flow. Furthermore, if a large pressure drop is acquired, the Knudsen number at the inlet is low and diffusion is restricted even further. On the other hand, we are neglecting any advective transport, which may not be appropriate to do. It seems worthwhile to investigate channel flow in more detail.

Colloquially, the term convective transport is often used for advective transport. Formally, convective transport includes diffusion and bulk movement of a fluid, while advective transport refers to bulk movement only. In this analysis the distinction plays an important role and we will use the formal nomenclature.

Literature on the subject of micro channel flow is quite abundant, see for example Agrawal (2011). Several models are available, notably one proposed by Livesey (2001) (see also Livesey (2004)). This particular model is quite broad in application as it can account for turbulence and channel entrance effects. For our purpose, though, this seems to unnecessarily complicate matters. We have also considered another model (Dongari, Sharma, & Durst, 2009; Dongari, Sambasivam, & Durst, 2009; Durst, Gomes, & Sambasivam, 2006; Dongari, Dadzie, Zhang, & Reese, 2011), developed specifically for flow in rectangular micro-channels. They introduce a mass diffusivity term in the Navier-Stokes (NS) equation to extend the validity of NS to higher Knudsen numbers. The model is fairly simple in use, but we couldn't reproduce the claimed excellent agreement when comparing predictions of the model to experimental data of Maurer, Tabeling, Joseph, and Willaime (2003) nor data of Ewart, Perrier, Graur, and Méolans (2007).

The method we use is based on adding an advective and a diffusive contribution to acquire the total flow rate. This is equivalent to a method outlined by Thompson and Owens (1975). The advective component needs to be valid up to high Knudsen numbers for which we use a second order slip flow model (Maurer et al., 2003). We follow Browne and John (1970) in multiplying the first order slip flow term in the advective transport equation by $(1+\mathrm{Kn})^{-1}$. This removes the unphysical contribution of viscous flow in the free molecular flow regime. Along the same line of reasoning we multiply the second order term by $(1+\mathrm{Kn})^{-2}$. The diffusive flow contribution needs to interpolate between bulk and free molecular diffusion. The proposed advective and diffusive flow rate contributions are:

$$Q_\mathrm{A} = -\frac{wh^3}{12\mu}\left(1+6A_1\frac{\mathrm{Kn}}{1+\mathrm{Kn}}+12A_2\left(\frac{\mathrm{Kn}}{1+\mathrm{Kn}}\right)^2\right)P\frac{\mathrm{d}P}{\mathrm{d}x} = -f_\mathrm{A}(P)\frac{\mathrm{d}P}{\mathrm{d}x} \tag{12}$$

$$Q_\mathrm{D} = -wh\left(\frac{D_\mathrm{b}D_\mathrm{fm}}{D_\mathrm{b}+D_\mathrm{fm}}\right)\frac{\mathrm{d}P}{\mathrm{d}x} = -f_\mathrm{D}(P)\frac{\mathrm{d}P}{\mathrm{d}x} \tag{13}$$

Note that the Knudsen number as function of the pressure is obtained by using Eq. (2). The constants of the slip-flow model are $A_1=1.2$, $A_2=0.23$ for helium (Maurer et al., 2003). These coefficients depend on the channel wall and gas species, but typically do not vary more than 10%.

The diffusive term is based on a inverse addition of the bulk diffusion coefficient D_b and the free molecular diffusion coefficient D_fm. In the limit of high Knudsen numbers, the diffusive contribution converges to a molecular flow expression while for low Knudsen numbers bulk diffusion is obtained. For intermediate Knudsen values the term is an interpolation between bulk and free molecular diffusion. The bulk diffusion coefficient from kinetic gas theory is $\frac{1}{3}\langle v\rangle\lambda$, which in terms of the Knudsen number reads

$$D_b = \tfrac{1}{3}\langle v\rangle \mathrm{Kn}h. \tag{14}$$

For free molecular flow, an expression for the flow conductance of rectangular channels is used (Thompson & Owens, 1975; Livesey, 1998), which is based on theory by Smoluchowski (1910). By employing Eq. (7) we obtain

$$\begin{aligned} D_{fm} = \frac{1}{4}\langle v\rangle \Bigg(& h\cdot\ln\left(\frac{w}{h}+\sqrt{1+\frac{w^2}{h^2}}\right)+w\cdot\ln\left(\frac{h}{w}+\sqrt{1+\frac{h^2}{w^2}}\right) \\ & -\frac{\left(h^2+w^2\right)^{3/2}}{3hw}+\frac{h^3+w^3}{3hw}\Bigg). \end{aligned} \tag{15}$$

When using the abbreviating functions f_A and f_D in Eqs. (12) and (13), the transport equation becomes

$$Q = -\left(f_A(P)+f_D(P)\right)\frac{dP}{dx}. \tag{16}$$

For calculating the flow rate we can employ that Eq. (16) is separable and obtain

$$Q = -\frac{1}{L}\int_{P(0)}^{P(L)}\left(f_A(P)+f_D(P)\right)dP. \tag{17}$$

If Q is known, we can solve Eq. (16) to obtain the pressure profile along the channel for a given upstream or downstream pressure. It is important to realize that the used Knudsen number is a local property in the channel. This means that the relative contributions of advective and diffusive transport can vary along the channel. Very large pressure differences in a channel are therefore acceptable in calculations.

Fig. 4 shows flow calculations using Eq. (17), compared to experimental data by Maurer et al. (2003); Ewart et al. (2007) and Arkilic, Schmidt, and Breuer (1995, 1997). These three data sets are acquired in different ways of varying the inlet and the outlet pressure, so the sets offer good diversity for testing the model. The resulting correspondence is about as good as the experimental variation. This is fortunate, given the relative simplicity of the model, notably the somewhat bold interpolation scheme for bulk and free molecular diffusion. Since our model is simple and reproduces experimental data well, it is a good choice for calculating channel flow.

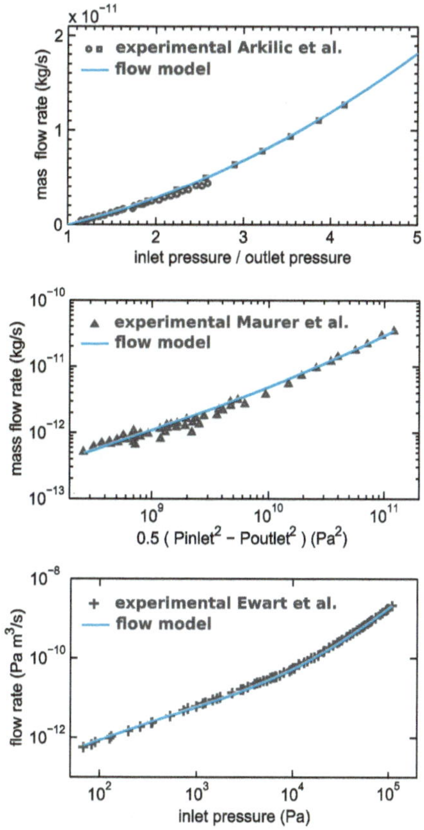

Figure 4 Comparison of the proposed model using Eq. (17) and experimental data shows excellent agreement. All measurements are performed using helium gas. No free parameter was adjusted to fit the data; the theoretical curves are obtained by directly using the reported channel dimensions and pressure settings. The channel dimensions of data by Arkilic et al. (1995, 1997) are $L \times w \times h = 7500 \times 52.25 \times 1.33$ µm. The outlet pressure was fixed at $P_{outlet} = 1 \times 10^5$ Pa. The data by Maurer et al. (2003) is obtained from a channel of $L \times w \times h = 10000 \times 200 \times 1.14$ µm. The inlet and outlet pressure were varied together linearly in the range $P_{inlet} = 0.26 \times 10^5 - 5 \times 10^5$ Pa and $P_{outlet} = 0.12 \times 10^5 - 1 \times 10^5$ Pa. Ewart et al. (2007) used a channel of $L \times w \times h = 9390 \times 492 \times 9.38$ µm and a fixed inlet to outlet pressure ratio of 5 was maintained.

Fig. 5 shows the calculated flow contributions and the pressure profile of an example channel. In this example, both advective as well as diffusive transport plays a role and their relative importance changes across the channel.

Fig. 6 shows the flow conductance for a variety of inlet pressures, as characterized by the local Knudsen number at the inlet. Since the effects of

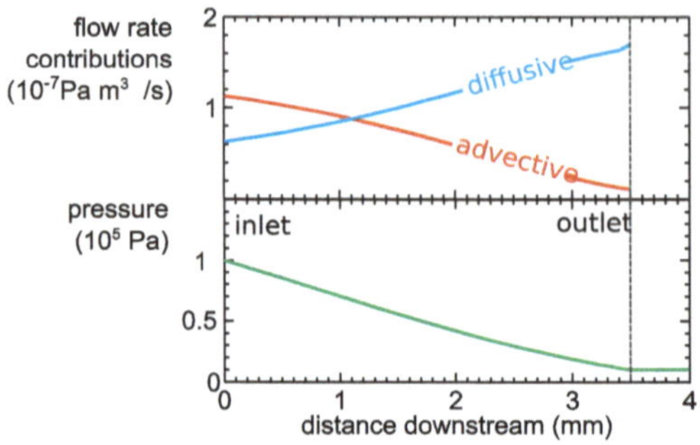

Figure 5 Flow calculation for a rectangular channel of $h \times w \times L = 500$ nm $\times$ 100 μm $\times$ 3.5 mm, supplied with an inlet pressure $P = 1 \times 10^5$ Pa and $P = 1 \times 10^4$ Pa at the outlet. Notice how the diffusive and advective contribution vary along the channel.

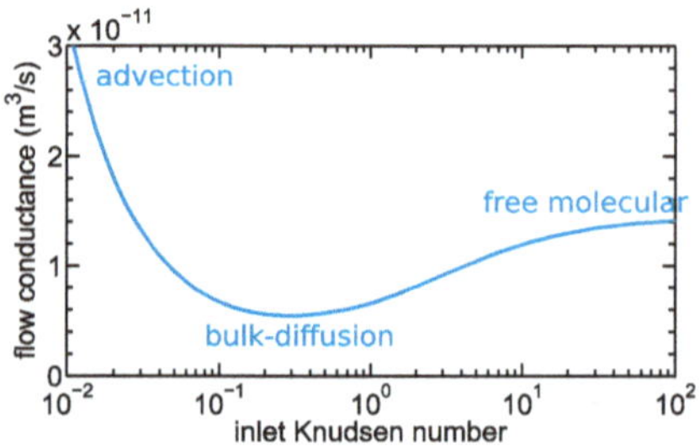

Figure 6 Flow calculations for a rectangular channel of $h \times w \times L = 500$ nm $\times$ 100 μm $\times$ 3.5 mm, supplied with different pressures at the inlet, and vacuum at the outlet. The Knudsen number varies along the channel and the given number is calculated at the channel entrance. The flow conductance is defined as the flow rate divided by the inlet pressure. The dominant flow mechanisms are indicated in the different regimes.

free molecular flow, molecular diffusion and advective flow are competing as a function of the Knudsen number, a minimum known as the Knudsen minimum or Knudsen paradox is obtained.

The reasonably shallow region around $\mathrm{Kn} = 1$ results in the free molecular flow assumption yielding a fair order of magnitude estimate, even down to $\mathrm{Kn} \approx 0.1$. The underlying transport mechanism is not free molecular flow though. For low Knudsen values ($\mathrm{Kn} < 0.1$), the advective contribution becomes very large and a free molecular approximation leads to underestimates of the flow conductance. In general, the shape of this curve depends on the pressures and geometry involved, so it is best to rely on Eq. (17) for accurate results.

2.4 Radial distribution around the apertures

In the close proximity of the double aperture structure we expect a radially symmetric pressure drop. The range of this radial symmetry is at most the width of the channel, typically some tens of micro meters. This distance is relatively short compared to the channel length of some mm's, however, the channel size decreases when approaching the double apertures, so an appreciable flow resistance may be obtained anyway.

The channel flow model from the previous section can be recycled when appropriately transformed to cylindrical symmetry. Since the model does not consider the sidewall effects, the sidewalls can be considered connected. The substitutions $w = 2\pi r$ and $x = r$ in Eqs. (12) and (13) lead to radial symmetry. The bulk diffusivity is not a geometrical property, so it can be used as is. The molecular flow diffusivity, however, was derived for the rectangular geometry and may not describe the radial symmetry very well. We follow Browne and John (1970) and use

$$D_{\mathrm{fm}} = \tfrac{2}{3}\langle v\rangle\, h. \tag{18}$$

Strictly speaking, this expression is theoretically not completely sound. It is based on Knudsen's method, which is known to be only fully correct for tubes (Thompson & Owens, 1975). No improved expression seems to be available and Browne and John (1970) showed good correspondence to experimental data.

When using the substitutions $w = 2\pi r$ and $x = r$, f_{A} and f_{D} become functions of r. Eq. (16) is no longer separable, but can still be solved numerically with relative ease. Fig. 7 shows the result of the radial pressure drop calculation for a few geometrical configurations. The pressure drop is defined as the pressure at $r = R = 40$ μm, divided by the pressure at the aperture position. The flow rate is calculated by Eq. (10), and the result has been used to solve Eq. (16) in combination with the aforementioned substitution.

From Fig. 7 we conclude that the effect of a radial pressure drop is not negligible. On the other hand, it is not so severe that it poses a fundamental problem. Secondly, we note that the dependence on the Knudsen number is fairly weak.

The dashed lines in the graphs show a different choice for the end of the radial regime of 40 um. The lines are very close to the 50 um line, showing that the result is very insensitive to the choice of R. This is not surprising, far away from the center the channel is relatively large and the contribution

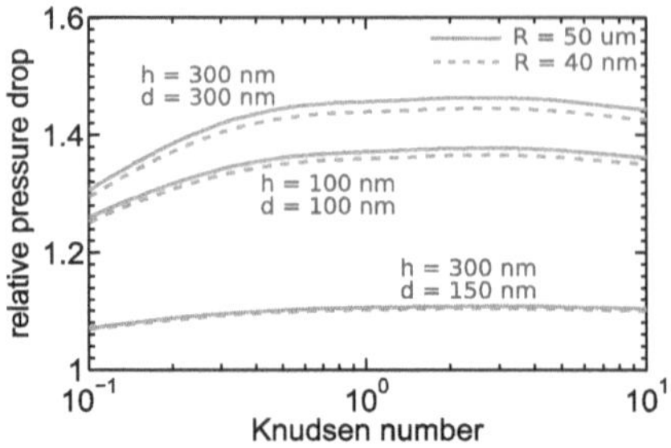

Figure 7 Calculations in the radial regime show a small relative pressure drop and a weak dependence on the Knudsen number. The relative pressure drop is defined as the pressure 50 um away from the apertures divided by the pressure between the two apertures. The Knudsen number is defined at the aperture location. The curves for $R = 40$ um shows that the result is insensitive to the choice of R.

to the flow resistance is only minor. This justifies the somewhat arbitrary choice of R. In all calculations Ma < 0.2.

2.5 Full system analysis

We have discussed the flow along the channel, the radial flow towards the aperture, and the outflow through the aperture. The combination of these gives us the behavior of the full system. The pressure between the apertures P_a is the independent variable, as it is the pressure we need to tune for NAIS operation. The dependent variable is the inlet pressure, as it is the pressure we will need to apply in an experiment.

For a given geometry and required aperture pressure, we can calculate the mass flow through the entire system using Eqs. (4) and (10). The aperture pressure is the boundary condition for Eq. (16) in the radial regime. The resulting pressure at R serves as the boundary condition for the calculation of Eq. (16) in the linear regime. It was shown in Fig. 7 that the end of the radial regime is not very important and we choose for R half the width of the channel ($R = 50$ μm). The channel length is specified as the distance between the inlet and the aperture center.

Fig. 8 shows the calculation result for the typical NAIS nano-fluidic channel. It is not obvious from an ion optical point of view what the ideal channel height and aperture diameter should be, so a variety of configurations is shown in one plot. We consider a fixed Knudsen number of Kn $= 1$. It is not possible to simply scale these curves, so for other values of the Knudsen number a new calculation is required.

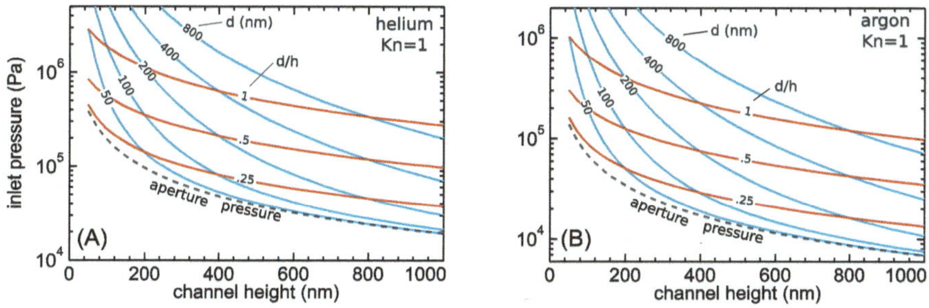

Figure 8 Calculated required inlet pressure for achieving an aperture pressure such that Kn = 1 at the aperture location. The calculation incorporates the outflow through two apertures, the drop in the radial regime, and the drop in the linear regime. The dashed line indicates the aperture pressure determined by the height and Kn = 1. The fixed channel dimensions used are $L = 3.5$ mm and $w = 100$ μm.

Perhaps most striking in Fig. 8 is the large pressure drop for various configurations. Without the proper remedies or a least awareness of the pressure loss, we can expect poor performance. More specifically, consider a channel of 100 nm high and 100 nm aperture. This configuration requires more than ten bar of pressure at the inlet. This is undesirable as it poses high requirements on the channel integrity.

The largest factor of this big pressure drop is the linear regime of the channel. Widening the channel is one way to improve and does not influence the ion optical properties, but we may need a very wide channel to achieve sufficiently low inlet pressures. A disadvantage of wide channels is the large membrane surface, which makes the chips more prone to membrane failure. Decreasing the channel length is always advantageous, but may not be possible due to practical limitations. A very effective way of lowering the pressure drop is increasing the channel height, but the ion optical consequences need to be considered.

3. Conclusion

A relatively simple model of gas flow for the NAIS nano-fluidic configuration was established. Good correspondence to experimental data of other researchers validates the presented method for calculating flow through a rectangular nano-channel. The flow is governed by an advective and a diffusive component and the relative contributions depends strongly on the Knudsen number.

Employing the model, we found a significant pressure drop between the inlet of the channel and near the double aperture structure for the typical dimensions and pressures we are aiming to operate the NAIS with. The main cause is the poor flow conductance of the relatively long nano-fluidic channel. The pressure drop locally around the aperture is also significant but less severe.

For ion-optical reasons a particular aperture size is chosen and preferably the channel is designed in accordance with this aperture. Simply choosing a large height is very effective in reducing the pressure drop, but if the height is constant across the entire device this influences the ion-optical properties as well. Widening or shortening the channel reduces the pressure drop too, though less effectively than increasing the channel height and has other practical disadvantages. An alternative channel design is a deep channel of say tens of microns, that is only confined very close to the aperture, if fabrication permits.

Fig. 8 shows the required inlet pressure for achieving Kn = 1 at the aperture. The figure reveals how serious the pressure drop can be in many configurations. An illustrative example is a 100 nm channel in combination with a 100 nm aperture diameter for argon. In this situation the pressure at the aperture needs to be 1.9×10^5 Pa. In order to achieve that, a 1.9×10^6 Pa inlet pressure is required. A 300 nm channel height in combination with a 100 nm aperture only requires 1.4×10^5 Pa at the inlet. If the aperture to channel height ratio needs to be fixed and larger than 0.5, a pressure drop of at least 3 times is unavoidable.

List of symbols

C_O	orifice flow conductance $\mathrm{m^3\,s^{-1}}$
C_{is}	isentropic flow cond. $\mathrm{m^3\,s^{-1}}$
C_{fm}	mol. flow cond. aperture $\mathrm{m^3\,s^{-1}}$
C_F	interpolated flow cond. $\mathrm{m^3\,s^{-1}}$
Q	gas flow rate $\mathrm{Pa\,m^3\,s^{-1}}$
Q_A	advective flow rate $\mathrm{Pa\,m^3\,s^{-1}}$
Q_D	diffusive flow rate $\mathrm{Pa\,m^3\,s^{-1}}$
D_b	bulk diffusivity $\mathrm{m^2\,s^{-1}}$
D_{fm}	free mol. diffusivity $\mathrm{m^2\,s^{-1}}$
α	transmission probability
Kn	Knudsen number
Re	Reynolds number
Ma	Mach number
l	typical dimension m
P_i	inlet pressure Pa
P_o	outlet pressure Pa

L	channel length	m
w	channel width	m
h	channel height	m
x	coordinate in flow direction	m
t	membrane thickness	m
r	radial position from aperture	m
R	end radial symmetric region	m
m	molecular mass	kg
ρ	mass density	$\mathrm{kg\,m^{-3}}$
μ	(dynamic) viscosity	Pa s
γ	ratio of heat capacities	
λ	molecular mean free path	m
v_s	speed of sound	$\mathrm{m\,s^{-1}}$
$\langle v \rangle$	average molecular velocity	$\mathrm{m\,s^{-1}}$
v	flow velocity	$\mathrm{m\,s^{-1}}$
T	temperature	K

Acknowledgment

I would like to thank Pieter Kruit for his guidance and contributions. This work is part of the Industrial Partnership Program (IPP) of the Stichting voor Fundamenteel Onderzoek der Materie (FOM), which is financially supported by the Nederlandse Organisatie voor Wetenschappelijk Onderzoek (NWO). The IPP program is co-financed by FEI Company. This work is based on the dissertation titled *The Nano-Aperture Ion Source*, by Leon van Kouwen.

References

Agrawal, A. (2011). A comprehensive review on gas flow in microchannels. *Int. J. Micronano Scale Transp.*, *2*(1), 1–40.

Arkilic, E. B., Schmidt, M. A., & Breuer, K. S. (1995). Slip flow in microchannels. In *Proceedings: Rarefied gas dynamics symposium 19* (pp. 57–65).

Arkilic, E. B., Schmidt, M. A., & Breuer, K. S. (1997). Gaseous slip flow in long microchannels. *J. Microelectromech. Syst.*, *6*(2), 167–178.

Browne, V. A., & John, J. E. A. (1970). Vacuum radial flow from the viscous through the free molecule regime. *Vacuum*, *20*(12), 525–533.

Colin, S. (2005). Rarefaction and compressibility effects on steady and transient gas flows in microchannels. *Microfluid. Nanofluid.*, *1*(3), 268–279.

Dongari, N., Dadzie, S. K., Zhang, Y., & Reese, J. M. (2011). Isothermal micro-channel gas flow using a hydrodynamic model with dissipative mass flux. *AIP Conf. Proc.*, *1333*(1), 718–723.

Dongari, N., Sambasivam, R., & Durst, F. (2009). Extended Navier-Stokes equations and treatments of micro-channel gas flows. *J. Fluid Sci. Technol.*, *4*(2), 454–467.

Dongari, N., Sharma, A., & Durst, F. (2009). Pressure-driven diffusive gas flows in micro-channels: From the Knudsen to the continuum regimes. *Microfluid. Nanofluid.*, *6*(5), 679–692.

Durst, F., Gomes, J., & Sambasivam, R. (2006). Thermofluiddynamics: Do we solve the right kind of equations? In *Proceedings of the international symposium on turbulence* (pp. 3–18).

Dushman, S. (1922). *Production and measurement of high vacuum*. General Electric review (Chapter 1).

Ewart, T., Perrier, P., Graur, I. A., & Méolans, J. G. (2007). Mass flow rate measurements in a microchannel, from hydrodynamic to near free molecular regimes. *J. Fluid Mech.*, *584*, 337–356.

Fujimoto, T., & Usami, M. (1984). Rarefied gas flow through a circular orifice and short tubes. *J. Fluids Eng.*, *106*(4), 367–373.

Gronych, T., Peksa, L., Prazak, D., Vicar, M., Repa, P., Krajicek, Z., … Tesar, J. (2010). Changes in the {NPL} orifice conductance on a transition from molecular gas flow to transitional flow. *Vacuum*, *84*(1), 280–282.

Henning, H. (1978). The approximate calculation of transmission probabilities for the conductance of tubulations in the molecular flow regime. *Vacuum*, *28*(3), 151–152.

Holman, J. P. (2002). *Heat transfer*. McGraw-Hill (p. 207).

Jitschin, W., Ronzheimer, M., & Khodabakhshi, S. (1999). Gas flow measurement by means of orifices and venturi tubes. *Vacuum*, *53*(1–2), 181–185.

Livesey, R. (1998). Flow of gases through tubes and orifices. In J. M. Lafferty (Ed.), *Foundations of vacuum science and technology*. Wiley (Chapter 2).

Livesey, R. (2001). Method for calculation of gas flow in the whole pressure regime through ducts of any length. *J. Vac. Sci. Technol. A*, *19*(4), 1674–1678.

Livesey, R. (2004). Solution methods for gas flow in ducts through the whole pressure regime. *Vacuum*, *76*(1), 101–107.

Maurer, J., Tabeling, P., Joseph, P., & Willaime, H. (2003). Second-order slip laws in microchannels for helium and nitrogen. *Phys. Fluids*, *15*(9), 2613–2621.

Santeler, D. (1986). New concepts in molecular gas flow. *J. Vac. Sci. Technol. A*, *4*(3), 338.

Sharipov, F. (1986). Rarefied gas flow into vacuum through thin orifice: Influence of boundary conditions. *AIAA J.*, *40*(10), 2006–2008.

Sharipov, F. (2004). Numerical simulation of rarefied gas flow through a thin orifice. *J. Fluid Mech.*, *518*, 35–60.

Sharipov, F., & Kozak, D. V. (2009). Rarefied gas flow through a thin slit into vacuum simulated by the Monte Carlo method over the whole range of the Knudsen number. *J. Vac. Sci. Technol. A*, *27*(3), 479–484.

Sharipov, F., & Seleznev, V. (1998). Data on internal rarefied gas flows. *J. Phys. Chem. Ref. Data*, *27*(3), 657–706.

Smoluchowski, M. (1910). Zur kinetischen Theorie der Transpiration und Diffusion verdünnter Gase. *Ann. Phys.*, *338*(16), 1559–1570.

Thompson, S. L., & Owens, W. R. (1975). A survey of flow at low pressures. *Vacuum*, *25*(4), 151–156.

van Essen, D., & Heerens, W. C. (1976). On the transmission probability for molecular gas flow through a tube. *J. Vac. Sci. Technol.*, *13*(6), 1183–1187.

Varoutis, S., Naris, S., Hauer, V., Day, C., & Valougeorgis, D. (2009). Computational and experimental study of gas flows through long channels of various cross sections in the whole range of the Knudsen number. *J. Vac. Sci. Technol. A*, *27*(1), 89–100.

Varoutis, S., Valougeorgis, D., Sazhin, O., & Sharipov, F. (2008). Rarefied gas flow through short tubes into vacuum. *J. Vac. Sci. Technol. A*, *26*(2), 228–238.

Varoutis, S., Valougeorgis, D., & Sharipov, F. (2009). Simulation of gas flow through tubes of finite length over the whole range of rarefaction for various pressure drop ratios. *J. Vac. Sci. Technol. A*, *27*(6), 1377–1391.

Weissler, G., & Carlson, R. (1979). *Vacuum physics and technology*. London: Academic Press Inc. (p. 16).

Weston, G. (1985). *Ultrahigh vacuum practice*. Butterworth and Co. (Chapter 1).

CHAPTER NINE

Optics of ion emission from the nano-aperture ion source

Leon van Kouwen
Department of Imaging Physics, Delft University of Technology, Delft, Netherlands
e-mail address: lvankouwen@gmail.com

Contents

Advances in Imaging and Electron Physics, Volume 212
ISSN 1076-5670
https://doi.org/10.1016/bs.aiep.2019.09.003

1. Introduction

A good ion source is crucial for high resolution ion optical systems, but such a system also needs a well designed optical column. Optimization of an optical column does not require knowing all the details of the charged particle source. More useful is the description of a particle source in terms of its characteristic quantities, notably brightness, energy spread, and virtual source size (such as an optimization by Hagen and Kruit (2009)).

The source properties of familiar sources such as the Schottky electron source (Bronsgeest, Barth, Swanson, & Kruit, 2008) and the liquid metal ion source (Swanson, 1983) have been studied quite well. For the nano-aperture ion source, the theory isn't as well established.

Studies on the source properties so far promised good performance; a brightness above 1×10^6 A/m^2 sr V and sub 1 eV energy spread (Jun, 2014). Some interesting trends regarding the voltage across the membrane and the aperture diameter were found. The modeling does not yet provide an understanding of how the gas and the electric fields in and around the chip relate to the optical properties. It furthermore doesn't include an analysis of the brightness dependence on the selected current, aberrations in the emission region, and an adequate analysis on the virtual source size and position.

In this chapter we study how the initial emission volume, the properties of the gas, and the electric fields establish the source properties. This enables appropriate optical column design and gives more insight for optimizing the source itself. Furthermore, experiments are easier when having a good idea what to expect under what conditions.

2. Source properties

The complicated emission region is usually replaced by a virtual source; an imaginary particle emitting object in a field-free region, instantaneously emitting at an energy equal to the acceleration voltage. This concept is visualized in Fig. 1. The resulting situation is optically equivalent to the real situation.

2.1 Virtual source size

The transverse size of the virtual source is important in probe forming systems because it determines the required magnification of the total system. The optical image of this object on the final target is called the geomet-

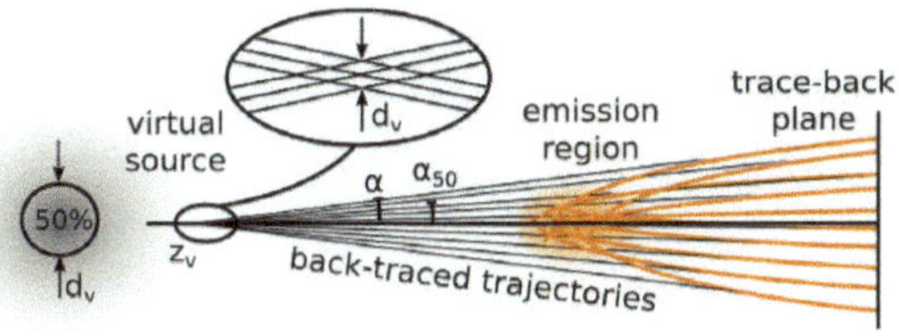

Figure 1 Particles are emitted from a source and the trajectories are determined by their initial conditions and the electrostatic fields. After acceleration up to a high voltage the trajectory tangent lines can be traced back to obtain the virtual properties.

rical contribution to the probe size. This contribution should always be as large as possible, without dominating the total probe size in order to get the most current in the probe. Said differently, any optimized system which is not limited by diffraction will have a significant geometrical contribution to its probe. For ions, the diffraction contribution is practically always negligible.

Most sources require a demagnification of a few times, but very large sources require very high demagnification, which may require several lenses and high alignment accuracy. On the other hand, very small sources such as carbon nano-tubes require magnification in order to get sufficient current in the probe (Kruit, Bezuijen, & Barth, 2006). This inherently means that the gun lens aberrations are magnified, which can cause challenging lens design. It is clear that the virtual source size is a crucial property of a particle source.

It is important to use a clear and appropriate measure to quantify the virtual source size. We use the full width 50% (fw50), which is the disk in the virtual object plane containing 50% of the total current. Barth and Kruit (1996) recommend the fw50 as it is fairly insensitive to the type of distribution and allows for easy addition of aberrations. Furthermore, it behaves well with respect to noise and outliers. Alternatives such as the RMS, full width half maximum or beam waist all score lower on robustness and applicability. Judging from the literature, not everyone is convinced of the fw50 superiority. In order to relate a fw50 to a perhaps more familiar measure Table 2 can be used.

The virtual source position is defined as the virtual plane at which we obtain the smallest fw50 disk and is essentially the object position.

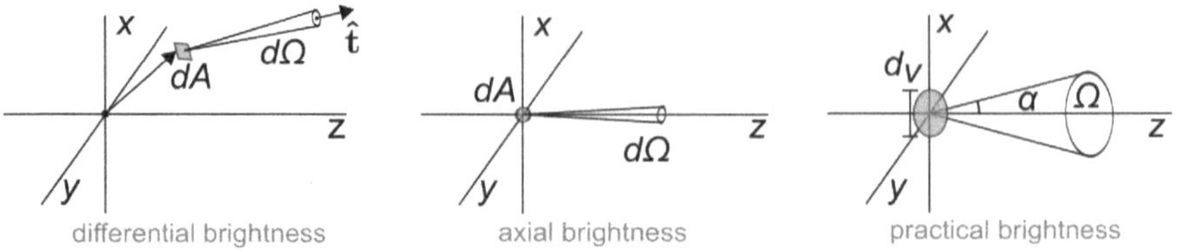

Figure 2 Three different flavors of the brightness. The differential brightness is the most general, where the position in space and the direction of the emitting cone are free variables. The axial brightness is the differential brightness at the origin and in the direction of the optical axis (z-axis). The practical brightness avoids using infinitesimal quantities and is the most useful brightness for column optimization and performance estimates. It is also typically the quantity acquired from measurements.

2.2 Brightness

One of the most important characteristics of a charged particle source is the brightness. In essence it is a measure of the beam intensity. The brightness comes in a few flavors and Fig. 2 indicates the three most important ones.

The definition of the differential brightness is

$$B_{\mathrm{r}}(\mathbf{r}, \hat{\mathbf{t}}) = \frac{\mathrm{d}^2 I}{\mathrm{d}\Omega \mathrm{d}A\Phi}. \tag{1}$$

It describes the amount of current that is confined in an infinitesimal solid angle $\mathrm{d}\Omega$, originating from the infinitesimal area $\mathrm{d}A$, aimed at direction $\hat{\mathbf{t}}$. The subscript "r" denotes reduced by the beam accelerating voltage Φ at position $\mathbf{r}$. No lenses, apertures nor acceleration or deceleration can change the brightness (Hawkes & Kasper, 1989) as a consequence of Liouville's theorem on the conservation of phase space density (Kardar, 2007). Related is the Helmholtz-Lagrange invariant (Szilagyi, 1988), which can be used to show the invariance of brightness as well. We should note that the invariance relies on the assumption of no internal interaction. This is often the situation, but when statistical coulomb interactions become significant this assumption fails and the brightness can deteriorate.

The differential brightness is very difficult to determine for every $\mathbf{r}$ and $\hat{\mathbf{t}}$ and one may question the purpose of such detailed information. For optical systems, the axial brightness is often more sensible to consider; brightness in the direction of the optical axis, originating from the optical axis. The definition is

$$B_{\mathrm{axial}} = B_{\mathrm{r}}(z, \hat{\mathbf{z}}). \tag{2}$$

Bronsgeest et al. (2008) explained that an even more useful version of the brightness is the practical brightness:

$$B_{\text{pract}} = \frac{I}{\frac{1}{4}\pi d_{\text{v}}^2 \pi \alpha^2 \Phi} \tag{3}$$

with d_v the fw50 of the virtual source size and α the half beam angle. It can conveniently be used to calculate the relation between current and spot size in an optical system. The axial and practical brightness are often closely related, for instance by a factor ln(2) for Schottky and thermionic sources. We can make the axial brightness also a bit more practical by using $J = 4I/\pi d_v^2$ instead of $\mathrm{d}I/\mathrm{d}A$.

The full beam current is rarely used in charged particle optical columns and more commonly a beam fraction is selected by an aperture. The most relevant brightness in such a configuration is the average practical brightness of the selected beam fraction. This quantity is found from

$$B(\Omega) = \frac{J}{\Omega\Phi}\int_0^{\Omega} \rho_\Omega \mathrm{d}\Omega \tag{4}$$

If we use $J = 4I/\pi d_v^2$ we make it a practical brightness as well. Quite often we work with the beam divergence angle α, which is related to the solid angle by $\Omega = 2\pi(1 - \cos\alpha) \approx \pi\alpha^2$.

2.3 Energy spread

Any charged particle source has an energy distribution associated with it. The range of energies in the distribution is much smaller than the final beam energy; typically 1 eV compared to 30 keV. Nevertheless, this small a-chromaticity can cause significant resolution loss due to the inability of electromagnetic lenses to focus the different energies at the same spot. The details of the distribution are usually summarized by a single number: the energy spread. For the same arguments as for the virtual source size, the use of a full width 50 is advocated.

Standard optical systems, so excluding time-dependent fields and energy filters, leave the energy distribution unaffected. This means that the energy spread is another constant source parameter. The energy spread is not truly invariant as the statistical coulomb interaction can make the spread larger, which is known as the Boersch effect. Most often this occurs right after emission and it is common to consider the Boersch effect as an inherent property of the source.

2.4 Source aberrations

Source aberration is blurring of the virtual source as a consequence of non-ideal imaging. Usually it is preferred to keep the virtual source properties and properties of the optical column, such as aberrations, separated. However, the optical effects of the initial acceleration stage and the virtual source formation are intimately related and cannot be considered individually.

We study the aberrations of the virtual source as attained during acceleration by looking at effects that cause a spread in the virtual source position. The size of the aberration disk can be estimated from $d_v \sim \Delta z_v \cdot \alpha$, with Δz_v a measure for the spread of the virtual source position. Analyzing a particular aberration boils down to finding an appropriate expression for Δz_v. The fw50 value of the aberration disk is given by

$$d_{50} = c_{50}\, \Delta z_v\, \alpha_{50}, \tag{5}$$

in which c_{50} is a dimensionless constant in the order of one. α_{50} is the angle that defines a cone that contains 50% of the current as selected by an aperture angle α. α_{50} can be found from

$$\int_0^{\Omega_{50}} \rho_\Omega\, \mathrm{d}\Omega = 0.5 \int_0^{\Omega} \rho_\Omega\, \mathrm{d}\Omega, \tag{6}$$

in combination with $\Omega_{50} = \pi\alpha_{50}^2$. The simplest approximation to α_{50} is assuming isotropic intensity, which is equivalent to a constant solid-angular probability density. It yields $\alpha_{50} = \frac{1}{\sqrt{2}}\alpha$.

2.5 Virtual source properties from ray tracing

Ray tracing is used extensively throughout this dissertation. We acquire the results as a set of position and velocities at a plane sufficiently far away from the chip such that all emission effects are included and the beam energy is much larger than any energy associated with the emission. This typically means acceleration up to 10 kV in 1 to 10 mm.

For determining the virtual source properties from our ray trace results we use a post-processing algorithm in MATLAB®. We trace N particles, but may only select a fraction of the beam for the analysis, effectively placing an aperture. We choose $N_{\text{sub}} \leq N$ and sort according to $\sin\alpha_i = v_{r,i}/v_{z,i}$. We relabel the set such that $\ldots < \sin\alpha_{i-1} < \sin\alpha_i < \sin\alpha_{i+1} < \ldots$ and select all particles for which $i \leq N_{\text{sub}}$.

We define trace-back position functions as:

$$x_i(z) = \frac{v_{x,i}}{v_{z,i}}(z - z_0) + x_{0,i}, \qquad y_i(z) = \frac{v_{y,i}}{v_{z,i}}(z - z_0) + y_{0,i}. \tag{7}$$

These functions are the black lines in Fig. 1 of the ith particle. The subscript 0 relates to the trace-back plane, also indicated in the figure. We define $r_i^2 = x_i^2 + y_i^2$ and for any requested value of z we relabel the set of r's by j such that ... $< r_{j-1} < r_j < r_{j+1} <$ .... We have $j_{50} = N_{\text{sub}}/2$, such that j_{50} is the index of the particle that defines a radius including 50% of the intensity. We use an optimizer to look for the value of z that yields the smallest r_{j50}. The obtained value for z is the virtual source position z_{virt} and the virtual source full width 50 diameter is

$$d_{\text{v}} = 2r_{j50}(z_{\text{virt}}). \tag{8}$$

The obtained virtual source size can be used as input for the computation of the practical brightness as defined by Eq. (4). Note that we do not actually need to solve the integral as the angular distribution is represented in particle counting. The integral is simply the number of particles within the subset divided by the total number of particles: N_{sub}/N. The appropriate Ω is found from the beam angle of the particle N_{sub}.

This method may be somewhat complex, but is as close as one can get to what will be measured in a real optical system. A perhaps more common alternative is first computing the (RMS) emittance and henceforth relating it to a brightness. This can be problematic though when incorporating the angular dependence of the brightness and is less resilient towards outliers. We note that the method used by van der Geer et al. (2007) using a 4D hyperelipsoid that just encompasses the sub-selected beam fraction before computing the emittance should give comparable results as our method.

The electric fields for complex geometries are computed with the FEM-based Electron Optical Design (SPOC, 2012). Due to the very large scale difference of 100 nm typical dimension to 1 mm, we stitch the FEM-computed field with an analytical expression for a uniform field at a position where the computed field is considered sufficiently uniform. This is typically after 1 um. Ray tracing is performed with the General Particle Tracer (PulsarPhysics, 2011) software, using 10 to 100k particles and high accuracy settings.

3. Emission in a uniform field

In essence the NAIS configuration consists of two membranes for confining the gas and two apertures to allow for electron entrance and ion emission. A bias voltage is applied across the membranes to extract the ions and an accelerating field right behind the exit membranes brings the beam to high energy, say 10 keV. The simplest manifestation of this is having infinitely thin membranes and equal electric fields between and outside of the membranes. The convenience of this configuration is that the electric field is uniform from the start of the emission region up to the end of the accelerating field. This allows for a fairly simple description of the optics. Such a simplification is not unrealistic; membranes of only tens of nanometers are not uncommon with modern day micro-machining techniques.

Since the ions are generated from within a gas, scattering of ions on neutral particles is inevitable. One aspect of the NAIS is operating it such that ion-neutral scattering does not deteriorate the performance. Previous work assumed that we can have about 1% of the electrons engaging in ionization in a gas with the gas still sufficiently dilute for neglecting ion neutral scattering (Jun, Kutchoukov, & Kruit, 2011).

In this chapter we focus on the rarefied limit as well, with the notion that higher particle density requires a more comprehensive approach, but may give a better performance. Ion-neutral scattering is the subject of the next chapter. We will use a slightly conservative ionization efficiency of $\eta < 0.5\%$. An electron column optimized for the NAIS may deliver 100 nA in a 50 nm (fw50) spot (Tondare, 2006; Jun, 2014) so we expect about 500 pA of generated ion current in the rarefied limit.

3.1 Initial particle distributions

For setting up an emission model, we will need to know the distributions for the initial conditions of the generated ions. We use these distributions to predict how the ions move after generation and use this to predict the optical properties of the source.

Upon ionization, there may be some momentum transfer from the impinging electron to the ion. Intuitively we can expect this to be negligible due to the large mass difference of the atom and the electron, as for example argued by Nanbu (Nanbu, 2000). We can try to back up this claim quantitatively. It is a textbook exercise to show that the maximum energy transfer ratio in a binary collision of an electron with mass m_e and an atom with mass m is $\frac{m_e}{m}\left(1+\frac{m_e}{m}\right)^{-2}$. For a 1 keV electron hitting a helium atom this gives 0.5 eV of energy transferred to the atom, which is much more than kT. This is of course a huge simplification and should be interpreted as an upper bound, but it does show that simply stating the mass-difference is insufficient argumentation for neglecting the ion recoil momentum.

Authors that have studied the kinematics of the outgoing electrons (electron-impact coincidence experiments, referred to as "(e, 2e)"), argue that the recoil ion momentum can be significant in ionizing collisions. See for example references (Durr et al., 2006; Ehrhardt, Fischer, & Jung, 1982; Catoire et al., 2006; Kheifets, Naja, Casagrande, & Lahmam-Bennani 2008, 2009) for mostly helium results, or (Lahmam-Bennani, Casagrande, & Naja, 2009; Naja et al., 2007) for some results on molecules. In the terminology of (e, 2e) experiments, a so-called recoil peak is identified in the differential cross section of the scattered electron. It is associated with a large momentum transfer to the ion. A second peak called the binary peak corresponds to a clean knock-out of the electron. Without going into the details, the explanation of the recoil peak is that after ionization, the electrons, mostly the slowest one, interact with the potential of the newly generated ion (Vriens, 1969).

It is difficult to use the work on (e, 2e) experiments for deciding upon the importance of ion recoil momentum in modeling the NAIS system. Typically only a specific configuration rather than the full cross section is computed or measured, for example the angular distribution of the resulting electrons for only a particular value of the ejected electron energy. For the NAIS we will inevitably have to deal with all possibilities and it is a matter of statistics whether or not the effect is important.

It may be possible to use the results for some representative particular configurations to estimate the average ion momentum, but fortunately something better is available. Jagutzki (Jagutzki et al., 1996) measured the ion recoil momentum in all three spatial dimensions of helium at 500 eV and 2000 eV initial electron energy. The most important result from their work for our purpose is that most collisions are associated with an ion momentum transfer in the order of 1 atomic unit of momentum. This corresponds to an energy of 1.8 meV for helium. Comparing this to a thermal energy of 25 meV we conclude that the ion recoil can at most cause a small deviation from the results based on purely thermal properties. For heavier gases, the effect will be even less important.

We will neglect the small momentum transfer of the impinging electron and assume a Maxwell-Boltzmann distribution of the gas. This also ignores the variations due to the confined geometry and the effect of temperature change caused by the expansion of the gas into a vacuum.

Some additional explanation is in place here as expanding gases are often associated with a big temperature decrease. An ideal gas will show no temperature change when it is freely expanding because it is not doing any work. When the van der Waals forces are important, the increase in inter-atomic distances requires work and the gas cools. A second mechanism is: when inter-atomic distances increase, fewer collisions occur, which means that on average the atoms spend less time high up in the repulsive part of the inter-atomic potential. This causes a temperature increase upon expansion.

The two competing effects are combined in the Joule-Thompson coefficient. For helium at room temperature and at 1 atm the coefficient is $-0.0616\ \mathrm{K\,atm^{-1}}$ (McCarty, 1973). So the temperature deviation caused by the expansion (increase in the case of helium) will typically be around 0.1 K. For argon and xenon the coefficients are $0.438\ \mathrm{K\,atm^{-1}}$ (Michels, Levelt, & Wolkers, 1958) and $1.87\ \mathrm{K\,atm^{-1}}$ (Michels, Wassenaar, Wolkers, & Dawson, 1956) respectively, so we acquire a typical temperature decrease in the order of 1 K; not very exciting either. We can safely neglect the Joule-Thompson effect for further purposes.

Under the aforementioned assumptions the probability density of a particular momentum vector is proportional to the Boltzmann factor: $\rho(\mathbf{p}) \sim e^{-p^2/2mkT}$. If we define $p_r^2 = p_x^2 + p_y^2$ with $\mathrm{d}p_x\mathrm{d}p_y = p_r\mathrm{d}p_\phi\mathrm{d}p_r$, and I is the current generated in some control volume, we obtain the radial momentum distribution

$$\rho_r(p_r) = \frac{1}{I}\frac{\mathrm{d}I}{\mathrm{d}p_r} = \frac{p_r}{mkT}e^{-p_r^2/2mkT}. \tag{9}$$

We can relate this to the more particle optical relevant angular probability density using $|\rho_r(p_r)\mathrm{d}p_r| = |\rho_\Omega(\Omega)\mathrm{d}\Omega|$, with $2me\Phi\Omega \approx \pi p_r^2$ and introducing the abbreviation $\gamma = e\Phi/\pi kT$, with $e\Phi$ being the final beam energy. Note that $2me\Phi \gg p_r^2$ was used. $d\Omega$ is the z-axis centered, annular shaped, differential solid angle. We obtain

$$\rho_\Omega(\Omega) = \frac{e\Phi}{\pi kT}e^{-e\Phi\Omega/\pi kT} = \gamma e^{-\gamma\Omega}. \tag{10}$$

We are often interested in the fraction of current in the selected beam cone. If we introduce η and integrate Eq. (10) up to Ω we obtain

$$\eta = \frac{I(\Omega)}{I_{\mathrm{tot}}} = \int_0^\Omega \rho_\Omega \mathrm{d}\Omega = 1 - e^{-\gamma\Omega}. \tag{11}$$

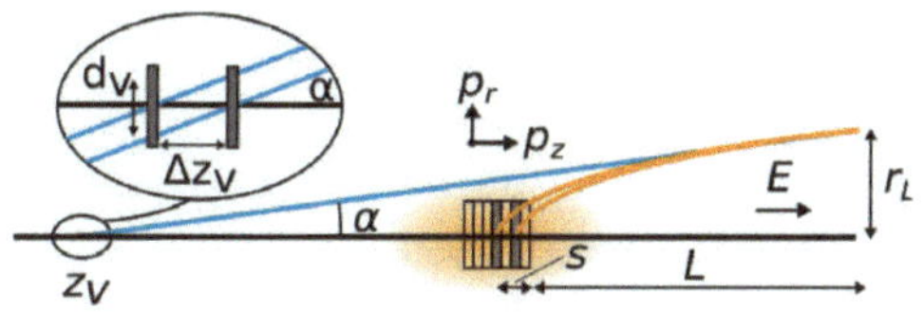

Figure 3 The concept of dividing the volumetric emission into slices and using test rays to obtain the virtual source properties and interpreting aberrations as a de-focus in the virtual plane. Volumetric aberration occurs because not all slices are imaged on to the same virtual source position. Normal energy aberration is a single emission slice not being mapped to the same virtual source position as a result of the initial energy variation. The rays start with radial momentum p_r and axial momentum p_z. L is length over which the particles are accelerated up to high voltage Φ.

With the angular distribution known we can find the 50% beam angle as a function of the accepted aperture angle as described by Eq. (6). Or, alternatively in terms of the selected current by using Eq. (11). By using Eq. (10) and Eq. (6) we get

$$\alpha_{50} = \sqrt{\frac{1}{\pi\gamma} \ln \frac{2}{e^{-\pi\gamma\alpha^2} + 1}} = \sqrt{\frac{1}{\pi\gamma} \ln \frac{2}{2 - \eta}}. \tag{12}$$

Furthermore, we would like to know the initial energy distribution. We will give the radial energy distribution ρ_{ϵ_r} and the normal energy distribution ρ_{ϵ_z}. Using conservation of differential probability: $|\rho_{\epsilon_r}(\epsilon_r) d\epsilon_r| = |\rho_r(p_r) dp_r|$ and by definition $2m\,\epsilon_r = p_r^2$, we find

$$\rho_{\epsilon_r}(\epsilon_r) = \frac{1}{kT} e^{-\epsilon_r/kT}. \tag{13}$$

Likewise we find the normal energy distribution,

$$\rho_{\epsilon_z}(\epsilon_z) = \frac{1}{\sqrt{\pi kT}} \frac{1}{\sqrt{\epsilon_z}} e^{-\epsilon_z/kT}. \tag{14}$$

3.2 Test-ray analysis

The emission region is a volume and in order to cope with this we divide the volume into thin slices, see Fig. 3. We study how one particular slice is virtually mapped and we study how the different slices are combined in the virtual plane. To do so, we use test ray analysis. The test ray starts on the axis at a distance s in front of $z = 0$ and travels across the acceleration length L.

Note that s is in the order of the gas region length (or membrane spacing), that is a few hundred nm, while L is typically some mm's; so $s \ll L$.

The test ray starts with initial momentum $\mathbf{p}_0 = p_r\hat{\mathbf{r}} + p_z\hat{\mathbf{z}}$ and we derive the virtual source position z_v from $z_v = L - r_L/\tan\alpha \approx L - r_L/\alpha$, with r_L the axial distance of the ray after an acceleration of length L. In Appendix A, a brief derivation of r_L is shown and the result is

$$r_L = 2(L+s)\frac{p_r}{p_{L,s}}(1-Z), \qquad 1-Z = \left(\sqrt{1+\left(\frac{p_z}{p_{L,s}}\right)^2} - \frac{p_z}{p_{L,s}}\right) \tag{15}$$

with $p_{L,s}^2 = 2meE(L+s)$ by definition. Note that the abbreviating term $Z \ll 1$ since the accelerating energy is in nearly any practical case much larger than the initial energy ($p_{L,s} \gg p_z$). The uniform field has no radial force so the final beam angle is simply $\alpha \approx \frac{p_r}{\sqrt{p_{L,s}^2+p_z^2}} \approx \frac{p_r}{p_{L,s}}$. We find for the axial intersection of the virtual ray

$$z_v = -L - 2s + 2(L+s)\cdot Z. \tag{16}$$

Interestingly, the virtual axis intersection only depends on the initial momentum in the axial direction and not on the radially directed momentum, which in turn solely governs the final beam angle.

3.3 First order properties

As a first step we look at a single emission slice. We set this plane in the origin so $s = 0$ and since the acceleration voltage is almost always much larger than the initial energy we can use $Z \approx 0$. So, to first approximation we observe from Eq. (16) that all the particles emitted from the origin appear to come from a single virtual point located at $z_v = -L$. Due to symmetry, non-axial rays have the same property only shifted in x and y. This means that at $-L$ a real image of the emission plane is made with magnification $M = 1$.

Almost inevitably, an accelerating electrode is apertured to allow the beam to pass through. This leads to an additional optical effect after uniform acceleration, which is well-know (Liebl, 2008). It will reposition the virtual source to $z_v = -L/3$ and the magnification becomes $M = 2/3$. For the rest of the presented analysis, this lens effect is deliberately excluded for simplicity, as it can always be added later. Of course, when an optical system is designed for a NAIS-based system, this lens effect should be incorporated.

What is described here shows similarity to a thermionic electron gun in the diode configuration (Liebl, 2008), however, very often the emission region is orders of magnitude larger. A real image is no longer the smallest possible back-traced image, but determined by the first lens. In that case it is easy to show that the virtual source is located at −3L and an estimate for the virtual source size is $d_v \approx 4L\sqrt{\frac{kT}{e\Phi}}$. This typically yields a virtual source size of at least several microns and is not applicable to the NAIS sub-micron sized emission region.

3.4 Chromatic aberration

Again, we consider a single slice in the gas emission region, but now we consider the distribution of initial momenta by looking at the small, but non-zero value of Z. From Eq. (16) we note that z_v is only dependent on p_z, so only a variation in the normal direction needs to be considered. The average normal momentum is simply zero, hence a ray with $p_z = 0$ can serve to define the Gaussian image plane. For the characteristic ray representing the distribution in p_z we use the momentum variation Δp_z. Following the idea explained in section 2.4 of describing the aberration by a variation in the virtual source position and using Eq. (16) we obtain

$$\begin{aligned}\Delta z_v &= 2L \cdot (Z(\Delta p) - Z(0)) \\ &= 2L\left(\frac{\Delta p_z}{p_L} - \sqrt{1 + \left(\frac{\Delta p_z}{p_L}\right)^2} + 1\right) \approx 2L\left(\frac{\Delta p_z}{p_L} - \frac{1}{2}\left(\frac{\Delta p_z}{p_L}\right)^2 + \ldots\right)\end{aligned} \tag{17}$$

This aberration can be interpreted as a chromatic effect and it is more customary to express it in terms of an energy spread. In this case we can simply use $\Delta p_z^2 = 2me\Delta\epsilon_z$ because the variation is defined with respect to zero energy. For this particular energy distribution (Eq. (14)) the most sensible definition of $\Delta\epsilon_z$ is the 50% point, such that 50% of intensity lies between zero and $\Delta\epsilon_z$. Integration of Eq. (14) up to 50% intensity gives $e\Delta\epsilon_z = 0.2275kT$.

We estimate the aberration disk in the virtual plane from Eq. (5), and by keeping only the first order term in Eq. (17) to obtain

$$d_{\text{n}} = 2c_{50}L\sqrt{\frac{\Delta\epsilon_z}{\Phi}}\alpha_{50}. \tag{18}$$

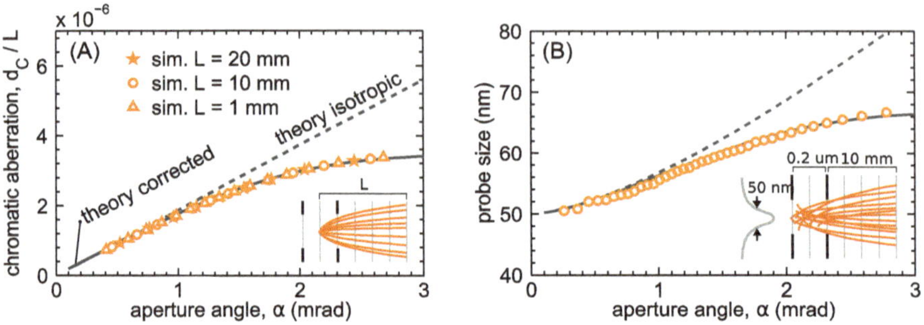

Figure 4 Ray tracing of 20k particles through a uniform field up to a final energy of 10 keV. Initial momenta are Maxwell-Boltzmann distributed. The dashed corrected theoretical curves uses Eq. (12) instead of an isotropic angular distribution. (A) All particles originate from the origin so the final fw50's represent a point spread function. The chromatic aberration is indeed inversely proportional to the acceleration field hence the reduction by the extractor length for constant final beam energy aligns the different simulation results on one curve. (B) Ray tracing particles originating from a realistic emission volume shows the combination of the geometrical contribution and the aberration disk.

For α_{50} we use the isotropic approximation $\alpha_{50} = \frac{1}{\sqrt{2}}\alpha$ to obtain a simple expression. From ray tracing we find $c_{50} = 1.75$ (see Fig. 4) and we get

$$d_n = 2.47\frac{\sqrt{\Delta\epsilon_z}}{E}\sqrt{\Phi}\alpha \tag{19}$$

This is a very practical form for the normal energy aberration disk. It is easily evaluated and we can see that the normal energy aberration can be minimized by using a high electric field. Note that keeping the field constant while changing the acceleration length has no influence as $\sqrt{\Phi}\,\alpha$ will remain constant.

Fig. 4 shows the result of a ray tracing simulation intended to get acquainted with the typical magnitude of this aberration and to compare to the theory. The particles start from a 293 K Maxwell-Boltzmann distribution, starting from the origin and being accelerated up to 10 keV of energy for a few different acceleration lengths (Fig. 4). A master curve is formed by dividing the obtained virtual source sizes by the acceleration length, showing indeed its proportionality. Also Eq. (19) is depicted.

An acceleration voltage of 1 kV/mm over 10 mm seems reasonable for a system that does not rely on field ionization. From Fig. 4A we can obtain that under these conditions the aberration disk nearly reaches 40 nm. This is considerable compared to a typically expected geometrical size of 50 nm. Nevertheless, when using an often employed engineering limit of

10 kV/mm of maximum field strength the aberration can be minimized to only a few nanometer.

The curves in Fig. 4 are not entirely linear. This can be explained by the decreasing angular density with increasing angle. We can account for this by using Eq. (12) instead of $\alpha_{50} = \frac{1}{\sqrt{2}}\alpha$ and henceforth the well-matching solid curve in Fig. 4A is obtained.

A second ray tracing simulation uses ions starting between z = −100 and 100 nm around the origin and a Gaussian radial intensity with a 50 nm fw50 to account for the expected electron beam profile. The acceleration length was 10 mm and the final energy 10 keV. The resulting virtual source size is shown in Fig. 4B.

In the figure we see that for angles approaching zero we find a 50 nm fw50, as expected since the magnification between emission and virtual plane is one. At higher aperture angles the normal energy aberration start to contribute. We find that a decent addition rule between the normal energy aberration and the geometrical spot size is

$$d_{50} = \left(d_g^{1.55} + d_n^{1.55}\right)^{1/1.55}. \tag{20}$$

3.5 Relation to low energy electron microscopy

The chromatic aberration discussed here shows resemblance to the aberrations encountered in low energy electron microscopy (LEEM), so it is interesting to compare the results. Aberrations in LEEM systems were first studied by Bauer (Bauer, 1984), but later also by others (Tromp, Wan, & Schramm, 2012; Rempfer & Griffith, 1989; Scheinfein, Qian, & Spence, 1993; Shao & Crewe, 1989). Both the NAIS as well as the LEEM system suffers from aberrations of a uniform field, yet, some essential differences can be identified.

For the NAIS, the initial energy and energy spread is in the order of kT and the normal and tangential energies are independent. Low energy electron microscopy deals with energies of typically a few to a few hundred eV with sub-eV energy spread and the tangential and normal energy distributions are not independent. Another, rather obvious, difference is for the NAIS the possibility of particles having an initial velocity vector in the negative z-direction, while in LEEM only forward emitted particles are possible.

Despite the differences we can try to relate the formalism of LEEM aberration to the formalism shown here. First we assume only forward emitted particles. Backward emitted particles have a considerably lower probability of escaping the gas confining membrane structure and so ignoring them all together is somewhat bold but not unreasonable. We describe the problem in terms of energy and realize that if no particles go backwards, the Gaussian image plane is defined by the average forward energy rather than zero energy. We use Eq. (16) to obtain

$$\begin{aligned}\Delta z_v &= 2L \cdot (Z_2(\epsilon_z + \Delta\epsilon_z) - Z_2(\epsilon_z)) \\ &= 2L\left(\sqrt{\frac{\epsilon_z + \Delta\epsilon_z}{\Phi}} - \sqrt{1 + \frac{\epsilon_z + \Delta\epsilon_z}{\Phi}} - \sqrt{\frac{\epsilon_z}{\Phi}} + \sqrt{1 + \frac{\epsilon_z}{\Phi}}\right)\end{aligned} \tag{21}$$

If we keep only the smallest terms we can fill in Eq. (5) to obtain the aberration disk as

$$d_n = 2L\alpha_{50}\left(\sqrt{\frac{\epsilon_z + \Delta\epsilon_z}{e\Phi}} - \sqrt{\frac{\epsilon_z}{e\Phi}}\right)\alpha_{50}. \tag{22}$$

This result, although slightly differently derived and formulated, was reported by Rempfer (Rempfer & Griffith, 1989). We can rewrite it in a more familiar form by expanding Eq. (21) over $\frac{\Delta\epsilon_z}{\epsilon_z}$:

$$d_n \sim L\sqrt{\frac{\epsilon_z}{\Phi}}\left(\frac{\Delta\epsilon_z}{\epsilon_z} \pm \frac{1}{4}\left(\frac{\Delta\epsilon_z}{\epsilon_z}\right)^2 + \ldots\right)\alpha. \tag{23}$$

If we only keep the lowest order term we get

$$d_n \sim C_c\frac{\Delta\epsilon_z}{\Phi}\alpha\,, \qquad C_c = L\sqrt{\frac{\Phi}{\epsilon_z}} \tag{24}$$

where C_c is the chromatic aberration coefficient. We have now obtained the more familiar expression that is linear with the relative energy spread and uses an aberration coefficient. This expression is very similar to what is found in low energy electron microscopy (Tromp et al., 2012; Rempfer & Griffith, 1989; Scheinfein et al., 1993; Shao & Crewe, 1989). We should note that the expansion over $\frac{\Delta\epsilon_z}{\epsilon_z}$ is not always appropriate because the condition $\Delta\epsilon_z \ll \epsilon_z$ is not always met. In fact, for a MB-distribution $\Delta\epsilon_z > \epsilon_z$. Eq. (19) does not rely on this condition and is therefore the preferred expression.

3.6 Volumetric emission aberration

The effect of volumetric emission compared to emission from a plane can be interpreted as an aberration because first order optical systems can only form sharp images of planes. We model this aberration by looking at the variation in the virtual source position of different slices.

We use two rays starting on the axis but separated by Δs, representing two characteristic slices. We ignore any chromatic effect because the two effects can be approached independently, just like conventional spherical and chromatic aberration for example. We can use Eq. (16) to obtain the difference in virtual position between the two rays and get $\Delta z_v = 2\Delta s$. As a first approximation we assume isotropic angular density and use Eq. (5) to get $d_V = \sqrt{2} c_{50} s\,\alpha$. With $c_{50} = 0.81$ found from ray tracing we obtain

$$d_V = 0.57 \Delta s\,\alpha. \tag{25}$$

To get a feeling for the magnitude we suppose that the aperture angle is such that it selects a decent fraction of the beam. Then, $\alpha \approx \sqrt{\frac{kT}{e\Phi}} = 1.6$ mrad. The typical membrane spacing is a few hundred of nm's, say 500 nm, so for a 10 keV acceleration voltage we find $d_v < 1$ nm. This is much smaller than the typical virtual source size of around 50 nm. We conclude from this estimate that emitting from a volume in a uniform field does not cause a significant aberration contribution.

3.7 Brightness

For the situation of particles emitted from a Maxwell-Boltzmann distributed gas we can use Eq. (4) and Eq. (10) to find

$$B(\Omega) = \frac{J}{\Omega \Phi}\left(1 - e^{-\gamma\Omega}\right) \approx \frac{eJ}{\pi k T}\left(1 - \tfrac{1}{2}\gamma\Omega + \ldots\right) \tag{26}$$

The axial brightness is obtained when selecting a vanishingly small cone of the beam ($\Omega \to 0$) and by doing so we obtain the familiar Langmuir (Langmuir, 1937) equation $B = \frac{eJ}{\pi kT}$.

An angular brightness dependence can be deceiving as simply changing the beam energy already changes the solid angle corresponding to an equal beam current. We can rewrite Eq. (26) in terms of the beam fraction η using Eq. (11) and obtain

$$B(\eta) = \frac{\eta}{\ln \frac{1}{1-\eta}} \cdot \frac{eJ}{\pi k T}. \tag{27}$$

This expression is the same everywhere in the optical system as long as the conditions for phase space conservation are met.

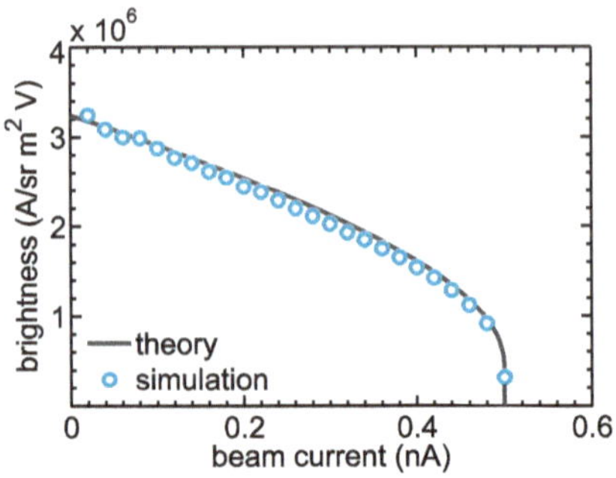

Figure 5 Ray tracing of particles starting uniformly from within a 200 nm distance in *z* and a Gaussian initial radial distribution with a 50 nm fw50. The gas temperature was 293 K and the initial velocities are Maxwell-Boltzmann distributed. The electric field strength was 10 kV/mm.

Fig. 5 shows a ray tracing simulation for comparison to Eq. (27). A high electric field of 10 kV/mm was applied to ensure negligible aberrations. We see that when nearly the full beam current is required, one will obtain a quite low brightness. For example 95% of the beam current relates to 32% of the axial brightness. A better figure of merit is the brightness of half the beam current, this gives 72% of the axial brightness.

4. Emission from a realistic geometry

The uniform field case is a good start for basic understanding, and may even be approachable. However, in practice this configuration is not so easily achieved. Restrictions to the applicable field across the membranes vary between chip designs, so it is important to study the influence of different field strengths in and outside the chip. It may even be desirable to have a different field in and outside. Furthermore, the finite membrane thickness and the blockage of ions by the membrane can be important. This section is concerned with the implications of a realistic geometry compared to the uniform case.

4.1 Chip exit aperture lens

We will try to extend our simple uniform model by the inclusion of a lens effect at the exit aperture. Again we use a test ray and determine its virtual z-position from which we derive the optical properties. In Fig. 6 the test ray and the symbols used are shown. The test ray starts at distance s from

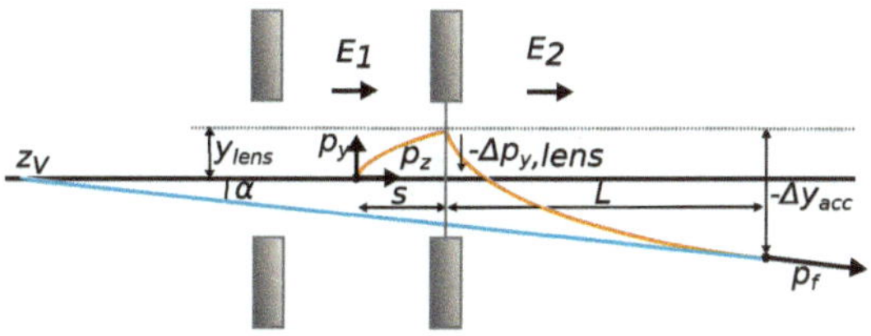

Figure 6 Symbol definitions for the presented theory.

the chip exit, has initial momentum p_x, p_z, and is accelerated over a length L up to a high voltage Φ. Variations in s and the initial momenta are related to the gaseous volume and its properties.

For the sake of obtaining a model that is simple enough to be useful we assume the lens to be thin, consider only the first order focusing effect, and neglect the lens effect from the back of the chip. It is known from earlier work that the lens effect from the back can influence the escape probability (Jun, 2014), so a complete model should include this as well. However, only a small fraction of the generated particles feel it so it can never be a dominant effect.

So, in the presented model the ions travel through a uniform field up to the membrane, feel a thin lens focusing effect, and again encounter a uniform electric field up to some high voltage. We use the fact that the thermal initial energy is much smaller than the energy acquired from the field in the chip, which is in turn much smaller than the final beam acceleration voltage ($kT \ll E_1 s \ll E_2 L$). We do take into account, and this is crucial, the difference in energy by which the particles pass through the lens. This is what makes the chip lens so extraordinary; the energy difference between particles is as big as the average energy.

Surely this model is not expected to describe all the involved optical effects in detail, however, it should encompass the most dominant effects. Further analytical modeling might become so complicated that its use can be questioned.

The focal distance of a lens created by a field difference separated by a thin electrode is (Liebl, 2008)

$$f = \frac{2V_a}{E_2 - E_1} = \frac{2E_1 s}{E_2 - E_1} = \frac{2ks}{1-k}, \qquad k = \frac{E_1}{E_2} \tag{28}$$

in which we have introduced the field ratio k. The momentum kick associated with this lens effect is

$$\Delta p_{\mathrm{y,lens}} = \frac{e(E_1 - E_2)}{2v_{\mathrm{z,lens}}} y_{\mathrm{lens}}. \tag{29}$$

The radial distance at the lens location is governed by the travel through the uniform field in the chip. For this purpose convenient equations are derived in Appendix A and the result is

$$y_{\mathrm{lens}} = 2s\frac{p_y}{p_s}(1 - Z_1), \qquad 1 - Z_1 = \left(\sqrt{1 + \left(\frac{p_z}{p_s}\right)^2} - \frac{p_z}{p_s}\right), \tag{30}$$

with $p_s = \sqrt{2meE_1 s}$. Since $v_{\mathrm{z,lens}} = \frac{p_s}{m}$ for $p_z \ll p_s$, which we have assumed, we now have the ingredients to derive the final beam angle. A little algebra gives

$$\sin\alpha = \frac{p_y + \Delta p_{\mathrm{y,lens}}}{p_{\mathrm{f}}} = \frac{p_y}{p_{\mathrm{f}}}\left(\frac{3}{2} - \frac{E_2}{2E_1} - Z_1\left(\frac{1}{2} - \frac{E_2}{2E_1}\right)\right). \tag{31}$$

We can derive the magnification from the Helmholtz invariant: $p_{\mathrm{i}}^2\, d_{\mathrm{i}}^2 \sin^2\alpha_{\mathrm{i}} = p_{\mathrm{f}}^2\, d_{\mathrm{f}}^2 \sin^2\alpha$ with i for initial and f for final. The initial angle is $\sin\alpha_{\mathrm{i}} = \frac{p_y}{p_{\mathrm{i}}}$, so

$$M = \frac{d_{\mathrm{f}}}{d_{\mathrm{i}}} = \frac{\sin\alpha_{\mathrm{i}}\, p_{\mathrm{i}}}{\sin\alpha\, p_{\mathrm{f}}} = \frac{p_y}{p_{\mathrm{i}}}\frac{1}{\sin\alpha}\frac{p_{\mathrm{i}}}{p_{\mathrm{f}}} = \left(\frac{3}{2} - \frac{E_2}{2E_1} - Z_1\left(\frac{1}{2} - \frac{E_2}{2E_1}\right)\right)^{-1}. \tag{32}$$

Now we can also see that $\sin\alpha = \frac{p_y}{p_{\mathrm{f}}} M^{-1}$.

We find the virtual source position from

$$z_v = L - \frac{y_{\mathrm{lens}} + \Delta y_{\mathrm{acc}}}{\tan\alpha} \tag{33}$$

We use the result from Appendix A again to obtain

$$\Delta y_{\mathrm{acc}} = 2L\frac{p_y + \Delta p_{\mathrm{y,lens}}}{p_L}(1 - Z_2), \qquad 1 - Z_2 = \left(\sqrt{1 + \left(\frac{p_s}{p_L}\right)^2} - \frac{p_s}{p_L}\right), \tag{34}$$

with $p_L = \sqrt{2meE_2 L}$. Since $p_y + \Delta p_{\mathrm{y,lens}} = p_{\mathrm{f}}\sin\alpha = p_y M^{-1}$ we obtain

$$\Delta y_{\mathrm{acc}} = 2L\frac{p_y}{p_L}M^{-1}(1 - Z_2) \tag{35}$$

Table 1 Different regimes for the chip exit lens are identified, depending on the relative field strengths inside and outside of the chip. The focal distance for equal field strengths is undefined as no lens effect is present.

Fields		f	M
$E_1 > E_2$	$k > 1$	$-$	$+$
$E_1 = E_2$	$k = 1$		1
$E_2 < 3E_1 < 3E_2$	$\frac{1}{3} < k < 1$	$+$	$+$
$3E_1 = E_2$	$k = \frac{1}{3}$	$+$	∞
$3E_1 < E_2$	$k < \frac{1}{3}$	$+$	$-$

We use $\tan\alpha \approx \alpha \approx \sin\alpha = M^{-1}\frac{p_y}{p_f}$, $p_f = \sqrt{2me(E_1 s + E_2 L)}$. After some tedious algebra of filling in Eq. (33) we get

$$z_v = L - 2L(1 - Z_2)\sqrt{1 + \frac{sk}{L}} - 2s(1 - Z_1)M\sqrt{1 + \frac{L}{sk}}. \tag{36}$$

4.2 First order properties

If we set $Z_1 = 0$ in Eq. (32) we obtain

$$M = \left(\frac{3}{2} - \frac{1}{2k}\right)^{-1} = \frac{2k}{3k - 1}. \tag{37}$$

Since a small variation in the magnification does not give rise to aberrations the small non-zero value of Z_1 is irrelevant. Note that the magnification is independent of s, meaning that the magnification is uniform with respect to the starting slice position. Table 1 summarizes the different operation regimes possible due to the chip exit aperture lens.

Firstly, if we look at the uniform field case ($k = 1$) we obtain $M = 1$. This result corresponds to what was found in the previous section; a good sanity check. Furthermore we obtain $z_v = -L - 2s$ from Eq. (36), also the result we obtained in the uniform field analysis.

The first order optical properties of the presented theory are shown in Fig. 7. We can see from Eq. (28) that $\frac{f}{s} = \frac{2k}{1-k}$. The quantity f/s is a quite general measure for the exit aperture lens strength. Note that this does not include the influence of the uniform field before and after the exit aperture. In contrast, the magnification and the virtual source position are quantities of the full imaging system. If $k = 1$ we see in all the graphs that the uniform properties are obtained. We note that the magnification shows asymptotic

Table 2 Ratio table for different distribution function metrics. The "fw50" is the diameter containing 50% of the intensity. The "fwhm" is the full width at half maximum. "25-75%" refers to a knife-edge rise distance; for 1D distributions this is also called the inter-quartile range. The conversions assume Gaussian beam profiles, although the conversion between fw50 and 25-75% in 2D is quite close for a wide range of distributions (Bronsgeest et al., 2008).

Dim	fw50	25–75%	fwhm	σ
1D	1	1	0.573	1.349
2D	1	1.746	1	2.355

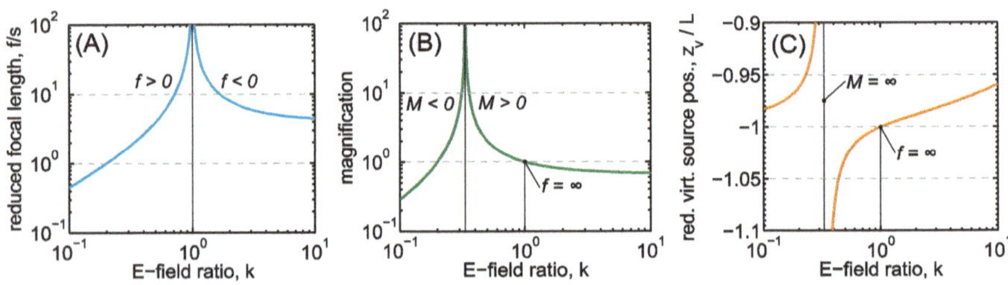

Figure 7 Theoretical predictions for the first order optical properties resulting from modeling the exit aperture as a thin lens and the fields before and after as uniform. (A) The focal distance f divided by the object to lens distance s is a general measure of the exit membrane lens effect potency. The lens is strongest for the lowest k-value. (B) The magnification can vary by an order of magnitude and shows asymptotic behavior at $k = 1/3$. (C) The reduced virtual source position is in general close to 1, but is weakly depending on s and L. In this example $L = 10$ mm and $s = 500$ nm.

behavior for $3E_1 = E_2$. Besides this remarkable asymptotic behavior, the magnification varies roughly between 0.3 and 3 for $k = 0.1$ to 10. Clearly the exit aperture lens effect is important.

In Fig. 8 we compare the prediction of the virtual source position to a realistic ray trace example of two different chip geometries. We compare to a situation with realistic starting conditions due to the neutral particle distribution, as well as the blockage of the membranes. We assume the radial emission profile to follow the ebeam intensity profile, which we assume to be Gaussian. A more simplified comparison is run too, in order to avoid any aberration as to study the first order properties. This is done by letting all particles start at the middle of chip, with a radial fw50 of 5 nm and a gas temperature of 5 K. We see in the figure that indeed the virtual source position is close to $-L$. The variations around $-L$ are reasonably

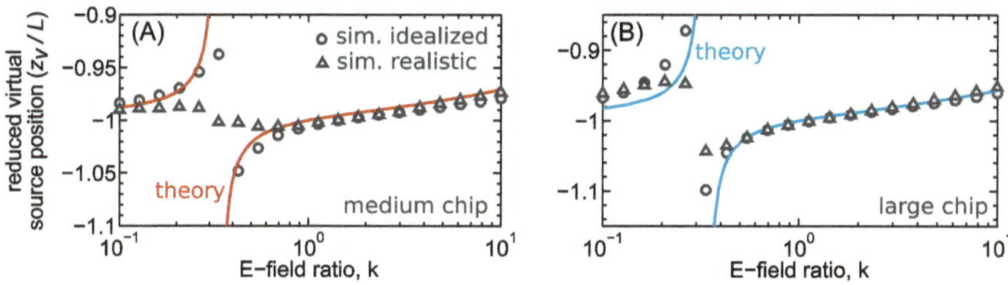

Figure 8 Theory compared to ray tracing using realistic ion starting conditions. The theory shows fair agreement, although the asymptote is not as explicit in ray tracing as it is theoretically. The idealized situation is a gas at $T = 5$ with particles starting from the plane in the center of the chip. (A) spacing × thickness × aperture = 300 × 100 × 150 nm, (B) spacing × thickness × aperture = 1000 × 100 × 500 nm.

well described by the theory, especially when $k > 0.4$. One aspect of the difference is the penetration of the E_2 field into the chip. The fact that the discrepancy is larger for the larger apertured chip supports this claim.

We do not see the predicted asymptotic behavior, but merely a tamed tendency of such. We note that the asymptotic behavior depends on a very precise balance of quantities, but in a realistic situation this balance is easily destroyed by small non-uniformities of the fields or the non-zero lens thickness. Furthermore, in the presented analysis it is assumed that the smallest back-traced image is a real image of the emission distribution. This is reasonable and was already shown to be valid in the uniform case. However, if the magnification becomes very large, the exit aperture becomes source size defining.

The virtual source size of the same ray trace simulation is shown in Fig. 10, but is better discussed in conjunction with the aberrations.

4.3 Aberrations

We can study the aberrations by looking at the variations in the virtual source positions according to Eq. (36). The normal energy aberration is found by looking at the variation in the initial normal momentum while keeping s fixed:

$$\Delta z_v = z_v(\Delta p_z) - z_v(0) = 2sM\sqrt{1 + \frac{L}{sk}}(Z_1(\Delta p_z) - Z_1(0)) \tag{38}$$

$$\approx 2M\frac{\sqrt{E_1 s + \Phi}}{E_1}\frac{\Delta p_z}{\sqrt{2m}} \approx 2M\frac{\sqrt{\Phi}}{E_1}\frac{\Delta p_z}{\sqrt{2m}} \tag{39}$$

Note that we have used $\Phi = E_2L$ and $E_1s \ll \Phi$, and expanded Z_1 up to first order. We would like to transform the aberration into terms of energy spread. Because Δp is defined with respect to zero we have $\Delta p_z^2 = 2me\Delta\epsilon$. The equation for the normal energy aberration disk with $d_n = c_{50}\Delta z_v \alpha$ becomes

$$d_n = 2c_{50}M\frac{\sqrt{\Delta\epsilon_z}}{E_1}\sqrt{\Phi}\alpha. \tag{40}$$

We see great resemblance to the uniform case (Eq. (19)). Since this more general case converges to the uniform case we can use $c_{50} = 1.75$, as found earlier. The relevant electric field is shown to be the in-chip field, so the normal energy aberration is effectively all generated right in the beginning of the trajectory.

Secondly we look at volumetric emission effects caused by the variations in s. It is easier to first look at the virtual source position for a particular s and taking the lowest order in the chromatic effects, so $Z_1 = 0$. From Eq. (36) we get

$$z_v = -L + 2L\sqrt{1+\frac{ks}{L}}\left(\sqrt{1+\frac{ks}{L}} - \sqrt{\frac{ks}{L}}\right) - 2s\sqrt{1+\frac{L}{sk}}M \tag{41}$$

We have used $p_s/p_L = \sqrt{\frac{E_1s}{E_2L}}$ to write Z_2 in terms of k, s and L. We can simplify by expanding over s/L, which is small, resulting after convenient rearrangement into

$$z_v = -L + 3\sqrt{L}\left(k^{\frac{1}{2}} - k^{-\frac{1}{2}}\right)M\sqrt{s} - 2ks. \tag{42}$$

We have kept the higher order term $-2ks$ just to show that if $k = 1$ the virtual source position is $z_v = -L - 2s$, exactly what was found in the uniform case. If k is only slightly non-zero, we find that this term is negligible compared to the others and is omitted for further use. The variation in the virtual source position is found from the variation in s and gives $\Delta z_v = 3\sqrt{L}\left(k^{\frac{1}{2}} - k^{-\frac{1}{2}}\right)M(\sqrt{s_+} - \sqrt{s_-})$ where $s\pm = s_0 \pm 0.5\Delta s$, such that s_0 is the center of the chip and the region in z determined by Δs should contain half the ion generation intensity. The full width 50 disk of the volumetric aberration becomes

$$d_V = 3c_{50}M\left(k^{\frac{1}{2}} - k^{-\frac{1}{2}}\right)E_2^{-\frac{1}{2}}\left(\sqrt{s_+} - \sqrt{s_-}\right)\sqrt{\Phi}\alpha, \tag{43}$$

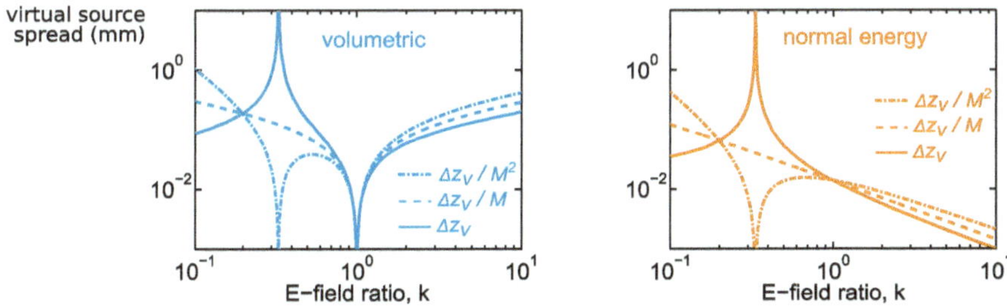

Figure 9 The spread in z_V is measure for the aberrations, however, because also a magnification is present in the system it is unclear what is the right figure of merit for the influence of these aberrations. Three different candidates are shown here. In this example: $E_2 = 1$ kV/mm, $T = 293$ K, $s_+ = 750$ nm, and $s_- = 250$ nm.

with $c_{50} = 0.57$. We observe that we can minimize the aberration by making the fields as equal as possible, and by making E_2 as high as possible.

In Fig. 9 we have visualized the normal energy and the volumetric aberrations by showing the spread in the virtual source position. It is important to compare the contribution of the aberrations to the geometrical probe size. If we simply look at Δz_v we neglect that a magnification can make the aberration more or less important. A better figure of merit would be $\Delta z_v/M$. What is then still neglected is that the beam angle for a particular current scales with $1/M$. So, purely theoretically, the influence on the virtual source size when selecting a particular current is measured by $\Delta z_v/M^2$. From this point of view, we see that two sweet spots exist for the volumetric aberration, namely $k = 1/3$ and $k = 1$. Since at $k = 1/3$ the normal energy aberration is also least influential, this analysis suggests that operating at $k = 1/3$ is most favorable for minimizing aberrations.

We compare the theory to ray tracing in Fig. 10. First we look at the ray trace curve with the circular marks. This curve is found from a cold gas (5 K) for which all ions are generated at the center of the chip. This configuration is in practice not achievable, but a good setup for comparing to the theory because aberrations are negligible. We compare to the theoretical magnification and see good correspondence down to $k = 0.4$. For the lowest k-values we see that the theory describes the trend, but lacks accuracy. This can be attributed to the field penetration making the particle's energy at the lens higher than is incorporated in the theory.

When looking at the more realistic simulation (triangular markers), we see that the combination of source magnification and aberrations can reasonably well explain the behavior of the virtual source size. At the lowest

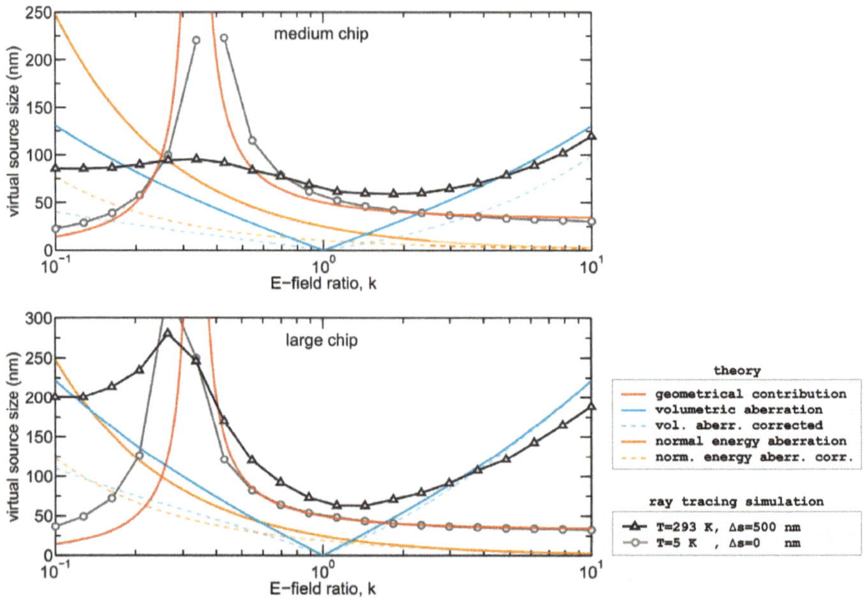

Figure 10 Virtual source size ray tracing results a medium sized chip (spacing × thickness × aperture = 300 × 100 × 150 nm) and a larger chip (spacing × thickness × aperture = 1000 × 100 × 500 nm) compared to theoretical predictions for different contributions. For further comparison an ideal situation at low temperature, with all particles originating from the center of the chip is included. The theory seems to predict the general trends well, although for $k < 0.3$ deviation occurs. The corrected theoretical curves take into account the limitation of the current as blocked by the exit aperture.

k-values, the aberrations seem to be an overestimate though, particularly for the medium sized chip. That is partially attributed to field penetration, but also because not all ions get extracted. The solid theoretical lines are based on the assumption that all current is emitted, while the dashed theoretical lines use the simulated emitted current to convert to an effective 50% beam angle according to Eq. (6). This correction seems to explain why the aberrations in the medium sized chip are not as large as predicted by the simplest theory for these low E_1 values.

An important difference between the two simulated chip geometries is the severity of the asymptotic behavior, which for the larger chip is much more pronounced. The theory does not take into account that the exit aperture can become the source size determining factor. The smaller chip runs quicker into this transition. This makes us doubt the idea that $\Delta z_v/M^2$ is the best figure of merit for the aberration for it assumes a fully developed

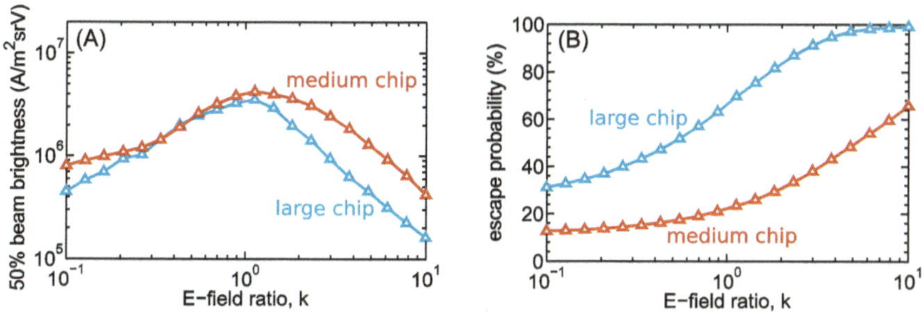

Figure 11 Ray trace results for a medium sized chip (spacing × thickness × aperture = 300 × 100 × 150 nm) and a larger chip (spacing × thickness × aperture = 1000 × 100 × 500 nm) with fixed $E_2 = 1 \times 10^6$ V/m. The total generated current was fixed at 500 pA. (A) The brightness is computed by using a back-traced fw50 probe size made out of the inner 50% current of the total beam cone. Clearly $k = 1$ is the optimum, where the volumetric aberration is lowest. (B) The escape probability is defined as the total emitted current to generated current ratio. In this case, k is varied by varying the in-chip field E_1, hence the large difference in escape probabilities.

asymptote. More practical is to use $\Delta z_v/M$ as a measure for the aberration influence.

In Eqs. (40) and (43) we see that both aberration disks are proportional to the magnification. The aforementioned analysis tells us then, that the influence of the aberrations is not very much affected by the first order properties of the exit aperture lens. The normal energy is minimized by the highest E_1. The volumetric aberration is lowest when E_2 is highest and when the fields are most equal. In general $k \approx 1$ is a good choice. The normal energy aberration is not necessarily lowest then, but we find it to be negligible for a wide range of configurations as long as E_1 is reasonably high, say > 3 kV/mm.

4.4 Brightness

We study the brightness by ray tracing using realistic starting conditions in two different chip geometries. The result is shown in Fig. 11A, the brightness shown here is formed by 50% of the beam cone. Immediately we notice the maximum at $k = 1$. The reduction of the brightness with respect to its maximum is caused by the blurring, mostly due to volumetric emission aberration.

Interestingly, there is no clear notion of the asymptotic behavior at $k = 1/3$. This is maybe not surprising because the asymptotic behavior only causes first order effects, which are inherently not affecting the brightness.

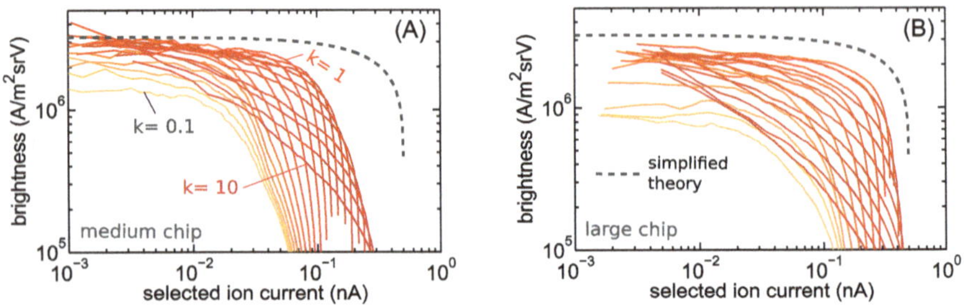

Figure 12 The computed average brightness of a selected beam current (Eq. (4)) from a realistic ray tracing setup. The theoretical curve is given by Eq. (27). Note how the theoretical curve serves as an upper-bound and the largest deviation occurs for the highest current. For very low currents, all but the lowest k-value configurations converge towards the theoretical prediction. (A) spacing × thickness × aperture = 300 × 100 × 150 nm, (B) spacing × thickness × aperture = 1000 × 100 × 500 nm.

It also shows that $\Delta z_v/M$ is indeed a good figure of merit for the aberration, for would it be Δz_v or $\Delta z_v/M^2$, we would see a peak or valley respectively in the brightness at $k = 1/3$.

Fig. 11B shows the escape probability due to the varying field E_1, by which we vary k. Obviously the larger apertured chip has a higher escape probability and for any chip a higher in-chip field minimizes membrane blockage. More subtle is the notion that a lower escape probability means that the 50% beam cone will have a smaller beam defining angle and so the aberrations are smaller then. This explains why the brightness reduction for $k < 1$ is less severe than for $k > 1$.

Note that the reduced escape probability is not directly related to brightness loss. The aperture mostly blocks particles with a high initial transverse energy and this group of particles ends up in the outer beam cone. Loosing them does lower the current, but also the 50% beam angle and the brightness is not very much affected.

Perhaps a clearer, or more fair way of looking at the brightness is by comparing the same currents. Fig. 12 does so by plotting the average of a particular selected current. The theoretical prediction assuming a uniform field and no chip blockage is plotted as well (Eq. (27)). The theoretical curve is never matched, except for the lowest currents where the ray trace results tend to converge towards the theory for all but the lowest k-values. For such small currents, the beam angle is vanishingly small such that the axial aberrations vanish as well. The fact that for the lowest k-values the curves do not converge towards the theoretical curve is rather unexpected.

Since it cannot be an axial aberration, and current directed towards the aperture can not be blocked the effect is not obvious.

If $k = 0.1$ the total energy acquired in the field is for the medium and large chip respectively 30 meV and 100 meV in the case of $E_2 = 1$ kV/mm. The typical thermal energy is 25 meV and we conclude that a reasonable amount of particles can exit the chip from the back. This can explain up to a factor of two because only half of the particles start in the backward direction. Looking quantitatively at the discrepancy, this can't be the full explanation. Throughout the theoretical analysis it was assumed that $eE_1s \gg kT$, however this assumption breaks down at these low k values. This gives rise to an additional variation in the lens strength and hence a variation in magnification. Furthermore, we may have off-axis effects, notably a distortion of the initial ion distribution profile when mapped on to the virtual source plane. The last two effects can both cause brightness decrease. A quantitative analysis of these plausible influences is omitted as it is tricky as well as of limited use. We remember that when eE_1s approaches kT, bad performance and bad understanding occurs.

4.5 Energy spread

The ion-energy distributions of the NAIS have been studied in quite some detail by Jun (Jun, 2014). Here we briefly discuss the most important results. Since we apply a bias voltage across the membranes, the ions are generated at different potential planes. The result is an energy spread in the beam in the order of the applied bias voltage. Fig. 13 shows simulated energy distributions and it can be seen that the width of the distributions indeed resemble the applied bias voltage. Note that the thermal energy is only 25 meV, typically much smaller than the applied bias voltage. The field penetration of the acceleration field outside of the chip adds extra spread. This is mostly noticeable when the applied bias voltage is low, and when the aperture is large, see Figs. 13A and E. The ions generated in front of the chip are responsible for a low energy tail.

One of the key insights resulting from Jun's work (Jun, 2014) is that the geometry and the field strengths determine the shape of the distributions. The field strength inside the chip determines which ions make it out of the double membrane structure. For example, a very weak field does not direct many ions from the back-side towards the exit aperture. This can be observed in the energy spectrum because not many high energy particles

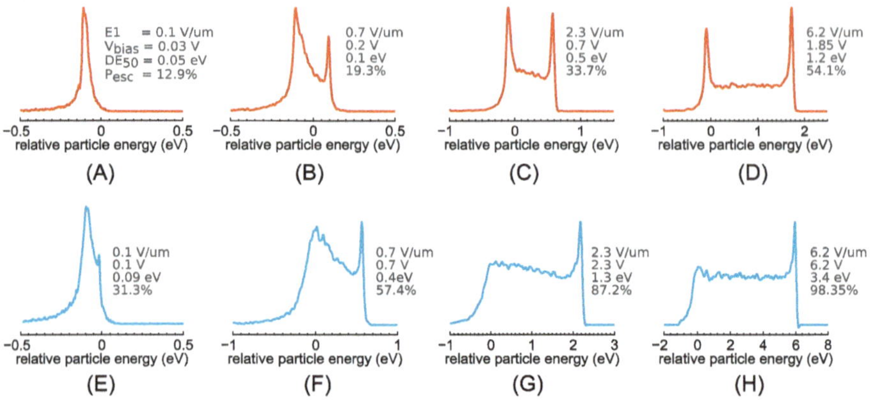

Figure 13 The energy distributions of scatter-free, coulomb interaction free, ray tracing simulations. The indicated energy is relative to an acceleration energy of 10 keV. (A–D) Medium sized chip, spacing × thickness × aperture = 300 × 100 × 150 nm. (E–H) A larger chip, spacing × thickness × aperture = 1000 × 100 × 500 nm.

become part of the beam. Notice how in Fig. 13A to D or E to H, the high-energy part of the spectrum becomes more dominant when increasing the field strength.

Jun also explains that a bi-modal distribution is obtained when the field strength is rather high. This is as a consequence of a lensing effect at the back, causing the back-side ions to have a higher probability of being emitted than some ions halfway. While we do not dispute this insight, it is not the complete story. Even if this lens effects was completely absent, the bimodality is obtained. This is because the field strength near and within the membranes is weaker, which results in a larger region at similar energies, hence an increased region in the energy spectrum. Furthermore, when increasing the field strength, ions make it out easier, simply because their trajectories are more parallel to the direction of the field, which is in the direction of the optical axis towards the exit aperture. The resulting distribution is a combination of these effects and the balance of the effects depends on the geometry and the field strengths. Notice how the low energy peak can be wider and less high compared to the high energy peak. This can be explained by the field penetration, increasing the local field strength and thereby causing the local spread to be wider than at the back.

5. Brightness from phase space evolution

For most sources, the theoretical brightness is derived by starting from the phase space density function,

$$f(\mathbf{r}, \mathbf{p}) = \text{phase-space density.} \tag{44}$$

It represents the number of particles in an infinitesimal 6D volume in real and momentum space combined. Alternatively, it is the particle density at position $\mathbf{r}$ for particles having momentum $\mathbf{p}$. We have already derived an equation for the brightness starting from the energy distributions in the gas. It is good to check if this different approach leads to the same conclusion, but it also gives some new insight. It may also be a better basis for further study, such as the inclusion of gas scattering.

We will assume a simplified configuration and later comment on the relaxation of the assumptions. The configuration is a top hat gas distribution with length L and uniform neutral particle density n with a constant electric field E applied in the positive z direction. Very much like the thin-membrane equal-field configuration discussed earlier in this chapter.

5.1 From phase space to brightness

We begin the derivation by giving the differential current from a phase space element in direction $\hat{\mathbf{t}}$. To acquire the current from an ion particle density n_I we use $dI = en_\mathrm{I}vdA$, with v the ion velocity. In this case, we also need to take the momentum space volume element into account. We obtain

$$\mathrm{d}^2 I = ef(\mathbf{r}, p\hat{\mathbf{t}})v\mathrm{d}V_p\mathrm{d}A. \tag{45}$$

The term $\mathrm{d}V_p$ represents the volume in momentum space and in terms of the solid angle it is $\mathrm{d}V_p = p^2\mathrm{d}\Omega\mathrm{d}p$. We aim to relate this differential current to the brightness as defined by Eq. (1). We can establish this by taking the derivative of the brightness with respect to the momentum in the $\hat{\mathbf{t}}$ direction to obtain

$$\frac{\mathrm{d}B(\mathbf{r}, \hat{\mathbf{t}})}{\mathrm{d}p} = \frac{\mathrm{d}^3 I}{\Phi\mathrm{d}\Omega\mathrm{d}A\mathrm{d}p} \tag{46}$$

If we plug Eq. (45) into Eq. (46) we find that the differential brightness can be found from the phase space density as

$$\mathrm{d}B(\mathbf{r}, \hat{\mathbf{t}}) = \frac{e}{m\Phi}f(\mathbf{r}, p\hat{\mathbf{t}})p^3\mathrm{d}p. \tag{47}$$

An integral form of this relation between phase space density and brightness, with integration over energies rather than momentum is found in the books of Hawkes and Kasper (Hawkes & Kasper, 1989). The derivation is different, but the result can be shown to be equivalent to Eq. (47).

5.2 Evolution of phase space

In order to calculate the brightness from the phase space density, we first need to have an expression for the phase space density. The evolution of phase space density in the case of non-interacting particles is described by the kinetic Boltzmann equation,

$$\frac{\partial f}{\partial t} = -\mathbf{v} \cdot \nabla f(\mathbf{r}, \mathbf{p}) - \mathbf{F} \cdot \nabla_p f(\mathbf{r}, \mathbf{p}) + S(\mathbf{r}, \mathbf{p}). \tag{48}$$

The term $S(\mathbf{r}, \mathbf{p})$ is the source term, representing the production of ions at position $\mathbf{r}$ having momentum $\mathbf{p}$. $\mathbf{F}$ is the local external force and $\mathbf{v}$ the particle velocity. ∇_p is the gradient with respect to momentum space.

Eq. (48) is a non-trivial partial differential equation and simplifications are required to solve it. We are looking for steady state solutions so $\frac{\partial f}{\partial t} = 0$. Furthermore, we will try to calculate the axial brightness so only the axial phase space needs to be found. The axial phase space density $f(0, 0, z, 0, 0, p)$ is formed only by particles starting on the axis, in the direction of the axis. So, we can simplify the 6D problem to a 2D problem because we only need to solve (48) on the optical axis.

The force term appearing in the kinetic Boltzmann equation is in this system simply a constant electric field E directed in the positive z-direction. The source term needs to be related to the electron beam that generates the current. We can write $eS(\mathbf{r}, \mathbf{p}) = \frac{\mathrm{d}I}{\mathrm{d}V} \rho(\mathbf{p})$, in which $\rho(\mathbf{p})$ is the momentum probability density function of a generated particle. The differential ion current as generated by the ebeam is $\mathrm{d}I = I_e n \sigma_+ dz = J_e n \sigma \,\mathrm{d}V$, so we obtain $eS(z, p) = J_e n \sigma_+ \rho(p)$.

We assume the particles to obey Boltzmann statistics, so every micro state (a particular momentum vector in this case) has a probability $\sim e^{-\epsilon/kT}$, thus $\rho = a_n e^{-p^2/2mkT}$. The constant a_n is found by integration over 3D momentum space and gives $a_n = 2(2\pi mkT)^{-3/2}$ (the factor 2 here is not obvious, which will be explained hereafter). The source term becomes:

$$eS(z, p) = J_e \sigma_+ n a_n e^{-\frac{p^2}{2mkT}}, \tag{49}$$

and the kinetic Boltzmann equation then reads

$$\frac{p}{meE}\frac{\partial f(z,p)}{\partial z}+\frac{\partial f(z,p)}{\partial p}=Ae^{-\beta p^2}, \tag{50}$$

with the abbreviations $A=\sigma_+ nJ_e a_n/e^2E$ and $\beta=\frac{1}{2}mkT$. Note that independence of the momentum probability density on z and independence of the particle generation rate on p is implicitly assumed. This essentially corresponds to uniform properties of the gas, and no influence of the neutral particle's momentum on the ionization probability.

This PDE can be solved by the method of characteristics (Haberman, 2004). The first step is defining the appropriate boundary conditions. It is easiest to consider only particles traveling in the positive z-direction. We can incorporate particles traveling in the opposite direction by realizing that they will be reflected forward. By energy conservation they act as if initially directed forward. Since ion production is isotropic we can simply use double the forward-directed ion production rate to account for backward propagating particles. This explains the factor 2 in a_n. Note that this only holds if no particles escape from the backside, which is typically negligible indeed.

In the case of only forward propagating particles we need to solve the equation in the positive z and positive p quadrant. The boundary condition at $z=0$ is rather trivial. No ions are produced for $z<0$ so we obtain $f(0,p)=0$. This condition is not sufficient as it still allows negative momentum, i.e. ions propagating in the negative z-direction. This is not allowed because we specified our system in the positive zp-quadrant. Note that an alternative description could allow negative momentum, but this will make the math more difficult. One problem is then, that we need to solve the PDE between z is minus infinity and zero as well, otherwise particles can escape from the solution domain. Then, to name something, the source term becomes a non-continuous function and things become pretty hairy pretty quickly.

What we need besides $f(0,p)=0$ is the condition $f(z,0)=0$. It is perhaps counter intuitive as $\rho(0)\neq 0$ and so one might expect $f(z,0)\neq 0$. However, the phase space density at any particular position z_{par} is a summation of all the infinitesimal contributions of ion production coming from $z<z_{\mathrm{par}}$. For $p=0$, only ions originating from that particular position can contribute, as particles from other positions must have higher energies when arriving at z due to the electric field. This means that only an infinitesimal amount of ions contributes to $f(z,0)$, hence $f(z,0)=0$.

We can use the method of characteristics to solve the PDE (see for example (Haberman, 2004) for an introduction to the method). We start from

$$\frac{\mathrm{d}f}{\mathrm{d}p} = \frac{\partial f}{\partial p} + \frac{\partial f}{\partial z}\frac{\partial z}{\partial p} \tag{51}$$

and conclude that if $\frac{\partial z}{\partial p} = \frac{p}{emE}$, looking at (50), we have

$$\frac{\mathrm{d}f}{\mathrm{d}p} = Ae^{-\beta p^2}. \tag{52}$$

Straightforward integration yields

$$f(z,p) = \sqrt{\frac{\pi}{4\beta}}A\,\mathrm{erf}\left(\sqrt{\beta}p\right) + C, \tag{53}$$

with C an unknown integration constant. Note that the solution is only valid on the characteristic curves defined by $\frac{\partial z}{\partial p} = \frac{p}{meE}$. Explicitly the curves are

$$z = \frac{p^2}{2meE} + z_0 = \frac{p^2 - p_0^2}{2meE}, \tag{54}$$

where $z = z_0$ if $p = 0$ and $p = p_0$ if $z = 0$. In the case of $z_0 > 0$, which is equivalent to $eEz > \frac{p^2}{2m}$, the characteristic curve is space-like and we need to use the boundary $f(0, z_0) = 0$ to obtain

$$f(0, z_0) = \sqrt{\frac{\pi}{4\beta}}A\,\mathrm{erf}\left(\sqrt{\beta}\cdot 0\right) + C = 0, \tag{55}$$

so $C = 0$. If $z_0 < 0$, so $eEz < \frac{p^2}{2m}$, we need to use the momentum-like boundary condition $f(p_0, 0) = 0$ and obtain

$$C + \sqrt{\frac{\pi}{4\beta}}A\,\mathrm{erf}\left(\sqrt{\beta}p_0\right) = 0. \tag{56}$$

We relate p_0 to z and p using Eq. (54) and get

$$C = -\sqrt{\frac{\pi}{4\beta}}A\,\mathrm{erf}\sqrt{\beta}\sqrt{p^2 - 2meEz}). \tag{57}$$

Combining the results for the two different regimes yields the full solution as

$$f(p,z)=\begin{cases}\sqrt{\frac{\pi}{4\beta}}A\,\mathrm{erf}\left(\sqrt{\beta}p\right), & \text{if} \quad eEz<\frac{p^2}{2m}\\ \sqrt{\frac{\pi}{4\beta}}A\left(\mathrm{erf}\left(\sqrt{\beta}p\right)-\mathrm{erf}\left(\sqrt{\beta}\sqrt{p^2-2meEz}\right)\right), & \text{if} \quad eEz>\frac{p^2}{2m}\end{cases}$$

Note that $z\in(0,s)$ and $p\in(0,\infty)$ with s the length of the gaseous region.

5.3 Computing the brightness

We now know f right after the ionization volume so we could calculate the brightness using Eq. (47). However, the brightness of a non-accelerated beam is not as meaningful and we prefer the brightness of an accelerated beam, up to a potential Φ, typically several kV. If we indicate the variables after acceleration by primes, we have from Eq. (47) $\mathrm{d}B=\frac{e}{m\Phi}f'(z',p')p'^3\mathrm{d}p'$. After the gaseous region there are no phase space sources so that by Liouville's theorem $f'(z',p')=f(s,p)$. From energy conservation we get $p^2+2me\Phi=p'^2$, and thereby $p\mathrm{d}p=p'\mathrm{d}p'$ to finally obtain

$$B_{\mathrm{axial}}=B(L,\hat{\mathbf{z}})=\frac{e}{m\Phi}\int_0^{\infty}\left(p^2+2me\Phi\right)f(s,p)p\mathrm{d}p. \tag{58}$$

The integral needs to be split for $p>\sqrt{2meEs}$ and $p<\sqrt{2meEs}$ because the phase space density function is different for these two intervals. The integral is tricky, but it turns out to be analytically solvable and the result is

$$B_{\mathrm{axial}}=\frac{eJ_e\sigma_+ns}{\pi kT}\left(1+\frac{Es}{2\Phi}+\frac{kT}{2e\Phi}\right). \tag{59}$$

If we look at $\Phi\gg Es$, $e\Phi\gg kT$, we obtain the familiar $B=\frac{eJ_e\sigma_+ns}{\pi kT}=\frac{eJ}{\pi kT}$. We conclude that with the phase space method we get the same result as working directly with energy distributions.

We have calculated the on-axis brightness, but since the E-field was assumed uniform, the analysis actually does not rely on being on-axis, so the result is valid for a practical brightness of vanishingly small acceptance angle. If the field is not uniform, it is still a very close estimate for the axial brightness because on the axis only a field in the z-direction is possible. Furthermore, on-axis current will never be blocked by a membrane.

Since it is rare that the axial and the practical brightness are very different (only a very spiky intensity distribution can do that), we expect that the small-angle practical brightness is quite independent of chip design. This is also what is observed when studying the chip lens effects. Only when the field ratio k is very low ($k < 0.3$) we see that the small-angle practical brightness is different from the brightness predicted here.

If the gaseous region was not a top hat but a more realistic smooth distribution, the math would be more complicated, but the final result won't change as long as $e\Phi \gg eEs$. The result here is only valid in the rarefied limit, where gas scattering is insignificant. In principle scattering can be included in the kinetic Boltzmann equation, however the tricky math already encountered in this simplified situation demonstrates that this is not a very promising direction.

6. Conclusion

In the rarefied limit of negligible ion-neutral interaction, it was found that the exit aperture of the chip can cause a significant lens effect, depending on the ratio of the electric fields inside and outside of the chip. The most direct effect is a magnification from the ion emission area to the virtual source area. The virtual source position is always close to one extractor length before the chip, excluding the extractor aperture lens.

Two aberrations were identified, namely chromatic and volumetric emission aberration. The chromatic aberration is caused by the intrinsic thermal energy spread of the ions in the direction of the optical axis. This spread leads to a variation in the virtual source position, hence can be interpreted as an aberration. The volumetric aberration originates from the ions being emitted from different positions along the optical axis, rather than from a plane. This aberration is affected strongly by the chip exit aperture lens. Poor choice of source configuration can cause these aberrations to lead to reduced performance. A good rule of thumb is that the electric fields inside and around the chip need to be approximately equal and above 3 kV/mm in order to suppress these aberrations sufficiently.

We note that the rarefied limit is not necessarily the best choice, as a higher gas density is expected to give better performance. However, the optical effects described here will be reminiscent in more complex situations and should be understood first. Furthermore, the analysis here can be very helpful for exploring alternative optimizations, such as ultra low energy spread, high current, low-end model (larger chip), or different e-beam properties.

Appendix A

Deriving relations of charged particles in a uniform field is a textbook exercise (Liebl, 2008), but it is shown here briefly anyhow to introduce the specific formalism. We start from a particle in the origin a $t = 0$, having initial momentum $\mathbf{p}_0 = p_r\hat{\mathbf{r}} + p_z\hat{\mathbf{z}}$. From newton's second law and a uniform field pointing in the z-direction we obtain

$$\mathbf{r}(t) = \left(\frac{1}{2}\frac{eE}{m}t^2 + \frac{p_z}{m}t\right)\hat{\mathbf{z}} + \left(\frac{p_r}{m}t\right)\hat{\mathbf{r}}. \tag{60}$$

Suppose the particle is accelerated across length u and has final coordinates $\mathbf{r}(t_u) = u\hat{\mathbf{z}} + r_u\hat{\mathbf{r}}$. We eliminate t and obtain

$$r_u^2 + 4u\frac{p_r p_z}{2emEu}r_u - 4u^2\frac{p_r^2}{2emEu} = 0, \tag{61}$$

and solve for r_u to finally get

$$r_u = 2u\frac{p_r}{p_u}\left(\sqrt{1+\left(\frac{p_z}{p_u}\right)^2} - \frac{p_z}{p_u}\right), \quad \text{with: } p_u = \sqrt{2meEu} \tag{62}$$

$$= 2u\frac{p_r}{p_u}(1 - Z_u) \approx 2u\frac{p_r}{p_u}\left(1 - \frac{p_z}{p_u} + \ldots\right). \tag{63}$$

Most often the energy obtained in the accelerating field is much larger than the initial energy, making $Z_u \ll 1$.

List of symbols

B, B_r	reduced brightness	A/m^2 sr V
B_{axial}	axial brightness	A/m^2 sr V
B_{pract}	practical brightness	A/m^2 sr V
$B(\Omega)$	angular dep. brightness	A/m^2 sr V
Ω	solid angle	sr
α	beam angle	rad
d_v	virtual source size	m
Φ	ion acceleration voltage	V
I	ion current	A
J	ion current density	$A\,m^{-2}$
L	acceleration length	m
ρ	probability density function	
ϵ_z, ϵ_r	normal energy, radial energy	eV
$\Delta\epsilon_z$	normal energy spread fw50	eV
z_{virt}	virtual source position	m

Δz_v	virt. source pos. variation	m
E	electric field strength	$\mathrm{V\,m^{-1}}$
E_1	inside chip field	$\mathrm{V\,m^{-1}}$
E_2	outside chip field	$\mathrm{V\,m^{-1}}$
γ	$e\Phi/\pi kT$	
η	beam fraction	
Δp_z	normal momentum spread	kg m/s
a_n	normalization constant	
S	phase space source	
r_L	radial distance at $z = L$	m
n_I	ion particle density	$\mathrm{m^{-3}}$
p_0	boundary momentum	kg m/s
k	Boltzmann constant or E_1/E_2	
s	distance to chip exit	m
Δs	ion emission spread	m
$s+, s-$	$s_0 \pm \frac{1}{2}\Delta s$	m
d_{50}	aberration disk fw50	m
c_{50}	constant for the fw50	
α_{50}	50% beam angle	rad
d_n	normal energy aberration	m
d_V	volumetric emission aberr.	m
d_i, d_f	initial, final object size	kg m/s
p_i, p_f	initial, final momentum	kg m/s
p_s, p_L	$\sqrt{2meE_1 s}, \sqrt{2meE_2 L}$	kg m/s
Z, Z_1, Z_2	abbreviation of expansion	
M	magnification	
f	phase-space density	m
f	focal distance	m
J_e	electron current density	$\mathrm{A\,m^{-2}}$
σ_+	ionization cross section	m
V_p	momentum-space volume	$\mathrm{kg^3\,m^3/s^3}$
β	$\frac{1}{2}mkT$	
A	$\sigma_+ n J_e a_n / e^2 E$	
$p_{L,s}$	$\sqrt{2meE(L+s)}$	kg m/s
z_0	boundary position	m
m	ion mass	kg
m_e	electron mass	kg
$\mathbf{F}$	external force	N
$\mathbf{v}$	ion velocity	m/s

Acknowledgment

I would like to thank Pieter Kruit for his guidance and contributions. This work is part of the Industrial Partnership Program (IPP) of the Stichting voor Fundamenteel Onderzoek der Materie (FOM), which is financially supported by the Nederlandse Organisatie voor Wetenschappelijk Onderzoek (NWO). The IPP program is co-financed by FEI Company. This work is based on the dissertation titled *The Nano-Aperture Ion Source*, by Leon van Kouwen.

References

Barth, J., & Kruit, P. (1996). Addition of different contributions to the charged particle probe size. *Optik*, *101*(3), 101.

Bauer, E. (1984). The resolution of the low energy electron reflection microscope. *Ultramicroscopy*, *17*(1), 51–56.

Bronsgeest, M. S., Barth, J. E., Swanson, L. W., & Kruit, P. (2008). Probe current, probe size, and the practical brightness for probe forming systems. *J. Vac. Sci. Technol. B*, *26*(3), 949.

Catoire, F., Staicu-Casagrande, E. M., Nekkab, M., Cappello, C. D., Bartschat, K., & Lahmam-Bennani, A. (2006). Investigation of the (e, 2e) single ionization of He and Ar at large energy loss close to minimum momentum transfer. *J. Phys. B, At. Mol. Opt. Phys.*, *39*(12), 2827.

Durr, M., Dimopoulou, C., Dorn, A., Najjari, B., Bray, I., Fursa, D. V., ... Ullrich, J. (2006). Single ionization of helium by 102 eV electron impact: Three-dimensional images for electron emission. *J. Phys. B, At. Mol. Opt. Phys.*, *39*(20), 4097–4111.

Ehrhardt, H., Fischer, M., & Jung, K. (1982). Triple differential cross section measurements for the electron impact ionization of helium: Comparison with a second born approximation. *Z. Phys. A*, *304*(2), 119–124.

Haberman, R. (2004). *Applied partial differential equations* (fourth edition). Pearson Prentice Hall (Chapter 12).

Hagen, C., & Kruit, P. (2009). Optimization of focused ion beam performance. *J. Vac. Sci. Technol. B*, *27*(6), 2654.

Hawkes, P., & Kasper, E. (1989). *Principles of electron optics, Vol. 2*. Academic Press (pp. 971–973).

Jagutzki, O., Spielberger, L., Dorner, R., Nuttgens, S., Mergel, V., Schmidt-Bocking, H., ... Buck, U. (1996). Recoil-ion momentum distribution for He(e, 2e) He+ - and He(e, e3) He++ - reactions. *Z. Phys. D*, *36*(1), 5–10.

Jun, D. (2014). *Development of the nano-aperture ion source (NAIS)* (Doctoral dissertation). Delft University of Technology. Retrieved from repository.tudelft.nl.

Jun, D. S., Kutchoukov, V. G., & Kruit, P. (2011). Ion beams in SEM: An experiment towards a high brightness low energy spread electron impact gas ion source. *J. Vac. Sci. Technol. B*, *29*(6), 06F603.

Kardar, M. (2007). *Statistical physics of particles*. Cambridge University Press (p. 59).

Kheifets, A. S., Naja, A., Casagrande, E. M. S., & Lahmam-Bennani, A. (2008). DWBA-G calculations of electron impact ionization of noble gas atoms. *J. Phys. B, At. Mol. Opt. Phys.*, *41*(14), 145201.

Kheifets, A. S., Naja, A., Casagrande, E. M. S., & Lahmam-Bennani, A. (2009). An energetic (e, 2e) reaction away from the Bethe ridge: Recoil versus binary. *J. Phys. B, At. Mol. Opt. Phys.*, *42*(16), 165204.

Kruit, P., Bezuijen, M., & Barth, J. E. (2006). Source brightness and useful beam current of carbon nanotubes and other very small emitters. *J. Appl. Phys.*, *99*(2), 024315.

Lahmam-Bennani, A., Casagrande, E. M. S., & Naja, A. (2009). Experimental investigation of the triple differential cross section for electron impact ionization of N2 and CO2 molecules at intermediate impact energy and large ion recoil momentum. *J. Phys. B, At. Mol. Opt. Phys.*, *42*(23).

Langmuir, D. (1937). Theoretical limitations of cathode-ray tubes. *Proc. Inst. Radio Eng.*, *2*(8), 977–991.
Liebl, H. (2008). *Applied charged particle optics*. Springer (Chapter 1).
McCarty, R. (1973). Thermodynamic properties of helium 4 from 2 to 1500 K at pressures up to 1e8 Pa. *J. Phys. Chem. Ref. Data*, *2*(4), 923.
Michels, A., Levelt, J., & Wolkers, G. (1958). Thermodynamic properties of argon at temperatures between 0 C and 140 C and at densities up to 640 Amagat (pressures up to 1050 atm.). *Physica*, *14*, 769–794.
Michels, A., Wassenaar, T., Wolkers, G., & Dawson, J. (1956). Thermodynamic properties of xenon as a function of density up to 520 Amagat and as a function of pressure up to 2800 atmospheres at temperatures between 0 C and 150 C. *Physica*, *12*(1–5), 17–21.
Naja, A., Staicu-Casagrande, E. M., Lahmam-Bennani, A., Nekkab, M., Mezdari, F., Joulakian, B., ... Madison, D. H. (2007). Triply differential (e, 2e) cross sections for ionization of the nitrogen molecule at large energy transfer. *J. Phys. B, At. Mol. Opt. Phys.*, *40*(18), 3775–3783.
Nanbu, K. (2000). Probability theory of electron-molecule, ion-molecule, molecule-molecule, and coulomb collisions for particle modeling of materials processing plasmas and gases. *IEEE Trans. Plasma Sci.*, *28*(3), 971–990.
PulsarPhysics (2011). *General particle tracer, version 3.10*. Retrieved from www.pulsar.nl/gpt.
Rempfer, G., & Griffith, O. (1989). The resolution of photoelectron microscopes with UV, X-ray, and synchrotron excitation sources. *Ultramicroscopy*, *27*(3), 273–300.
Scheinfein, M., Qian, W., & Spence, J. (1993). Aberrations of emission cathodes: Nanometer diameter field-emission electron sources. *J. Appl. Phys.*, *37*(5), 2057.
Shao, Z., & Crewe, A. (1989). On the resolution of the low-energy electron reflection microscope based on wave electron optics. *Ultramicroscopy*, *31*(2), 199–204.
SPOC (2012). *Electron optical design, version 3.072*. Retrieved from www.lencova.com.
Swanson, L. (1983). Liquid metal ion sources: Mechanism and applications. *Nucl. Instrum. Methods Phys. Res.*, *218*(1–3), 347–353.
Szilagyi, M. (1988). *Electron and ion optics*. New York and London: Plenum Press.
Tondare, V. (2006). *Towards a high brightness, monochromatic electron impact gas ion source* (Doctoral dissertation). Delft University of Technology. Retrieved from repository.tudelft.nl.
Tromp, R., Wan, W., & Schramm, S. (2012). Aberrations of the cathode objective lens up to fifth order. *Ultramicroscopy*, *119*, 33–39.
van der Geer, S. B., Reijnders, M. P., de Loos, M. J., Vredenbregt, E. J. D., Mutsaers, P. H. A., & Luiten, O. J. (2007). Simulated performance of an ultracold ion source. *J. Appl. Phys.*, *102*(11), 094312.
Vriens, L. (1969). Angular correlation of scattered and ejected electrons in ionizing collisions. *Physica*, *45*(3), 400–406.

CHAPTER TEN

A model for ion-neutral scattering in the nano-aperture source

Leon van Kouwen
Department of Imaging Physics, Delft University of Technology, Delft, Netherlands
e-mail address: lvankouwen@gmail.com

Contents

Advances in Imaging and Electron Physics, Volume 212
ISSN 1076-5670
https://doi.org/10.1016/bs.aiep.2019.09.004

1. Introduction

When ions are generated between the NAIS apertures, they are inevitably surrounded by neutral gas particles. On the way out of the ionization volume, an ion may interact with one or more of these particles. In this chapter, a model is presented that is used to investigate ion-neutral interaction in the nano-aperture ion source.

Previously, a hard-sphere molecular diameter was used to estimate a pressure such that the mean free path of the ions equals the membrane spacing (Tondare, 2006; Jun, 2014). Ion-neutral scattering is then assumed to be almost negligible. One problem is that the molecular diameter is a rather ambiguous concept, but moreover this approach is problematic for the following two reasons. Firstly, although a molecular diameter model may have been successful in describing neutral-neutral scattering, it is inappropriate for ion-neutral interaction. Secondly, one may find better performance in a regime where ion-neutral scattering is more abundant. In fact, it is evident that gas scattering is a significant phenomenon any optimized NAIS system. If the particle density is sufficiently low to have negligible gas scattering, an optimum is not reached as the brightness can be increased by simply increasing the particle density.

2. Ion-neutral interaction

2.1 Momentum transfer and charge exchange

Qualitatively we can describe ion-neutral interaction as depicted in Fig. 1. Upon encounter, the ion polarizes the neutral atom, creating a strong interaction potential, when compared to neutral-neutral interaction. Hence, scattering with large momentum transfer is more likely for ions than for neutrals. The interaction potential acts over a long range such that a high probability of weak scattering is present as well. We furthermore recognize hitting the rigid core of the interaction potential, which resembles a hard-sphere collision.

In addition to momentum transfer, the ion and the neutral can exchange an electron fairly easily because of the symmetry in the electron energy level. Rapp and Francis (1962) showed that the probability of exchanging charge is a rapidly oscillating function of the impact parameter up until a particular length, whereafter it attenuates quickly. Most often, the detailed oscillating behavior is neglected and an average probability of 0.5 is used for ions passing a neutral within a particular distance. For energetic

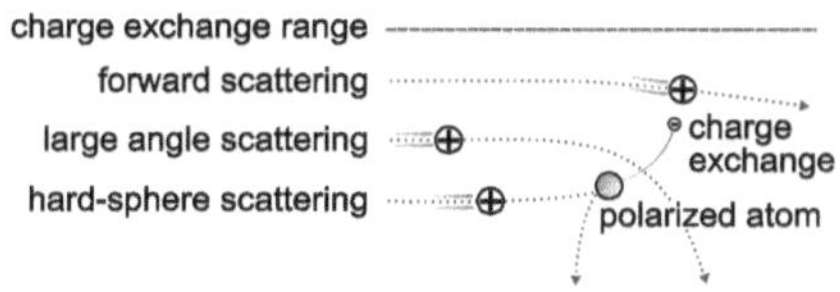

Figure 1 Ion-neutral encounter. The ion trajectory may be hardly affected, follow a curved path closely passing the neutral, or scatter on the rigid core of the atom. In addition one electron may be exchanged.

ions ($\epsilon > 1$ eV), the inter-atomic distance over which charge exchange can occur is much larger than the distances over which significant momentum transfer occurs. In that case, the identity of the ion and the neutral essentially change without any deflection. When the impact parameter is small though, charge exchange still occurs with probability of about 0.5, but an identity switch is no longer a sufficient description of the collision.

2.2 Cross sections for ion-neutral scattering

A hard-sphere model for ion-atom scattering using double the molecular cross section for neutral particles in combination with an identity switch of probability 0.5 has been used (Nanbu, 2000). This gave reasonable results for the drift velocity under certain conditions, but gives considerable errors in for example the transverse diffusion of ions in a neutral gas.

A very accurate approach to ion-neutral interaction would be using an energy dependent differential cross section. Such an approach can take into account momentum transfer as well as charge exchange, when combined with energy and momentum conservation. In the end, it is just an atom and an ion going in and the same going out. Whether the outgoing ion was originally an atom or an ion doesn't matter for the post-collision momentum vector calculation. The disadvantage of a differential cross section approach is the scarcity of data and the computational complexity. It also diminishes intuitive understanding to some extent.

Phelps (1994) argues that the differential cross section in the center-of-mass frame can be separated (artificially), into an isotropic part, a forward scattering peak, and a back-scatter peak. One can recognize the isotropic part as a variable hard sphere collision with or without charge exchange. For isotropic scatter events, charge containment and charge exchange are indistinguishable. It can be shown that the back-scatter peak in the center-of-mass frame is identical to an ion-neutral identity switch in the lab-frame. This also called idealized charge exchange, meaning that charge is ex-

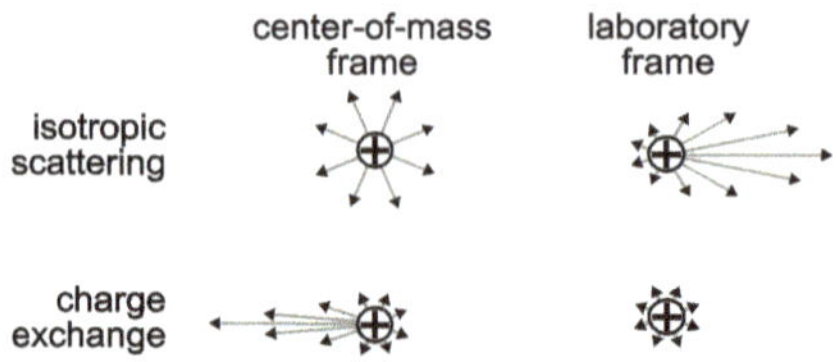

Figure 2 A sketch of the post-collision ion velocity distribution. In this example the initial ion velocity is much larger than the thermal neutral. The term isotropic collision is claimed due to the isotropy in the center-of-mass frame, but in the laboratory frame the distribution is forward-skewed due to the conservation of the ion momentum.

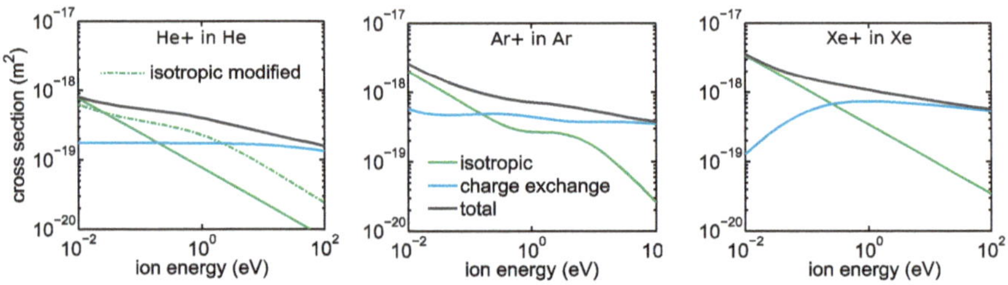

Figure 3 Ion-neutral scattering cross sections of argon (Phelps, 1994), helium, and xenon (Phelps, 2014) ions in their parent gas on an energy domain relevant for the NAIS.

changed without any exchange of momentum. The velocity distributions of isotropic and charge exchange collisions are sketched in Fig. 2. The forward scatter peak corresponds to ions being affected very weakly by the polarizing potential. This peak is neglected because it doesn't influence the kinematics very much. The cross sections as suggested by Phelps (1994) are shown in Fig. 3.

A word of caution is in place here. In the literature one can find total charge exchange cross sections, see for example Pullins and Dressler (2000). However, using such cross sections and assume the charge exchange to be ideal, leads to an underestimate of the momentum exchange for low energies. In the Phelps (1994) model, the portion of charge exchange events with considerable momentum transfer is included in the isotropic cross section and excluded from the idealized charge exchange cross section. This improves the accuracy for describing the anisotropy at low energies.

It should be mentioned that yet a different model was proposed by Nanbu and Kitatani (1995), but this is more computationally demanding and quite frankly a bit non-transparent. A simplified version of this model is proposed by Nanbu and Wakayama (1999) and it shows fair agreement with experimental data, but not as good as the model by (Phelps, 1994) does.

Phelps (1994) described his model for argon initially and gave fits for the cross sections. The cross sections are given as a function of the relative energy between the ion and the neutral $\epsilon = \frac{1}{2}m|\mathbf{u}_\mathrm{i} - \mathbf{u}_\mathrm{n}|^2$ in units of electron volt with the cross section in m^2. The fit for the isotropic part of the cross section is

$$\sigma_{\mathrm{iso,Ar}}(\epsilon) = \frac{2 \times 10^{-19}}{\epsilon^{0.5}(1+\epsilon)} + \frac{3 \times 10^{-19}\epsilon}{(1+\epsilon/3)^2}. \tag{1}$$

The idealized charge exchange cross section is found indirectly through the use of a total momentum transfer cross section. This cross section is associated with the average momentum transferred in a collision. The total momentum transfer cross section is found to follow the fit

$$\sigma_{\mathrm{m,Ar}}(\epsilon) = \frac{1.15 \times 10^{-18}(1+0.015/\epsilon)^{0.6}}{\epsilon^{0.1}}. \tag{2}$$

Phelps explains that back-scatter events on average transfer double the momentum compared to isotropic collisions. So, a total cross section describing the net momentum transfer is $\sigma_\mathrm{m} = \sigma_\mathrm{iso} + 2\sigma_\mathrm{cx}$. Hence, the idealized charge exchange cross section is given by

$$\sigma_{\mathrm{ctr,Ar}}(\epsilon) = \tfrac{1}{2}(\sigma_{\mathrm{m,Ar}} - \sigma_{\mathrm{iso,Ar}}). \tag{3}$$

Note that the total cross section describing the probability of any event happening, is simply $\sigma_\mathrm{tot} = \sigma_\mathrm{iso} + \sigma_\mathrm{cx}$.

Elsewhere, fits for other noble gases are reported (Phelps, 2014), but these are never officially published and presumably not as thoroughly verified as the ones for argon. For xenon the cross sections are

$$\sigma_{\mathrm{iso,Xe}}(\epsilon) = \frac{33.9 \times 10^{-20}}{\epsilon^{0.5}} \tag{4}$$

$$\sigma_{\mathrm{cx,Xe}}(\epsilon) = \frac{36 \times 10^{-20}\left(1+100\epsilon^2\right)^{0.2}}{\epsilon^{0.42}\left(1+0.09/\epsilon\right)^{1.3}\left(1+0.001\epsilon\right)^{0.25}}. \tag{5}$$

The isotropic cross section is essentially a so-called spiraling cross section (Langevin, 1905) (see for example Chanin and Biondi (1957) for more on this relation).

For helium we found it necessary to make changes to the cross sections from the fits proposed by Phelps (2014). We will state the result here and motivate the changes in the section on model validation.

$$\sigma_{\mathrm{iso,He}}(\epsilon) = \frac{4.5 \times 10^{-20}\left(1 + 8\epsilon^{0.65}\right)}{\epsilon^{0.5}(1 + 1.3\epsilon)^{0.7}} \tag{6}$$

$$\sigma_{\mathrm{cx,He}}(\epsilon) = \frac{1 \times 10^{-19}}{(0.001\epsilon)^{0.15}(1 + 5/\epsilon)^{0.15}(1 + 0.001\epsilon)^{0.25}}. \tag{7}$$

Surely there is a lot more complexity to ion-neutral interaction than this model encompasses (Barata, 2007), but these details are not relevant for estimating the properties of the NAIS. The limitation of the Phelps model is found mainly in the accuracy of approximating the differential cross section by an isotropic and a back-scatter part. The model has been used by various others for example Jovanovic, Vrhovac, and Petrovic (2002), Robertson and Sternovsky (2003) and more recently Ristivojevic and Petrovic (2012). All authors concluded good agreement with the experimental data.

3. Monte-Carlo implementation

We intend to use the interaction model discussed in the previous section in a Monte-Carlo ray tracing scheme. The details of the Monte-Carlo implementation are elaborated in this section. The method is similar, though not identical to what is used elsewhere (Jovanovic et al., 2002; Robertson & Sternovsky, 2003; Nanbu, 2000; Ristivojevic & Petrovic, 2012).

3.1 Collision kinematics

When ray tracing, we are interested in the change in the velocity vector of the ion in the case of a scatter event. It is a textbook exercise to calculate the post collision velocities of isotropically scattering equal-mass particles. The result (Nanbu, 2000) is

$$\mathbf{u}_{\mathrm{i}}' = \tfrac{1}{2}(\mathbf{u}_{\mathrm{i}} + \mathbf{u}_{\mathrm{n}}) + \tfrac{1}{2}|\mathbf{u}_{\mathrm{i}} - \mathbf{u}_{\mathrm{n}}|\hat{\mathbf{R}} \tag{8}$$

$$\mathbf{u}_{\mathrm{n}}' = \tfrac{1}{2}(\mathbf{u}_{\mathrm{i}} + \mathbf{u}_{\mathrm{n}}) - \tfrac{1}{2}|\mathbf{u}_{\mathrm{i}} - \mathbf{u}_{\mathrm{n}}|\hat{\mathbf{R}}. \tag{9}$$

A primed vector indicates post-collision and $\hat{\mathbf{R}}$ is a randomly oriented unit vector. The pre- and post-collision velocity vectors here are given in the lab-frame and can be used right away in a simulation. We use i for ion, n for neutral atom and $\mathbf{u}$ for velocity.

Idealized charge exchange is essentially an identity switch and the post collision velocities are given by:

$$\mathbf{u}_i' = \mathbf{u}_n \tag{10}$$

$$\mathbf{u}_n' = \mathbf{u}_i. \tag{11}$$

3.2 The scattering frequency

In the Monte-Carlo ray-tracing scheme we need to determine every time step which ions should change their velocity due to a scatter event. Therefore, we define a scattering frequency as the number of scatter events per unit of time:

$$\nu(\mathbf{u}_i) = \text{scattering frequency}. \tag{12}$$

The use of this quantity is finding the probability of an event in one time step. In the simple case that the medium is stationary, we can find it from $\nu = n\sigma(u_i)u_i$. If, however, the medium is not static, the contribution to the collision frequency should be integrated over all possible velocity vectors of the neutral atoms. The generalized scatter frequency becomes

$$\nu(\mathbf{u}_i) = n \iiint \sigma(\mathbf{u}_i - \mathbf{u}_n)|\mathbf{u}_i - \mathbf{u}_n| f(\mathbf{u}_n) d\mathbf{u}_n \tag{13}$$

with $f(\mathbf{u}_n)$ the probability density function for the velocity of neutral particles.

We simplify the system by assuming the neutrals to be in thermal equilibrium. This means that the neutral velocities are Maxwell-Boltzmann distributed:

$$f(\mathbf{u}_n) = \left(\frac{m}{2\pi kT}\right)^{3/2} e^{-\frac{m|\mathbf{u}_n|^2}{2kT}}. \tag{14}$$

For neutrals that are expanding into the vacuum this assumption becomes doubtful. However, in Section 6 it is explained that the density outside the chip is very low, so inaccurate modeling at these locations is not problematic.

Eq. (13) is quite problematic from a practical point of view. Suppose using it as-is, then, every time step one would need to calculate the full 3D integral for every ion. This would be unacceptably slow. We could generate a 3D look-up table prior to the simulation, but luckily we have a better trick.

In Eq. (13), the ion velocity vector can be chosen in any direction due to the rotational symmetry of the neutral velocity distribution $f(\mathbf{u}_n)$.

Essentially, we employ the observation that $\nu(\mathbf{u}_i) = \nu(|\mathbf{u}_i|\hat{\mathbf{x}})$, for rotationally symmetric f. The advantage of using this symmetry is that the collision frequency becomes a function of the ion velocity magnitude only. All of a sudden, we can suffice with a 1D look-up table, which is quite convenient indeed.

The look-up table we pre-calculate is not exactly Eq. (13). Since the scattering frequency is proportional to the particle density we can tabulate $\nu(u_i)/n$ and at run-time multiply by the local particle density.

3.3 Event probability and time step control

For deciding if an event occurs in a time step, we need to transfer the scattering frequency to a scattering probability. The scattering frequency is time-independent so the number of collisions in a time interval Δt is described by a Poisson distribution: $P_k = (k!)^{-1}\mu^k e^{-\mu}$ with $\mu = \nu\Delta t$.

Our simulation is incapable of handling more than one collision per time-step so a Poisson distribution only makes sense if the criterion $P_{k>1} \ll P_1$ is met. This simply states that the probability of multiple scatter events needs to be negligible and this is equivalent to $\nu\Delta t \ll 1$.

An alternative to the Poisson distribution is

$$P_0 = 1 - \nu\Delta t \tag{15}$$

$$P_1 = \nu\Delta t \tag{16}$$

$$P_{k>1} = 0 \tag{17}$$

The advantage is that now we only require $\nu\Delta t < 1$ as $P_{k>1} \ll P_1$ is automatically met. The number of events over a long time T is clearly just νT, indeed what it should be, even if $\nu\Delta t$ is close to 1. A second advantage is that this distribution is computationally less costly.

The only error made here is the unjustified spatial separation of events. This boils down to an inaccuracy in the location of the scatter events. The criterion for a time step size thus becomes $\Delta t < \Delta x u_i$, with Δx the maximum allowed positional inaccuracy. The required time step depends on the specific simulation, but for example for 1 eV argon ions and a required positional accuracy of 2 nm, we require time-steps of about 1 ps.

More generally we try to relate the positional accuracy to the mean free path. Suppose we have an estimate for the mean free path λ in a particular configuration (see Section 5), then, we can estimate the maximum ion velocity from

$$u_{\mathrm{i}} = \sqrt{\frac{2eE\lambda}{m}} + u_{\mathrm{t}}, \tag{18}$$

with E the electric field strength. Note that the addition of the thermal velocity u_{t} is needed to ensure a small enough time step in the case of a weak field or a dense gas. The maximum allowed time step is given by

$$\Delta t < \frac{\Delta x}{\lambda}\left(\sqrt{\frac{2eE}{m\lambda}} + \frac{u_{\mathrm{t}}}{\lambda}\right)^{-1} \tag{19}$$

Setting $\frac{\Delta x}{\lambda} = 0.1$ was tested to be sufficient for most NAIS simulations.

The scatter probability depends on the ion velocity, the neutral particle density, and the time step and should therefore be calculated every time step for every particle. This is costly so an improvement here is effective. The null collision method, as explained for instance by Vahedi and Surendra (1995) introduces a constant probability that is larger than any possible real scattering probability. The null collision probability is $P_{\mathrm{null}} = \max(n\nu_{\mathrm{n}}\Delta t)$. Only in case that a random number between 0 and 1 is smaller than this maximum probability do we calculate the real scattering probabilities.

3.4 Sampling a neutral atom

In the case of a scatter event, we need to sample a neutral atom such that we can calculate the kinematics of the ion. As pointed out by Robertson and Sternovsky (2003) and by Ristivojevic and Petrovic (2012), sampling a neutral atom should not be done using a Maxwell-Boltzmann distribution, even though the neutral gas is assumed to be Maxwell-Boltzmann distributed. The probability of sampling a particular neutral velocity vector should be proportional to the amount it contributes to the collision frequency. This is simply the integrand of Eq. (13). So the probability density function for sampling a neutral atom $g(\mathbf{u}_{\mathrm{i}}, \mathbf{u}_{\mathrm{n}})$, is given by

$$g(\mathbf{u}_{\mathrm{i}}, \mathbf{u}_{\mathrm{n}}) = \frac{\sigma(\mathbf{u}_{\mathrm{i}} - \mathbf{u}_{\mathrm{n}})|\mathbf{u}_{\mathrm{i}} - \mathbf{u}_{\mathrm{n}}|f(\mathbf{u}_{\mathrm{n}})}{\nu(\mathbf{u}_{\mathrm{i}})/n}. \tag{20}$$

When sampling from a distribution function, inverse transform sampling is often used. This requires the inverse of the cumulative probability density function, which is not attainable in this case. An alternative is the acceptance-rejection method. One generates uniformly distributed random variables over a pre-defined domain and any sample under the curve of the probability distribution is usable. This method requires the peak value of the probability density function. We can use the symmetry of the Maxwell-Boltzmann function again to simplify the task, as $\max g(\mathbf{u}_i, \mathbf{u}_n) = \max g(|\mathbf{u}_i|\hat{\mathbf{x}}, \mathbf{u}_n)$. A straightforward numerical optimization routine is used to find the maximum for a broad range of ion velocities. It turns out, a look-up table is not needed because the peak value of the distribution is fairly constant, so we can use a single number. For helium, argon and xenon we find that the values $1.4 \times 10^{-10}\ \mathrm{s^3\,m^{-3}}$, $4.2 \times 10^{-9}\ \mathrm{s^3\,m^{-3}}$, and $2.6 \times 10^{-8}\ \mathrm{s^3\,m^{-3}}$ are at least $\max g(\mathbf{u}_i, \mathbf{u}_n)$. The values work for both isotropic as well as charge exchange collisions.

If $u_i \gg u_n$ then $g(\mathbf{u}_i, \mathbf{u}_n) \approx f(\mathbf{u}_n)$, so in many cases sampling from a Maxwell-Boltzmann distribution will give only a limited inaccuracy. Using the distribution described here, however, ensures proper sampling in any case. In particular this could be important in weakly accelerating situations with ion energies not much larger than the thermal energy.

3.5 Software

A useful model of the NAIS does, besides ion-neutral scattering, include the electrostatic field inside the chip, the acceleration field outside the chip, and the coulomb interactions between the ions. Therefore, it is chosen to implement the ion neutral scattering model into the commercial software (PulsarPhysics, 2011). This software takes care of the ray tracing including coulomb interactions. In this way, all relevant physics can be simulated simultaneously. We calculate the electric fields using another commercial software package (SPOC, 2012).

4. Model validation

4.1 Method of validation

To check if the simulation implementation is sound, a simple configuration was set up and tested. Ions starting in the origin are traced through a region of constant particle density $n = 3.535 \times 10^{22}\ \mathrm{m^{-3}}$, constant temperature of 293 K. Fig. 4 shows an example of some traced trajectories. After traveling for several thousands of mean free path-lengths, two quantities often encountered in plasma physics are computed and compared to literature values: the drift velocity and the reduced transverse diffusion coefficient. This approach is similar to what is done by Robertson and Sternovsky

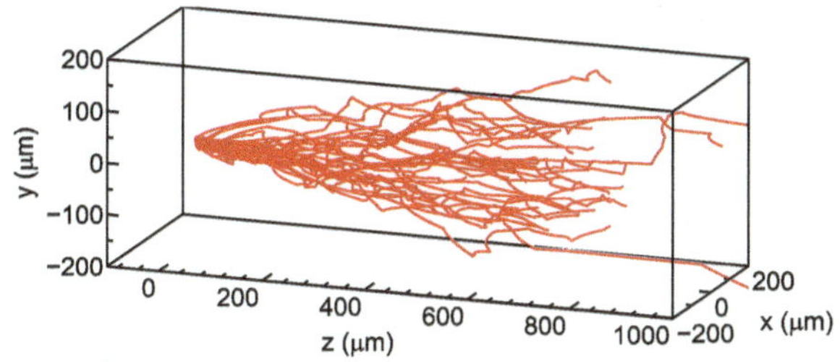

Figure 4 Example of tracing Ar+ ions through argon gas with particle density 3.535×10^{22} m^{-3} and a temperature of 293 K. An electric field of 10 V/mm was applied in the positive z-direction. Such tracing is used to determine the drift velocity and the transverse diffusion coefficient for testing the simulation. Shown here is a set of 30 particles each having around 25 scattering events, but in a typical model validation simulation 3000 particles are traced each undergoing around 5000 scattering events.

(2003). The computed quantities are

$$u_{\text{drift}} = \frac{\langle z \rangle}{t_{\text{sim}}}, \qquad D_{\text{t}} = \frac{1}{4}\frac{\langle r^2 \rangle}{t_{\text{sim}}}, \qquad \mu = \frac{u_{\text{drift}}}{E}. \tag{21}$$

Particularly the reduced transverse diffusion coefficient (D_t/μ) is infamous for its sensitivity to errors in the model or inaccurate handling of the anisotropy in the cross sections (Jovanovic et al., 2002).

The maximum simulation time step was determined by Eq. (19) using a very accurate setting: $\Delta x/\lambda_{\text{eff}} = 1/200$ with mean free path estimates are taken from Section 5. The simulation time t_{sim} is adjusted accordingly to run for about 5000 mean free paths lengths. A total of 3000 particles are used.

4.2 Argon

Fig. 5 shows the ray tracing result for argon. The correspondence with experimental drift velocities is nearly perfect and the match for the transverse diffusion coefficient is also quite good. This result confirms a correct implementation of the model and shows that the cross sections suggested by Phelps are describing the ion-neutral interaction quite well.

For the sake of comparison, a fixed-size hard-sphere simulation was run, with twice the molecular cross section of neutral-neutral interaction, as suggested by Nanbu (2000). The molecular diameter used was 369 pm (Roth, 1990). We see that for both quantities the values are quite far off and

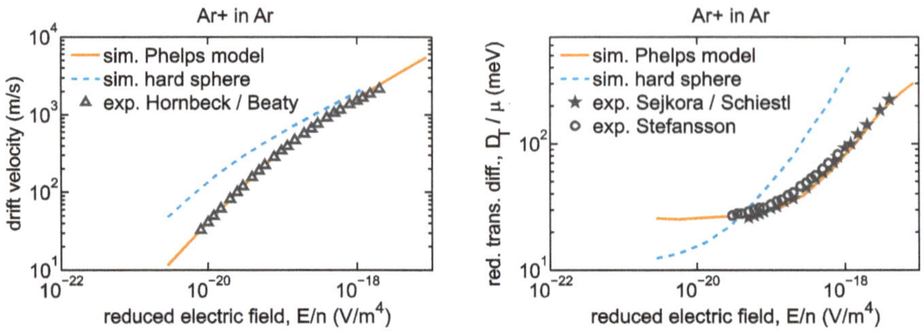

Figure 5 Comparison of the simulation implementation used in this work to experimental data. Settings: $n = 3.535 \times 10^{22}\ m^{-3}$ and $T = 293$ K, the electric field was varied. The experimental values for the drift velocity are taken from tabular data composed by Ellis, Pai, McDaniel, Mason, and Viehland (1976), which is based on measurements performed by Hornbeck and Molnar (1951) and by Beaty and Patterson (1965). The transverse diffusion coefficient data is taken from measurements by Stefansson and Skullerud (1999) and also from tabulated values composed by Viehland and Mason (1995), which is based on measurements from Sejkora, Girstmair, Bryant, and Mark (1984) and Schiestl, Sejkora, Lezius, Foltin, and Mark (1993).

one can not expect accurate results from this, perhaps a little too simple, model.

4.3 Helium

The cross sections proposed by (Phelps, 2014) for helium were tested, but didn't show such a great accordance with measured transverse diffusion data as can be seen by dashed curve in Fig. 6. An adjustment to the cross sections was made, notably at energies around 1 eV the isotropic cross section is made a little larger than the spiraling cross section (Langevin, 1905). The same type of adjustment can be seen in the isotropic cross section for argon. The result of the adjustment is an excellent match between the simulation and the experimental data. Eq. (6) gives the adjusted cross sections.

4.4 Xenon

Cross sections for xenon (Phelps, 2014) were also tested. In the literature (Larsen & Elford, 1986) we find for the drift velocity a distinction between the ionic ground state $P_{3/2}$ and the metastable state $P_{1/2}$. The energy difference between these states is 1.31 eV and the lifetime of the metastable state is 48 ms (Barata, 2005). Phelps does not elaborate on different excitations of the ionized states for his cross sections and we can only assume the cross

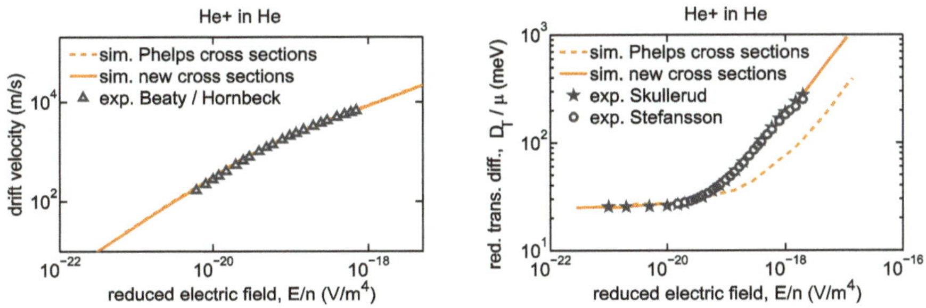

Figure 6 Comparison of the simulation implementation used in this work to experimental data. Settings: $n = 3.535 \times 10^{22}$ m^{-3} and $T = 293$ K, the electric field was varied. The experimental values for the drift velocity are taken from tabular data composed by Ellis et al. (1976), which is based on measurements performed by Hornbeck and Molnar (1951) and by Beaty and Patterson (1965). The transverse diffusion coefficient data is taken from measurements by Skullerud and Larsen (1990) and by Stefansson, Berge, Lausund, and Skullerud (1988).

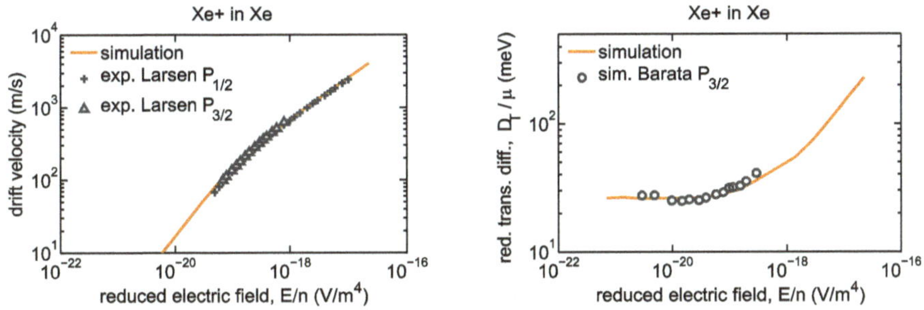

Figure 7 Comparison of the simulation implementation used in this work to experimental data. Settings: $n = 3.535 \times 10^{22}$ m^{-3} and $T = 293$ K while the electric field was varied. The drift velocity measurements are taken from Larsen and Elford (1986) and the distinction between the ground state $P_{1/2}$ and the excited metastable state $P_{3/2}$ is made. No experimental data for the transverse diffusion coefficient was available and a comparison was made to a simulation performed by Barata (2005).

sections to represent a combination of unknown proportion. The proportion of both states depends mainly on the means and details of ionization and will differ from experiment to experiment. Despite this ignorance, we can boldly state that knowing the exact proportion is not so relevant as it gives nearly the same scattering properties. We can see in Fig. 7 that the experimental data of drift velocity for the different excitations is hardly distinguishable. Note that this applies to argon as well (Barata, 2007).

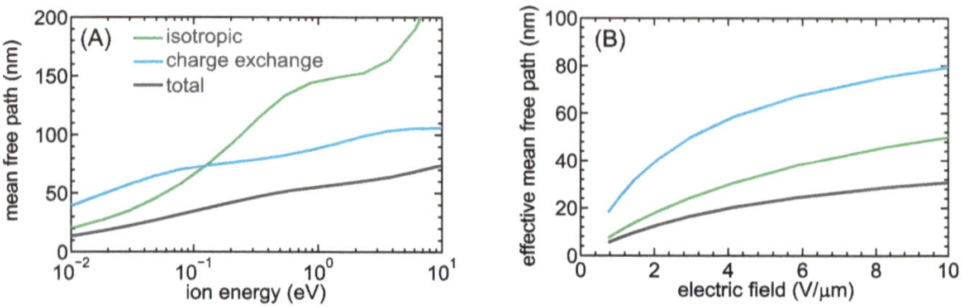

Figure 8 Ionic mean free path calculations at atmospheric pressure of argon ions through argon gas, based on the cross sections discussed in Section 2.2 at $T = 293$ K. (A) The argon ions travel through their parent gas at constant velocity. (B) Ions are accelerated in an electric field and the effective mean free path is calculated according to Eq. (27).

We could not find experimental data for the reduced transverse diffusion coefficient for xenon, so the next best thing was a comparison to other calculations executed by Barata (2005). These calculations were based on ion-atom interaction potential models by Tang and Toennies (1984) and have no direct relation to the cross sections proposed by Phelps. We find a good correspondence between the work by Barata and this work, at least in the range for which calculations were performed.

5. An effective mean free path

5.1 MFP for constant velocity

A useful and intuitive way of describing scattering is via a mean free path (MFP). For a constant cross section, a MFP is fairly simply given by $\lambda_{\text{eff}} = (\sigma n)^{-1}$. However, when the cross section is energy dependent, and the medium is non-stationary as well, the calculation is more tricky. We can calculate the MFP as the inverse of the collision density (number of collisions per unit length) and get

$$\lambda^{-1} = \frac{\nu(u_{\text{i}})}{u_{\text{i}}} = \frac{n}{u_{\text{i}}} \iiint \sigma(|\mathbf{u}_{\text{i}} - \mathbf{u}_{\text{n}}|)|\mathbf{u}_{\text{i}} - \mathbf{u}_{\text{n}}| f(\mathbf{u}_{\text{n}}) d\mathbf{u}_{\text{n}}. \tag{22}$$

It is interesting to relate the cross sections discussed in the previous section to a mean free path using this equation. Numerical evaluation is performed for a 1 bar pressure and the result is shown in Fig. 8A. Note that the MFP is inversely proportional to the particle density (or pressure) so the presented curve can be scaled accordingly.

5.2 MFP in an electric field

The mean free path calculation of the previous section gives some insight, however does not offer great understanding for the NAIS. It is rather unclear what ion energy value is appropriate. In fact, the ions are accelerated through the gas, starting with thermal energy, up to say 1 eV, hence there is not really one particular energy assignable. Nevertheless, a relevant quantity that is unambiguous can be identified, namely the electric field strength. Furthermore, in a NAIS system we are interested mostly in unscattered particles, as these are responsible for the high beam quality. Therefore, we define a mean free path as follows: the average unscattered travel distance for ions being accelerated in an electric field. To be clear, this MFP relates to the first scatter event after generation.

A mean free path is normally non-directional, but for the purpose of the NAIS, we are more interested in how far the particle travels towards the end of the gas region. Therefore we define the mean free path as the average traveled distance along the optical axis; not as the average length of the trajectory. Note that the particle starts with an initial thermal energy in the transverse directions. A most comprehensive analysis integrates over the initial thermal velocity vectors, but this leads to a numerical challenge, so this is avoided. Instead, we pick a typical initial ion, having kT of energy in the transverse direction. We assign no initial energy to the z-direction because the particle can travel forward or backward, and setting the initial velocity to zero approximately balances this out.

The mean free path as described above can be calculated from

$$\lambda_{\text{eff}} = \int_0^\infty z(t)\rho(t)dt, \tag{23}$$

with $\rho(t)$ the probability density function of an ion scattering between t and $t + dt$. The calculation is performed in the time domain to keep the equations and the numerical computation later-on the easiest. We find ρ from an intensity attenuation analysis. If I is the current of an initial ion beam, then the loss in current by scattering is described by

$$dI = -I\nu dt \;\rightarrow\; I = I_0 e^{-\int_0^t \nu dt'}. \tag{24}$$

This is similar to a Lambert-Beer law, but the integral in the solution can't be expressed explicitly because the collision frequency is not an explicit function; it is the result of the volume integral of Eq. (13). We can relate the current to a probability density according to

$$\rho(t)dt = -\frac{dI}{I_0} = -d(e^{-\int_o^t \nu dt'}) = \nu e^{-\int_o^t \nu dt'} dt. \tag{25}$$

We also need the ion position and velocity as a function of time, while being accelerated in an electric field:

$$z(t) = \frac{eE}{2m}t^2 \quad , \quad |\mathbf{v}(t)| = \sqrt{\frac{2kT}{m} + \left(\frac{eE}{m}t\right)^2}. \tag{26}$$

Combining Eqs. (23), (25), and (26), the effective mean free path to a first scatter event in an electric field is given by

$$\lambda_{\text{eff}} = \frac{eE}{2m} \int_0^\infty t^2 \nu(t) e^{-\int_0^t \nu(t')dt'} dt. \tag{27}$$

The numerical evaluation of this integral for atmospheric pressure as a function of the electric field is shown in Fig. 8B. Note that, in contrast to the energy dependent MFP, this MFP does not scale simply with the particle density.

To get a feeling for the relative influence of the isotropic and charge exchange events to the MFP, we compute an effective MFP for both type of events according to

$$\lambda_{\text{iso}}^{-1} = \lambda_{\text{eff}}^{-1} \int_0^\infty \nu_{\text{iso}}(t) e^{-\int_0^t \nu(t')dt'} dt \tag{28}$$

$$\lambda_{\text{cx}}^{-1} = \lambda_{\text{eff}}^{-1} \int_0^\infty \nu_{\text{cx}}(t) e^{-\int_0^t \nu(t')dt'} dt. \tag{29}$$

Note that the non-sub-scripted ν represents the total collision frequency. These MFP's interpolate to the total MFP by inverse addition; exactly what is common for mean free paths. The integrals in these equations essentially represent the contributing fraction of the particular event-type to the full MFP calculation.

In Fig. 8B we can see that at a field of 4 kV/mm, the mean free path is about 20 nm. Rather small, compared to 70 nm derived from a molecular diameter model. The graph shows that at this particle density, isotropic collisions are more important than charge exchange events. Charge exchange events are not negligible though. Since the kinematics of the two event types are not the same, a molecular diameter model will have an additional error in describing the collisions, besides being of the wrong magnitude. Interestingly, the mean free path can be tuned between 10 to 30 nm by changing the electric field. The quick explanation is: a higher field brings the ions up to higher energy quicker, which makes the polarization potential less influential.

The calculations shown in Fig. 8 are only valid for atmospheric pressure. We would like to have a more general result, but the difficulty is that the mean free path does not simply scale with the particle density due to the energy dependence of the cross sections. However, if the particle density and the electric field are scaled simultaneously, all possible ions trajectories are scaled proportionally in terms of energy at the moment of a collision. This means that we can write $\lambda_{\text{eff}} = f\left(\frac{E}{n}\right)\frac{1}{n}$, with f some unknown function. E/n is the reduced electric field and the quantity $n\lambda_{\text{eff}}$ is a function of this variable only; a more general result indeed.

For the convenience, reasonable fits for $f\left(\frac{E}{n}\right)$ were found, allowing us to summarize a lot of information:

$$n\lambda_{\text{eff}} = f\left(\frac{E}{n}\right) \approx a \cdot \ln\left(b\frac{E}{n}\right) \quad \text{or} \quad \left(a\frac{E}{n}\right)^b. \tag{30}$$

We can use this equation to estimate the MFP for a particular combination of E and n. This circumvents the need for the complicated evaluation of Eq. (27) (the evaluation of (27) was of course needed to acquire the fits in the first place). The fitting results are given in Table 1. Note that sometimes a power law was more appropriate than the natural log to describe the curve; both have no physical basis.

5.3 Unscattered ionization efficiency

The fit functions (Eq. (30) and Table 1) for $n\lambda_{\text{eff}}$ are very useful for calculations, but $n\lambda_{\text{eff}}$ is somewhat of an indirect quantity in itself. More instructive is the efficiency of electron to ion current. For the moment, we simplify the gas density by a top-hat distribution of length l and particle density n. The ion current J_{I} as generated by electron impact ionization is then given by $J_{\text{i}} = J_{\text{e}} n \sigma_{\text{I}} l$, with J_{e} the electron current density.

Suppose the length l is at least a few MFP's long. The generated ions from the back won't make it out unscattered, but the ones generated within one or two MFP's from the exit aperture are quite likely to make it out. A reasonable simplification is to say that all ions within one MFP from the end make it out unscattered, and no other. This gives as an estimate for the

Table 1 Fit parameters for Eq. (30). Outside the specified domain the quality of the fits is not assessed. 'ln' indicates the natural logarithm fit and 'p' denotes power law. The units in the fit functions are meter and Volt.

	Fit	a	b	Fit domain E/n V m^2	Mean err. %	Max err. %
Helium						
Total	ln	6.93×10^{17}	5.39×10^{19}	$3.7 \times 10^{-20} - 5.0 \times 10^{-18}$	4.0	13.0
Isotropic	p	4.84×10^{56}	0.484		3.9	8.9
Char. Exch.	ln	6.46×10^{17}	4.59×10^{21}		3.7	13.2
Argon						
Total	ln	2.50×10^{17}	5.56×10^{19}	$2.3 \times 10^{-19} - 2.0 \times 10^{-17}$	1.2	5.0
Isotropic	ln	5.58×10^{17}	2.24×10^{19}		2.6	4.8
Char. Exch.	ln	3.97×10^{17}	4.35×10^{20}		3.0	8.6
Xenon						
Total	ln	2.22×10^{17}	2.32×10^{19}	$3.1 \times 10^{-19} - 2.4 \times 10^{-17}$	0.2	0.3
Isotropic	p	6.24×10^{48}	0.591		2.0	10.0
Char. Exch.	ln	1.02×10^{17}	3.04×10^{23}		0.9	1.7

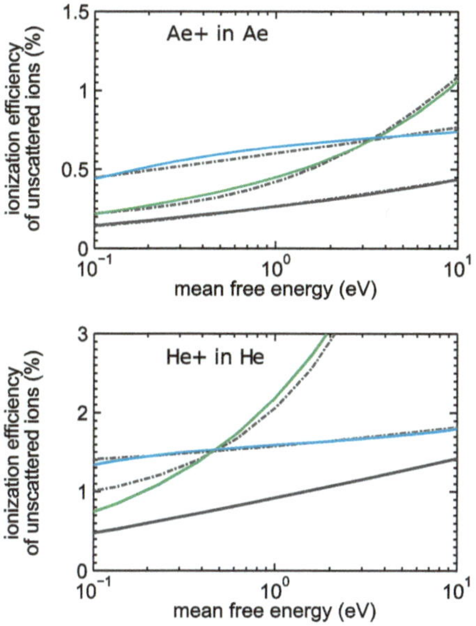

Figure 9 Calculated conversion ratio of electron current to unscattered ion current for a 1 keV electron beam. The ions are accelerated in a gas with uniform particle density and uniform electric field. Cross section data from Rejoub, Lindsay, and Stebbings (2002). The dashed gray lines are calculated from the fit functions given by Eq. (30).

ionization efficiency of unscattered particles:

$$\eta_{\text{unscat}} \approx n\sigma_{\text{I}}\lambda_{\text{eff}}. \tag{31}$$

We could plot the ionization efficiency versus E/n however E/n doesn't tell a whole lot either. We introduce a quantity with units of energy as

$$\epsilon_{\text{E}} = eE\lambda_{\text{eff}} = \text{mean free energy}. \tag{32}$$

This quantity represents the average energy an ion gains before it scatters. Since we can write it as $\epsilon_{\text{E}} = n\lambda_{\text{eff}}\frac{E}{n} = f\left(\frac{E}{n}\right)\frac{E}{n}$, we can combine it with Eq. (30) and obtain $n\lambda_{\text{eff}}$ as a function of ϵ_{E} only. So, ϵ_{E} is an alternative independent variable to E/n.

Fig. 9 shows the result of the unscattered ionization efficiency as a function the mean free energy ϵ_{E}. In this example we have used the ionization cross section of a 1 keV electron beam. Perhaps the most important message here is that we can expect in the order of 1% of unscattered ionization efficiency for typical NAIS configurations. The unscattered ionization efficiency may be somewhat tunable, but this requires orders of magnitude difference in the acquired energy. A high mean free energy translates to a high energy spread. We may get a higher total ionization efficiency, but the excess generated current will scatter on neutral particles and does not contribute as much to the brightness as unscattered ions do.

5.4 The ionic Knudsen number

So far in this section we have tried to develop an intuitive assessment of the NAIS emission region. We have introduced an effective mean free path and in the previous subsection we have simplified the particle density distribution by a top-hat distribution of length l. Both the mean free path and the gaseous length only have meaning relative to each other, and here we introduce a more general concept for characterizing the operation condition of a particular geometry. In the case of neutral particles, characterizing confined geometries is often done by the Knudsen number. We can extend that idea by introducing the ionic Knudsen number. It is the effective MFP, divided by an appropriate length measure for the gaseous region:

$$\text{iKn} = \text{ionic Knudsen number} = \frac{\text{ion mean free path}}{\text{length of gas region}}. \tag{33}$$

Arguably, Knudsen defined his number inconveniently. It is actually better to work with the inverse ionic Knudsen, as will become apparent. To avoid confusion, however, we keep the definition in line with the original Knudsen number.

The inverse ionic Knudsen number, iKn^{-1}, indicates the average number of scatter events for particles starting from the back of the confined gas region. If $\text{iKn}^{-1} > 1$, this is strictly speaking not entirely correct because the MFP used was calculated for single scatter events and not for multiple scatter events. In practice iKn^{-1} is a good indicator of the expected number of scatter events anyhow.

An inverse ionic Knudsen number $\text{iKn}^{-1} = 10$ means that most generated ions will scatter several times while a value of 0.2 is practically a scatter free situation. $\text{iKn}^{-1} = 1$ essentially marks the transition from scatter free to scatter rich systems.

The ionic Knudsen number can be found from the fit function of Eq. (30) as

$$\text{iKn} = \frac{1}{nl} f\left(\frac{E}{n}\right). \tag{34}$$

The total and the unscattered ionization efficiencies are related by

$$\eta_{\text{unscat}} \approx \text{iKn} \cdot \eta_{\text{tot}}. \tag{35}$$

The mean free energy can be calculated from

$$\epsilon_{\text{E}} = \text{iKn} \cdot eEl = \tfrac{eE}{n} f\left(\frac{E}{n}\right). \tag{36}$$

We give two examples of how to use the ionionic Knudsen number. We begin by setting iKn^{-1} to 3. Thus, the ions from the back will scatter about three times on average. Furthermore we choose $E = 3 \times 10^6$ V μm^{-1} and $l = 600$ nm, so the bias voltage across the membranes will be 1.8 V.

We calculate the mean free energy from Eq. (36) and get $\epsilon_{\text{E}} = 0.6$ eV. Because ϵ_{E} is smaller than the membrane bias, some energy redistribution will be present. From $l = 600$ nm and $\text{iKn}^{-1} = 3$ we have directly $\lambda_{\text{eff}} = 200$ nm. The required particle density can be obtained from Eq. (30) and gives $n = 4.5 \times 10^{24}$ m^{-3}, or 180 mbar. We calculate the total generated current efficiency and obtain 2.1%. The total unscattered efficiency is estimated to be 0.70%, so for a 100 nA ebeam we get 700 pA of high quality current. If the spatial fw50 of the ebeam is 100 nm, the brightness is about 2×10^6 A/m^2 sr V. Fairly quickly we have determined the configuration of a decent NAIS.

Suppose we would like to generate a large current, say 25% of the e-beam engaging in ionization. Furthermore, we want to keep the pressure below 1 bar ($n = 2.5 \times 10^{25}$ m^{-3} at room temperature). This leads to $l > 1$ μm. If we choose a field of 3×10^6 V μm^{-1} again, the mean free path is only 19 nm and $\text{iKn}^{-1} = 68$ for $l = 1.3$ μm, so heavy scattering will be present.

Inevitably, part of the generated current is blocked by the membranes. The escape probability is typically 50%, but the value depends strongly on the geometry and field strengths. In the simplified analysis here we can not take this into account quantitatively.

6. Neutral particle-density distribution

For simulating ion trajectories in the NAIS we need the neutral density distribution. This distribution dictates the starting positions of the ions and determines the ion-neutral scatter probability.

6.1 Choice of the model

We intend to operate the NAIS with $iKn^{-1} > 1$. Since the mean free path of neutral atoms in their parent gas is of the same order of magnitude as that of the ions, we expect to work with a Knudsen number Kn of around 1 (do not confuse the Knudsen number Kn for neutral particles with the ionic Knudsen number iKn). The situation around $Kn = 1$ is known as the transition regime. In this regime, inter-atomic collisions as well as wall effects can be relevant. Rademaker (2014) compared three models: free molecular flow, variable hard-sphere and variable soft-sphere. The free molecular flow model does not consider any inter-atomic interactions while the other two models do. At $Kn = 0.7$ he found no noteworthy difference in the final particle distribution for the different models. The simulated geometry was 100 nm spacing, 100 nm aperture diameter and 100 nm membrane thickness. In this configuration inter-atomic collisions are abundant, but did not manifest themselves in significant neutral particle density variation. Combined simulation and experimental work by Peatross and Meyerhofer (1993) is in agreement with the findings of Rademaker (2014).

We assume for other geometries that the difference in neutral particle density, given that Kn is not much smaller than 0.7, is also well described by free molecular flow. In addition, we recognize that the performance of the NAIS is not very sensitive to the details of the neutral particle density. This means that even at lower Knudsen numbers, where we do start to expect some inaccuracies, free molecular flow modeling can still serve as a reasonable approximation for NAIS simulations.

6.2 Free molecular flow

A major advantage of free molecular flow simulations is that the neutral particle density is scalable. So, we have to simulate every geometry only once. This is not possible for models that include particle-particle interactions. There are mainly two options for free molecular flow: Monte-Carlo particle tracking and the surface integral method. We need to know the density around the axis, which is a weak spot for a Monte-Carlo approach; not many particles contribute to the statistics close to the axis. The surface integral method does not suffer from such weakness. A commercial implementation of the method is offered by COMSOL (2013), which is used in this work.

Fig. 10 shows the result of a free molecular flow simulation of a particular chip. We see that the density decreases very rapidly from in between the membrane towards the vacuum region. Within the first micron, the

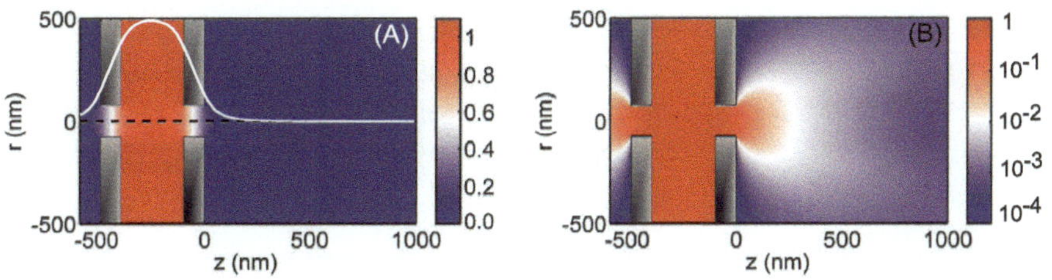

Figure 10 Normalized neutral particle density of NAIS chip geometry based on free molecular flow. (A) Linear color scale. The white curve represents the axial density profile. (B) A log color scale is used to reveal the distribution outside the chip better.

density is decreased more than three orders of magnitude. Because of this rapid decrease we expect limited gas scattering outside the chip.

It is also interesting to note that the radial dependence is not very strong. This can be seen from the absence of curvature in the color gradient inside the aperture. When comparing the density profile in the z-direction on-axis and off-axis more closely, only minimal differences were observed indeed. Therefore, we only take the z-dependence into account in the ray trace simulations. On the right hand side of Fig. 10, for r larger than say 100 nm, we see radial dependence outside the chip. However, practically no ions come there and the density is very low anyway.

6.3 Effective length

The length of the gas region is a crucial aspect of the NAIS. In first approximation this is equal to the membrane spacing. However, the aperture allows density outside the chip and a more appropriate measure of the gas length is the effective length:

$$l_{\mathrm{eff}} = \frac{1}{n_{\max}} \int n(z)dz. \tag{37}$$

This essentially simplifies the axial particle density profile into a top hat distribution, with the top hat height equal to the maximum particle density of the true axial profile. That is, the particle density in the center of the chip. Note that when loosely talking about the pressure or particle density, the maximum in the center of the chip is meant.

With this definition of the particle density, the total generated ion current can be calculated with

$$I = I_e \sigma_+ n_{\max} l_{\mathrm{eff}}. \tag{38}$$

Table 2 Simulation configuration.

Φ_I	10	kV	Φ_e	1.0	kV	spacing	300	nm
E_1	3×10^6	V/m	I_e	100	nA	thickness	100	nm
E_2	3×10^6	V/m	d_e	50	nm	aperture	150⌀	nm
T	293	K	shape	Gaussian				
N	20 000							

For simulations, this identity is used to find the relation between the particle density scaling factor n_{max}, and the current I, for a particular chip geometry characterized by l_{eff}.

7. Monte-Carlo ray tracing

The simulations described in this section aim to explore the effects of gas scattering on the optical performance of the NAIS. We ignore any coulomb-interactions, but keep in mind that accurate performance predictions require the inclusion of coulomb interactions as well.

We consider a quite typical configuration and the details are listed in Table 2. We keep the fields in and outside equal and reasonably high as to minimize optical aberrations.

7.1 Trajectory visualization

In Fig. 11, a few particle trajectories are plotted. We can see how the trajectories of many particles in the $\text{iKn}^{-1} = 25$ plot are fully dominated by scattering. Rather striking is that despite the scattering, still most particles are directed towards the aperture. Not in free flight, but in a drift-wise fashion.

7.2 Brightness and energy spread

More quantitative are the results shown in Fig. 12. The influence of gas scattering becomes clear when comparing the black-to-green brightness curves to the scatter-free gray dashed curves. For low iKn^{-1} there is no difference, but at higher values, the gas-scatter-including brightness curves are significantly reduced.

We can identify the responsible mechanisms by looking more quantitatively at the brightness reduction. For example, at $n = 1.6 \times 10^{25}$ m^{-3} ($\text{iKn}^{-1} = 8.2$) the brightness at 50% of the beam current is 6.5 times lower than the (non-physical) scatter-free situation. The virtual source area has

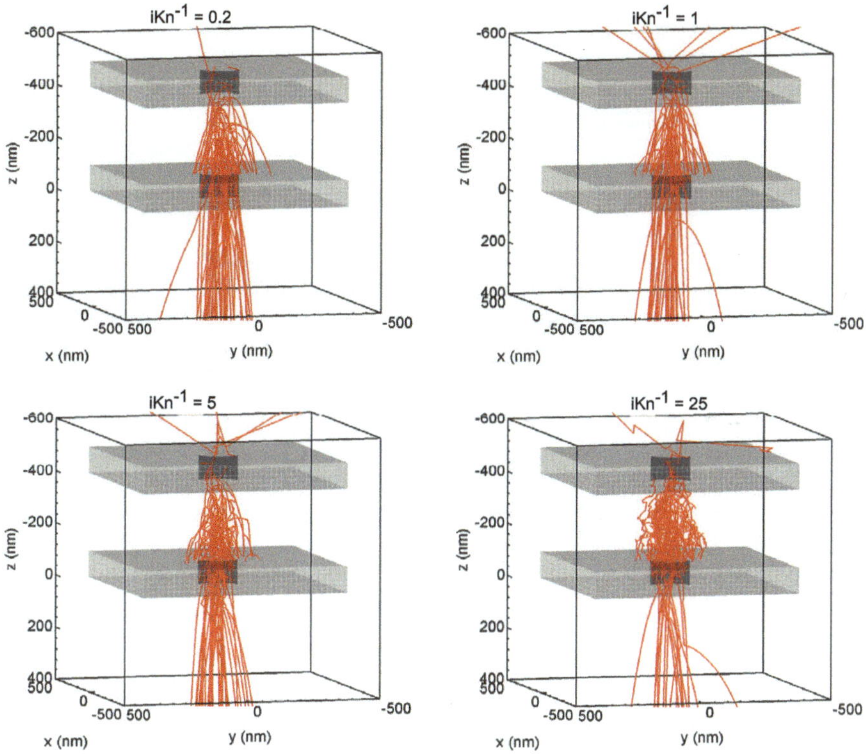

Figure 11 100 simulated ion trajectories for a typical NAIS configuration. Four cases of increasing neutral particle density are shown, characterized by the inverse ionic Knudsen number. At $iKn^{-1} = 0.2$ and also at $iKn^{-1} = 1$, gas scattering is limited, while at $iKn^{-1} = 5$ and $iKn^{-1} = 25$, the effect becomes rather clear.

grown by a factor of 1.7 and the total emitted current is reduced by a factor 1.6. The remaining factor is 2.4 and we can conclude that a large influence on the brightness loss must be due to induced transverse energy. So, some of the scattered particles don't make it out due to scattering, but also quite some do exit the chip after experiencing one or more scatter events.

If we simply look at the obtained values for the brightness, we see that $B > 1 \times 10^6$ A m^{-2} sr^{-1} V^{-1} in combination with several nA's of total current is not jeopardized by ion-neutral scattering.

The energy distributions depicted in Fig. 12B show for low iKn^{-1} a bimodal distribution. This is also reported by Jun (2014) and explained in the introduction chapter of this publication. For increasing particle density, at first only the right hand side of the distribution is influenced. This

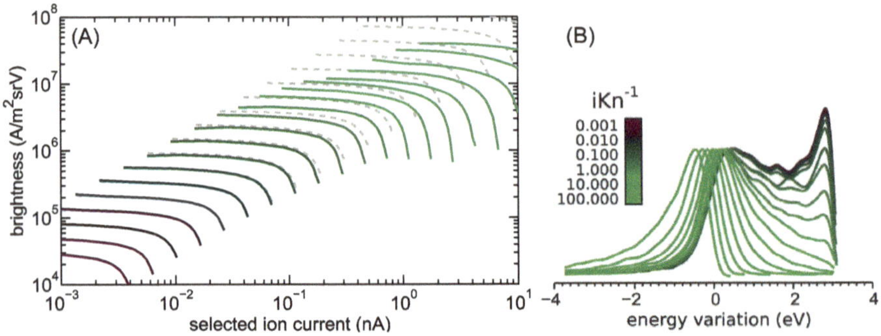

Figure 12 Monte Carlo ray trace result of a typical NAIS configuration using argon ions including electrostatic fields and ion-neutral scattering, but no coulomb interactions. (A) The brightness as function of the post-acceleration selected beam current. The light gray dashed lines indicate the brightness assuming no gas scattering. (B) Energy distribution functions of the final beam, shifted towards zero by subtracting the acceleration voltage. The distribution is a combination of initial positions in the potential landscape and the redistribution due to gas scattering. The distributions are arbitrarily scaled such that the shapes can be compared.

part of the distribution originates from the back of the chip and travels longest through the gas, hence these are most prone to ion-neutral scattering. These particles become less likely to escape, but also transfer energy in collisions and end up elsewhere in the final distribution. Interestingly, this effectively lowers the energy spread of the final beam. If the pressure is further increased the intensity on the left hand side starts to increase. This relates to increased scattering and increased ion generation in front of the chip.

8. Conclusion

A crucial aspect of NAIS operation is the inevitable scattering of ions on neutral atoms. A validation study showed that a hard sphere collision model gives inaccurate results for two quantities used for model validation; the drift velocities and the transverse diffusion coefficient. A more advanced model gave excellent results and is used to model ion-neutral scattering in the NAIS. In this model a charge exchange event and an isotropic event are considered, with energy dependent cross sections.

The effective mean free path and the ionic Knudsen number are shown to be useful concepts for characterizing the NAIS. Reasonably simple fit functions allow easy use of these quantities. Evaluation of the unscattered ionization efficiency indicates that not much more than 1% of the electron beam current of unscattered ions can be generated in a typical NAIS configuration.

A Monte Carlo ray trace simulation of a nominal NAIS geometry showed promising performance. In a fairly moderate regime, $\mathrm{iKn}^{-1} = 1-10$, we observe a brightness larger than 1×10^{6} $\mathrm{A\,m^{-2}\,sr^{-1}\,V^{-1}}$, combined with sub 1 eV energy spread. We conclude that ion-neutral scattering is not a show-stopper for achieving competitive ion source performance.

An important result is the notion that the current and the brightness tend to keep increasing with increasing particle density, despite more ion-neutral scattering. We do point out that at the highest particle densities simulated, the assumption in the modeling, notably the molecular flow assumption, become endangered and the results become inaccurate. It is expected, however, that the observed trends are not affected by a somewhat different neutral particle density distribution.

A fortunate influence of gas scattering is a reduction of the energy spread. Ions originating from the back of the chip can be slowed down by transferring energy to neutral atoms. In addition, ions from the back may scattering away from the aperture towards the membrane. This effectively yields energy filtering.

List of symbols

σ_{iso}	isotropic cross section m
σ_{CX}	charge exchange cross section m
E_1	inside-chip electric field $\mathrm{V\,m^{-1}}$
E_2	outside-chip electric field $\mathrm{V\,m^{-1}}$
E	electric field $\mathrm{V\,m^{-1}}$
f	Maxwell-Boltzmann distribution $\mathrm{s\,m^{-1}}$
g	neutral sample distribution $\mathrm{s\,m^{-1}}$
σ_{I}	ionization cross section m
ϵ	ion energy eV
$\mathbf{u}_{\mathrm{i}}$	ion velocity $\mathrm{m\,s^{-1}}$
$\mathbf{u}_{\mathrm{n}}$	neutral velocity $\mathrm{m\,s^{-1}}$
$\hat{\mathbf{R}}$	random unit vector
ν	scattering frequency $\mathrm{s^{-1}}$
u_{drift}	drift velocity $\mathrm{m\,s^{-1}}$
t_{sim}	total simulation time s
u_{t}	thermal velocity $\mathrm{m\,s^{-1}}$
μ	ion mobility $\mathrm{m\,V^{-1}\,s^{-1}}$
D_{t}	transverse diffusion coefficient $\mathrm{m^2\,s^{-1}}$
λ	mean free path m
λ_{eff}	effective mean free path m
λ_{iso}	isotropic mean free path m

λ_{cx}	charge exchange mfp	m
$\rho(t)$	probability density	s^{-1}
J_e	electron current density	$A\,m^{-2}$
J_I	ion current density	$A\,m^{-2}$
n	particle density	m^{-3}
n_{max}	max. particle density	m^{-3}
l_{eff}	effective gas length	m
l	gas length	m
iKn^{-1}	ionic Knudsen number	
$\eta_{unscatt}$	unscatt. ionization efficiency	%
η_{tot}	ionization efficiency	%
I	ion beam current	A
I_e	electron beam current	A
Φ	ion acceleration voltage	V
Φ_e	electron landing voltage	V

Acknowledgment

I would like to thank Pieter Kruit for his guidance and contributions. This work is part of the Industrial Partnership Program (IPP) of the Stichting voor Fundamenteel Onderzoek der Materie (FOM), which is financially supported by the Nederlandse Organisatie voor Wetenschappelijk Onderzoek (NWO). The IPP program is co-financed by FEI Company. This work is based on the dissertation titled *The Nano-Aperture Ion Source*, by Leon van Kouwen.

References

Barata, J. (2005). Calculation of drift velocities and diffusion coefficients of Xe+ ions in gaseous xenon. *IEEE Trans. Nucl. Sci.*, *52*(6), 2889.

Barata, J. (2007). Integral and differential elastic collision cross-sections for low-energy Ar+ ions with neutral Ar atoms. *Nucl. Instrum. Methods Phys. Res., Sect. A*, *580*(1), 14–17.

Beaty, E. C., & Patterson, P. L. (1965). Mobilities and reaction rates of ions in helium. *Phys. Rev. A*, *137*(2), A346.

Chanin, L., & Biondi, M. A. (1957). Temperature dependence of ion mobilities in helium, neon, and argon. *Phys. Rev.*, *106*(3), 473.

COMSOL (2013). *Comsol multiphysics 4.4 - molecular flow module.* Retrieved from www.comsol.com/molecular-flow-module.

Ellis, H., Pai, R. Y., McDaniel, E. W., Mason, E. A., & Viehland, L. A. (1976). Transport properties of gaseous ions over a wide energy range. *At. Data Nucl. Data Tables*, *17*(3), 177–210.

Hornbeck, J. A., & Molnar, J. P. (1951). Mass spectrometric studies of molecular ions in the noble gases. *Phys. Rev.*, *84*(4), 621.

Jovanovic, J., Vrhovac, S., & Petrovic, Z. (2002). Momentum transfer theory of ion transport under the influence of resonant charge transfer collisions: The case of argon and neon ions in parent gases. *Eur. Phys. J. D*, *21*(3), 335–342.

Jun, D. (2014). *Development of the nano-aperture ion source (NAIS)* (Doctoral dissertation). Delft University of Technology. Retrieved from repository.tudelft.nl.

Langevin, P. (1905). A fundamental formula of kinetic theory. *Ann. Chim. Phys., 5*, 245–288.

Larsen, P. H., & Elford, M. T. (1986). The mobilities of xenon ions in xenon and the derived charge transfer cross section for Xe+(2P3,2) ions in xenon. *J. Phys. B, At. Mol. Opt. Phys., 19*(4), 449–461.

Nanbu, K. (2000). Probability theory of electron-molecule, ion-molecule, molecule-molecule, and coulomb collisions for particle modeling of materials processing plasmas and gases. *IEEE Trans. Plasma Sci., 28*(3), 971–990.

Nanbu, K., & Kitatani, Y. (1995). An ion-neutral species collision model for particle simulation of glow discharge. *Jpn. J. Appl. Phys., 28*(2), 324–330.

Nanbu, K., & Wakayama, G. (1999). A simple model for Ar+ - Ar, He+ - He, Ne+ - Ne and Kr+ - Kr collisions. *Jpn. J. Appl. Phys., 38*(1(10)), 6097–6099.

Peatross, J., & Meyerhofer, D. D. (1993). Novel gas target for use in harmonic generation. *Rev. Sci. Instrum., 64*(11), 3066.

Phelps, A. V. (1994). The application of scattering cross sections to ion flux models in discharge sheaths. *J. Appl. Phys., 76*(2), 747.

Phelps, A.V. (2014). *Online collection of cross sections*. Formerly available online, JILA, Boulder, Colorado. Available on request: lvankouwen@gmail.com.

Pullins, S. H., & Dressler, R. A. (2000). Guided-ion beam measurements of Ar+ + Ar symmetric charge-transfer cross sections at ion energies ranging from 0.2 to 300 eV. *Z. Phys. Chem., 214*(9), 1279–1297.

PulsarPhysics (2011). *General particle tracer, version 3.10*. Retrieved from www.pulsar.nl/gpt.

Rademaker, G. (2014). *Fabrication and analysis of a freestanding nanofluidic channel* (Master's thesis). Delft University of Technology. Retrieved from repository.tudelft.nl.

Rapp, D., & Francis, W. E. (1962). Charge exchange between gaseous ions and atoms. *J. Chem. Phys., 37*(11), 2631.

Rejoub, R., Lindsay, B. G., & Stebbings, R. F. (2002). Determination of the absolute partial and total cross sections for electron-impact ionization of the rare gases. *Phys. Rev. A, 65*(4), 042713.

Ristivojevic, Z., & Petrovic, Z. L. (2012). A Monte Carlo simulation of ion transport at finite temperatures. *Plasma Sources Sci. Technol., 21*(3), 035001.

Robertson, S., & Sternovsky, Z. (2003). Monte Carlo model of ion mobility and diffusion for low and high electric fields. *Phys. Rev., 67*(4(2)), 046405.

Roth, A. (1990). *Vacuum technology* (second edition). New York: Elsevier.

Schiestl, B., Sejkora, G., Lezius, M., Foltin, M. P. S., & Mark, T. (1993). Interaction of rare gas ions with rare gas monomers and clusters: 1. Transverse diffusion coefficient DT/K for Ar+ in argon, and 2: Dissociation Xe+ ion capture by Ne and Ar clusters. In *Proc. 8th int. seminar on electron and ion swarm*.

Sejkora, G., Girstmair, P., Bryant, H. C., & Mark, T. D. (1984). Transverse diffusion of Ar+ and Ar2+ in Ar. *Phys. Rev. A, 29*(6), 3379.

Skullerud, H. R., & Larsen, P. (1990). Mobility and diffusion of atomic helium and neon ions in their parent gases. *J. Phys. B, At. Mol. Opt. Phys., 23*(6), 1017.

SPOC (2012). *Electron optical design, version 3.072*. Retrieved from www.lencova.com.

Stefansson, T., Berge, T., Lausund, R., & Skullerud, H. R. (1988). Measurements of transport coefficients for lithium ions in argon and helium ions in helium with a drift-tube mass spectrometer. *J. Phys. D, Appl. Phys., 21*(9), 1359.

Stefansson, T., & Skullerud, H. R. (1999). Measurements of the ratio between the transverse diffusion coefficient and the mobility for argon ions in argon. *J. Phys. B, At. Mol. Opt. Phys., 32*(5), 1057–1066.

Tang, K., & Toennies, J. (1984). An improved simple model for the van der Waals potential based on universal damping functions for the dispersion coefficients. *J. Chem. Phys.*, *80*(8), 3726.

Tondare, V. (2006). *Towards a high brightness, monochromatic electron impact gas ion source* (Doctoral dissertation). Delft University of Technology. Retrieved from repository.tudelft.nl.

Vahedi, V., & Surendra, M. (1995). A Monte Carlo collision model for the particle-in-cell method applications to argon and oxygen discharges. *Comput. Phys. Commun.*, *87*(1–2), 179–198.

Viehland, L., & Mason, E. (1995). Transport properties of gaseous ions over a wide energy range, IV. *At. Data Nucl. Data Tables*, *60*(1), 37.

CHAPTER ELEVEN

Ion emission simulations of the nano-aperture ion source

Leon van Kouwen
Department of Imaging Physics, Delft University of Technology, Delft, Netherlands
e-mail address: lvankouwen@gmail.com

Contents

1. Introduction

We apply electric and magnetic fields to tailor ion trajectories to our needs, however, the ions in the beam influence each other as well. Usually, in an undesired way. The so-called coulomb interactions have been an essential aspect of operation for many ions sources. Studies on the gallium LMIS showed that the source performance is limited by these interactions (Hagen, Fokkema, & Kruit, 2008; Radlicka & Lencova, 2008). For the ultra cold ion source, the coulomb interactions play a dominant role as well

Advances in Imaging and Electron Physics, Volume 212
ISSN 1076-5670
https://doi.org/10.1016/bs.aiep.2019.09.006

(ten Haaf, Wouters, van der Geer, Vredenbregt, & Mutsaers, 2014; van der Geer et al., 2007; Steele, Knuffman, & McClelland, 2011). One may expect negligible coulomb interaction effect as the NAIS operates with a relatively low current compared to a LMIS while the initial source size is larger. However, the extraction field in a LMIS type source is much larger than found in a NAIS configuration, so the influence can not be neglected a-priori.

Besides free-space coulomb interactions, we consider the influence of surface induced charge. It is well known that a charged particle in proximity of a conductor induces charge on the surface of the conductor. This charge can influence the trajectory of the free space particle. One example is the deflection of low energy electrons in a pinhole (Birkhoff et al., 1980). In most cases though, this effect is very weak. For the emission region of the NAIS, the particles are traveling only at about 1 eV of kinetic energy, through an aperture of the order 100 nm. Neglecting this effect may not be appropriate.

A second unconventional aspect that is studied, is the influence of the electron beam on the ion beam. Due to the large mass difference, an electron scattering on an ion does not alter the ions trajectory very much. However, since we have a very high electron current (100 nA typically), many electrons engage in scattering and a more elaborate analysis is required to determine the influence.

The regime where coulomb interactions become relevant is overlapping with the regime where ion-neutral scattering is present. Therefore, the model for ion-neutral scattering discussed in the previous chapter is included in the simulations presented here.

2. Ion-induced surface charge

A charged particle in the proximity of conducting surfaces will induce surface charge. A charged particle can not feel its own field, however, it does feel the effect of the induced surface charge. The electric field strength acting on a particle close to a conducting infinite plate in the x-y plane is simple to calculate using the method of images (Griffiths, 1999). The result is

$$\mathbf{E}_{\mathrm{q}} = -\kappa \frac{q}{4h^2} \hat{\mathbf{z}} \tag{1}$$

with h the distance between the charge and the plate and $\kappa = 8.99 \times 10^9$ $\mathrm{N\,m^2\,C^{-2}}$. This solution is useful to get acquainted with the order of magnitude of the effect. The electric field strength at 25 nm distance is

0.6 kV/mm, which is not negligible. At 5 nm it even reaches 14 kV/mm. The field strength can become equal or even exceed the typically applied electrostatic field, so the surface induced charge can certainly not be ignored a-priori.

The method of images is great for simple geometries, but is not a method that works in general. Interestingly, the method is so popular, that surface induced charge effects are often referred to as image charge.

2.1 Numerical approach

We will now consider a brute-force way to compute the self-field of a charged particle in any geometry. We assume a quasi-static situation, which means in this context that the charge in the conductor reassembles instantaneously upon ion movement. We will ignore the influence of surface charge, induced by one ion, acting on another. This assumption is justified later.

If the ion is at position $\boldsymbol{r}_{\mathrm{q}}$, the electrostatic potential in all space is described by the Poisson equation as

$$\epsilon_0 \nabla^2 \Phi(\boldsymbol{r}) = q\delta(\mathbf{r} - \boldsymbol{r}_{\mathrm{q}}), \qquad \Phi(\boldsymbol{r}_{\mathrm{s}}) = 0. \tag{2}$$

The term $q\delta(\boldsymbol{r} - \boldsymbol{r}_{\mathrm{q}})$ represents the ion as a point charge and $\Phi(\boldsymbol{r}_{\mathrm{s}})$ is the potential on the electrode surfaces. Note that even in a rotationally symmetric electrode configuration a 3D solution is required, for the point charge breaks the symmetry.

The field $\Phi(\boldsymbol{r}_{\mathrm{s}})$ is not directly useful. In fact, it is singular at the only position we would like to know it, that is, at $\boldsymbol{r}_{\mathrm{q}}$. It is, however, not this field that affects the motion of the ion. The ion is only affected by the charge it has induced on the conducting surfaces. Locally, we can compute the surface charge density from $\sigma(\boldsymbol{r}_{\mathrm{s}}) = -\epsilon_0 \left(\hat{\mathbf{a}} \cdot \nabla\Phi(\boldsymbol{r}_{\mathrm{s}})\right)$. We can use this result to calculate the self-field by summing the contributions of all surface elements at position $\boldsymbol{r}_{\mathrm{s}}$ of surface area $\mathrm{d}a$, using Coulombs law. Fig. 1A shows a sketch of the problem. The vector from a particular surface charge element to the position of the ion is $\boldsymbol{r}_{\mathrm{sq}} = \boldsymbol{r}_{\mathrm{q}} - \boldsymbol{r}_{\mathrm{s}}$, so we get

$$\mathbf{E}_{\mathrm{q}}(\boldsymbol{r}_{\mathrm{q}}) = \kappa \iint_S \frac{\boldsymbol{r}_{\mathrm{q}} - \boldsymbol{r}_{\mathrm{s}}}{|\boldsymbol{r}_{\mathrm{q}} - \boldsymbol{r}_{\mathrm{s}}|^3} \sigma(\boldsymbol{r}_{\mathrm{s}}) \mathrm{d}a. \tag{3}$$

The field obtained in $\boldsymbol{r}_{\mathrm{q}}$ space can be used to describe the motion $\boldsymbol{r}_{\mathrm{q}}(t)$ of an ion in charge-free region of space confined by electrodes. The field is a little peculiar though, as we haven't acquired it as a solution of Maxwell's

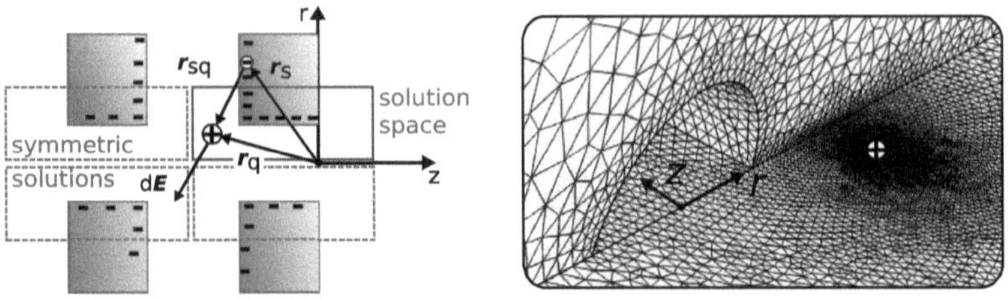

(A) (B)

Figure 1 Finite element method (FEM) for surface induced self-field effects. (A) NAIS electrode configuration cross section, showing the surface charge located at r_S giving a contribution **dE** to the free space charge at position r_q. (B) Cross sectional view of a 3D mesh for the FEM solution, showing the refinement at the location of the free charge.

equations directly. In fact $\nabla \cdot \mathbf{E}_q(\boldsymbol{r}_q) \neq 0$, so it can't be a solution of Maxwell's equations in charge-free space (Laplace equation). We will therefore call it a virtual field, as this field never physically exists. To clarify the point, if an ion is at position $\boldsymbol{r}_q$, we can not use Eq. (3) to calculate the field for some other ion at location $\boldsymbol{r}'_q$. This would neglect the direct ion-to-ion coulomb interaction, which is a topic later on in this chapter. The field at some other location can be found from $-\nabla\Phi(\boldsymbol{r}'_q)$, but we do not use this. Here we only consider surface induced self-fields because these are far more relevant than surface induced effects on neighboring ions.

We can not a priori assume that this field can be written as a gradient, i.e. $\mathbf{E}_q(\boldsymbol{r}_q) = -\nabla\Phi_q(\boldsymbol{r}_q)$. However it is quite easy to show using Leibniz' rule that

$$\Phi_q(\boldsymbol{r}_q) = -\kappa \iint_S \frac{\sigma(\boldsymbol{r}_s)}{|\boldsymbol{r}_q - \boldsymbol{r}_s|} da. \tag{4}$$

The interpretation of this potential is the work required to escape from $\boldsymbol{r}_q$ to infinity. We also conclude that $\mathbf{E}_q$ is, as expected, a conservative vector field.

Interestingly, the result of Eq. (3) is a rotationally symmetric field, if the electrode configuration is rotationally symmetric, even though the solution to Eq. (2) is not rotationally symmetric. Numerically this means we need to perform a 3D calculation of Eq. (2), for only a 2D set of values for $\boldsymbol{r}_q$. This is still computationally heavy but manageable, while a 3D set for $\boldsymbol{r}_q$ would be very unpleasant. We identify some more symmetry in our situation, which is shown in Fig. 1A.

For the numerical evaluation of $\mathbf{E}_q(\boldsymbol{r}_q)$, we have used commercial software COMSOL (2013). We perform a finite-element calculation of the

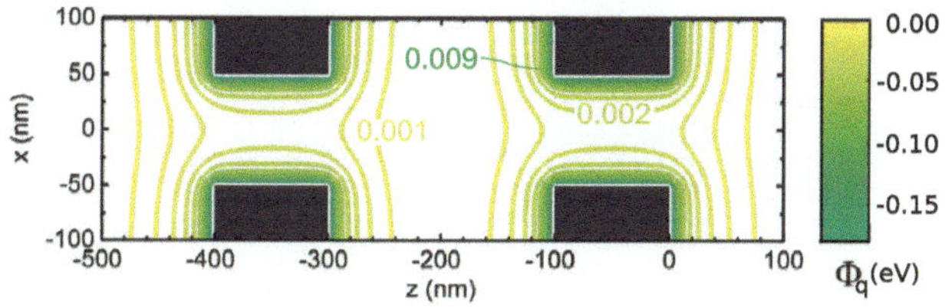

Figure 2 Electrostatic potential due to surface induced charge. The gradient of this field is the electrostatic self-field. Note that the field is virtual as it is not a solution to the Laplace equation, it is rather a look-up table for the self-induced field.

3D field for a set of 400 values of $\boldsymbol{r}_q$. The chip geometry considered has 100 nm membrane thickness, 100 nm aperture diameter and 200 nm membrane spacing. We found that it was absolutely crucial to heavily refine the mesh around the point charge. Such mesh refinement is shown in Fig. 1B. Because the free charge location is parameterized, we need to recalculate a new mesh for every solution. In total, such a field calculation took about one day on a decent desktop PC (CPU: Intel Xeon E5-1620). The resulting virtual electrostatic potential field is shown in Fig. 2.

We acknowledge that a boundary element method would be more natural to solve the surface induced self-field, as it directly gives the surface charge and doesn't suffer from the singularity at the charge location. We have used a finite element method (FEM) solver because it was more easily available to us and can do the job as well. We have tested the method by simulating a particle-plate configuration. We found excellent agreement between the numerical solution and the well-known analytical result for this situation. In the NAIS geometry, further mesh refinement did not yield noticeable change in the field hence we are confident the results are accurate.

2.2 Analytical approximation: method of images

The FEM method described here is accurate, but not very convenient. It is tedious to set up and it takes a day to accurately compute the field of one particular geometry. It is therefore worthwhile to consider alternatives. In the center of the chip, the surface induced charge effects mostly cancel out. We noticed that all relevant effect occurs in the exiting aperture, so maybe the effect can be simplified by only considering the effect in a cylindrical geometry. Analytical approaches to the point-charge and cylinder system are available, (Hernandes & Assis, 2005; Bouwkamp & De Bruijn, 1947), but these are not at all simple.

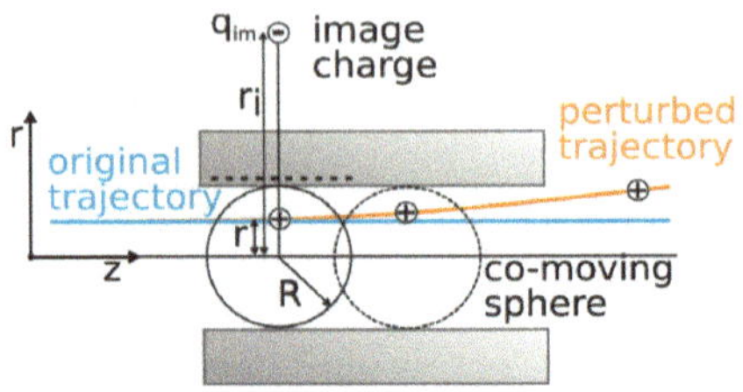

Figure 3 Approximate image charge configuration for the point charge in a cylinder. The image charge solution for a point charge and a sphere is moving along with the charge through the cylinder. This gives a good approximation because the forces in the z-direction are in both cases canceled out.

We may try to find an image charge configuration to solve the situation of a point charge in a cylinder. It can be shown, unfortunately, that no such image charge equivalent situation exists. It seems unlikely that a multiple-charge image can be found either.

We can, however, boldly try to employ the well-known solution for a sphere and a point charge. A sphere that just fits inside a cylinder, positioned at the same z-coordinate as the charge will attain quite similar surface induced charge forces. See Fig. 3 for the visualization of this concept. Of course the sphere deviates from the cylinder, however the force in the z-direction cancels out in both cases so this a-symmetry is partially irrelevant.

If we place a virtual charge of magnitude q_{im} at r_{im} according to

$$q_{\mathrm{im}} = -q\frac{R}{r_{\mathrm{q}}}, \qquad r_{\mathrm{im}} = \frac{R^2}{r_{\mathrm{q}}} \tag{5}$$

we obtain the image charge solution of a point charge inside a sphere (Griffiths, 1999). The acquired self-field is then

$$\mathbf{E}(\boldsymbol{r}_{\mathrm{q}}) = \kappa q \frac{R r_{\mathrm{q}}}{\left(R^2 - r_{\mathrm{q}}^2\right)^2}\hat{\boldsymbol{r}}. \tag{6}$$

We have acquired an extraordinary simple equation when compared to the full FEM solution method. It is of course not clear how good of an approximation this equation is. If $r_{\mathrm{q}} \to R$, the cylinder and the sphere case both converge to the infinite plate case. On-axis, the field is zero in both cases. At least the limiting cases are equal.

In Fig. 4, a ray trace example comparing the full FEM calculation method and the simple analytical model given here is shown. In this ex-

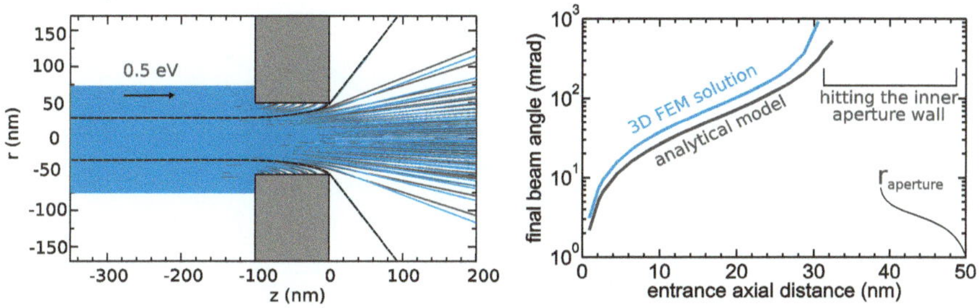

Figure 4 Test case for inspecting surface induced charge effects. A parallel beam with 0.5 eV kinetic energy fills a, perfectly conducting, aperture. A computational model and an analytical approximation are compared by using identical starting conditions for both ray traces. No other fields are applied. The dashed lines indicate that trajectories that are on the verge of escaping.

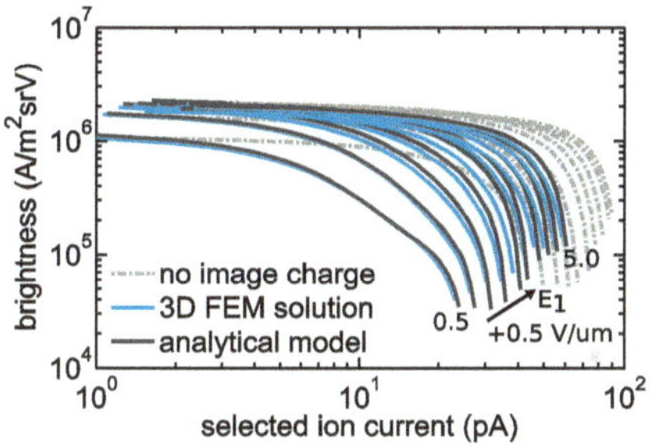

Figure 5 A comparison of two models for surface induced charge effects reveal good agreement between the models. This simulation corresponds to a spacing × thickness × aperture = 200 × 100 × 100 nm and inside and outside chip fields of $E_1 = E_2 = 1$ kV/mm.

ample a uniform, 0.5 eV mono-energetic argon beam travels through a 100 × 100 nm aperture. The correspondence can be considered quite good upon visual inspection in Fig. 4A; both the black and the blue rays behave similarly. The obvious notion for both models is that the beam is getting divergent and that the trajectories close to the aperture walls are drawn towards the wall. In Fig. 4B, we see more quantitatively that the analytical model predicts a weaker final beam angle. Nevertheless, the range in the axial entrance distance from which the ions get attracted onto the inner aperture wall is remarkably close.

A more direct way of testing the suitability of the analytical model is simulating a full NAIS configuration and compare what we are after in the end: optical properties. Fig. 5 shows the result of ray-tracing and we

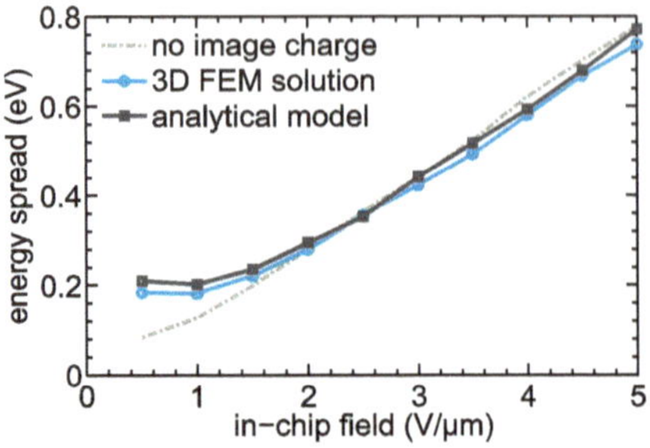

Figure 6 The comparison of the energy spread in the beam reveals that hardly any energy spread is introduced due to the surface induced charge, except for the very lowest in-chip fields. No noteworthy difference between the two models is observed. This simulation corresponds to a spacing-thickness-aperture combination of 200 × 100 × 100 nm and inside and outside chip fields of $E_1 = E_2 = 1$ kV/mm.

conclude that both models correspond very well to each other in terms of brightness. Also included is a simulation without any model for surface induced charge effects. Clearly, the effect is important for an accurate description of the ion emission. The brightness at low current is not very much affected, though.

The work required to escape from a NAIS chip is described by Eq. (4). Fig. 2 already shows that the effect of this should be rather limited. Only ions starting very close to the walls require more energy than a few tens of meV, but these are unlikely to escape the chip geometry anyway. We can inspect the chromatic effect of the image charge directly by looking at the fw50 energy spread of the ray trace simulation in Fig. 6. The energy spread contribution is negligible, except at the lowest bias voltage, where in this configuration it gave a 200 meV contribution. Again, we see no difference between the methods. Since the analytical model only considers the aperture, this energy spread contribution must all be due to particles generated in the aperture.

Now we have gained confidence in the applicability of the analytical model we can investigate it a bit further. We expand Eq. (6) and get

$$\mathbf{E}(r_\mathrm{q}) \approx \kappa \frac{q}{R^2}\left(\frac{r_\mathrm{q}}{R} + \frac{2r_\mathrm{q}^3}{R^3} + \frac{3r_\mathrm{q}^5}{R^5} + O\left(r_\mathrm{q}^7\right)\right)\hat{\boldsymbol{r}} \tag{7}$$

We can see essentially a first order lens effect and higher order aberrations. Note that the higher order terms are not decreasing in magnitude very rapidly. At $r_\mathrm{q} = \frac{1}{2}\sqrt{2}R$, the third order is already as large as the first order.

The image charge effects influences the center of the beam only slightly, while strongly attracting the outer beam part close to the aperture walls.

We have learned that the surface induced charge can lead to significant displacement of the ion trajectories. It should therefore be included in further simulation of the NAIS. The full calculation is a comprehensive process, but the analytical approximation gives good agreement for the brightness and the energy spread and can therefore be used in further study. Making the aperture a little bigger than 100 nm rapidly leads to less influence of the effect. A higher in-chip field and thinner membranes help to suppress the effect as well. Note that the particle mass does not influence the trajectory in an electrostatic field, so the effect is equal in magnitude for all ionic species.

3. Monte-Carlo ray tracing

3.1 Ion-ion interaction

One convenient way to calculate the ion-ion coulomb interactions is using the extended-two-particle approximation developed by Jansen (1988) (see also Kruit and Jansen (2008)). In situations where the beam envelope and beam voltage are not constant, an extension to the theory called the slice-method is required (Jiang, 1996). Unfortunately, the emission region of the NAIS does not fulfill the requirements for the extended-two-particle approximation to be valid. The angular current density needs to be uniform, which is not what we get out of a nano-aperture ion source. Furthermore, we can not take into account the details of the chip geometry such as volumetric emission or membrane blockage.

The most direct method of calculating coulomb interactions between particles in a beam, is summing over all the pair-wise coulomb forces:

$$\mathbf{F}_i = \kappa q_i \sum_{j=1}^{N} q_j \frac{\boldsymbol{r}_i - \boldsymbol{r}_j}{|\boldsymbol{r}_i - \boldsymbol{r}_j|^3}. \tag{8}$$

This can be used directly in a ray-tracing scheme. One big disadvantage is the inherent slowness of such a calculation; the number of calculations scales with the square of the number of particles involved. In addition, one needs quite some particles for decent statistics as the degrees of freedom of starting conditions is much larger than for non-interacting particles.

One method of reducing the calculation time is grouping particles such that only their net effect is used in the force calculation. This is called the Barnes-Hut algorithm (BH). We have used commercial software (PulsarPhysics, 2011), which offers both the full pair-wise calculation and the BH simplification. The grouped particles are approximated up to the fourth order multipole moment and a variable θ indicates the crudeness of particle bunching for which 1 is very crude and 0 is equal to the full pair-wise calculation.

We have found that in NAIS simulations the BH approximation gives identical results as the full calculation even for $\theta > 0.5$. To be on the safe side we have used $\theta = 0.25$. The total emission current of the NAIS is relatively low, so the coulomb interaction is mostly caused by the nearest neighboring particle, hence the BH approximation is very natural.

The method of acquiring the optical properties of the beam after ray tracing is described in the chapter on Optics of ion emission from a volume, found in this publication. Gas scattering is included in ray tracing as well, and the details of the implemented model can be found in the previous chapter. The total number of particles traced for a typical set of parameters is 20k, which gives enough statistics to also study a fraction of the beam.

It is important to keep the time step smaller than the average time between the generation of particles to avoid unphysical particle bunching. We typically set $\Delta t < 0.3\frac{e}{I}$. Note that unphysical bunching is partially prevented automatically because of the volumetric emission. This time-step requirement is checked in addition to the time step requirements discussed in the previous chapter.

3.2 Nominal geometry using argon

For understanding the main optical effects due to coulomb interactions we simulate a chip geometry with spacing × thickness × aperture $= 300 \times 100 \times 150$ nm, using argon ions. This is considered a production feasible, ion optically decent, geometry. In order to avoid trouble with lens effects, we would like to have high and equal fields. We set $E_1 = E_2 = 10$ kV/mm. The ebeam is assumed to be Gaussian shaped in the xy-plane with a 50 nm fw50 and 100 nA of current. The e-beam divergence and the influence of ionization on the e-beam current density is ignored. We interpret this simulation configuration as what is realistically achievable if the full system is well under control and up to spec. See Table 1 for a summary of the settings.

Ultimately we are interested in the most realistic simulation. This needs to include the electric fields, ion-neutral scattering, ion-ion interaction

Table 1 Simulation settings for Monte-Carlo ray tracing as used throughout this chapter unless otherwise stated.

Φ_I	10	kV	Φ_e	1.0	kV	spacing	300	nm
E_1	1×10^7	V/m	I_e	100	nA	thickness	100	nm
E_2	1×10^7	V/m	d_e	50	nm	aperture	150⊘	nm
T	293	K	shape	Gaussian		species	argon	
N	20 000							

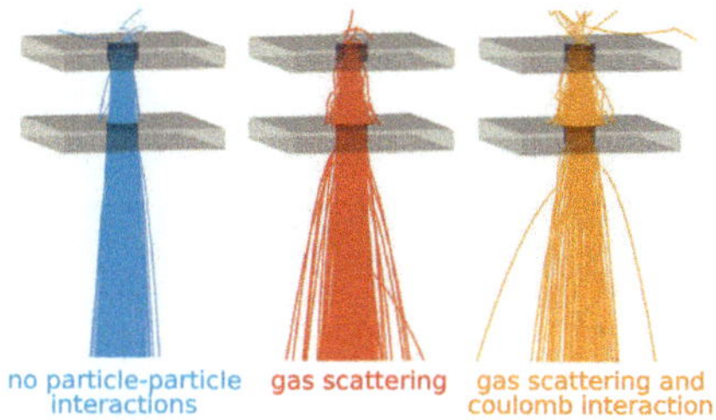

Figure 7 Trajectories of argon ions for three cases. No interaction yields only smooth trajectories and a narrow emitted beam. Including gas scattering broadens the beam and more particles are blocked. When including coulomb interactions, further broadening of the beam is observed.

and surface induced coulomb interaction simultaneously. For better understanding, we compare such a complete model to a situation where no particle interaction are included and to one where only gas scattering is present. Fig. 7 shows what a typical set of simulated trajectories looks like.

Fig. 8 shows the simulated practical brightness as formed by the inner 50% of the beam. Different emission currents are obtained by changing the particle density in the chip. A dry-run ignoring all particle-particle interactions shows the expected result, that is, the brightness scales linearly with the current indefinitely. In the low current regime, this dry run coincides with the particle interaction simulations and it is possible to increase the current and the brightness for free by increasing the particle density. The ion-neutral interaction included simulation shows a reduction of the brightness if the current of the beam is increased even further. It is very interesting to see that the ion-neutral interaction alone does not seem to form a true limit of the brightness as increasing the beam current keeps on giving a higher brightness. It is the coulomb interactions that pose the limit to the brightness ultimately. Note that both the ion-neutral and ion-ion interactions begin to become important more or less at the same current,

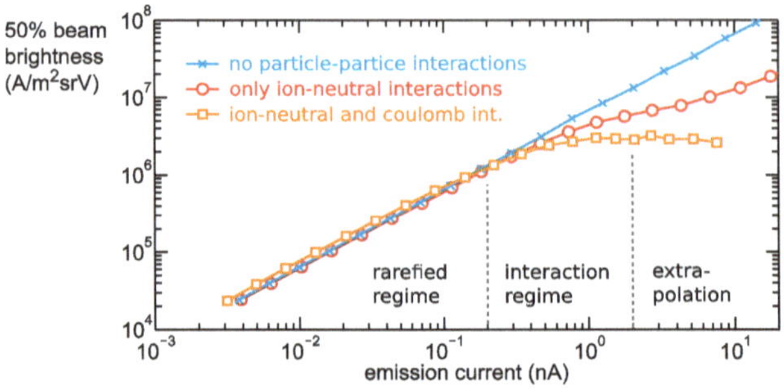

Figure 8 Monte-Carlo ray tracing result of 20k argon ions. The current in the beam is varied by using different particle densities in the chip. The 95% confidence interval for the statistical precision of the simulation is smaller than the marker sizes.

which is truly a coincidence. We can reach 1×10^6 $\mathrm{A\,m^{-2}\,sr^{-1}\,V^{-1}}$ without being influenced very much by any particle interaction.

The effect of surface induced charge is best seen in the low current regime as particle-particle interactions are negligible here, while the surface induced charge effects do not depend on the particle density. The coulomb interaction included curve is not exactly converging towards the other curves. We observe a reduction of total emission current, however, without deteriorating the brightness. This means that the current was mostly scraped off from the outer part of the beam, which we can expect by looking at Fig. 4. It seems as if the surface induced charge effect is actually beneficial, as the brightness is higher in the lower current regime. This would be a misconception because ion-neutral scattering occurs at a lower current as well, so the peak brightness is not higher. This confusion would be removed by plotting the brightness versus iKn^{-1}, but here we would like to have a direct beam property on the axis.

Currents above 2 nA in this simulation should be considered an extrapolation of the model. At such high currents, the approximation for a uniform e-beam in z becomes inaccurate and the effects of electron multiplication and redirection during electron impact ionization should be taken into account for more accurate results. The background neutral particle density is also no longer well described by molecular flow. Furthermore, the required pressure for obtaining > 2 nA emission current for this chip is in the several bar range, which may pose a danger for chip integrity and may also give a too high vacuum load.

Fig. 9 shows three building blocks of the brightness: the virtual source size, the divergence angle and the beam current. The escape probability

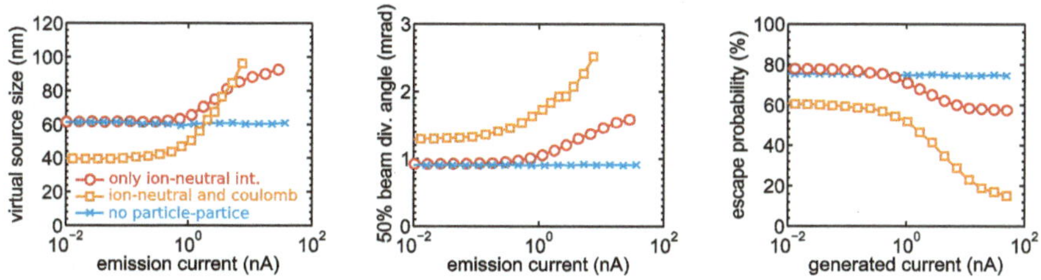

Figure 9 The virtual source size and the beam divergence angle characterize the shape of the beam. We can see the influence of gas scattering and coulomb interactions as we increase the emission current by increasing the particle density. The escape probability indicates what fraction of the generated current makes it out of the chip.

represents the fraction of current that can escape the chip and is plotted versus the total amount of generated current. In the lower current regime, the surface induced charge is responsible for lowering the virtual source size and increasing the divergence angle; essentially a lens effect. It furthermore lowers the escape probability, yielding a lower emission current. At higher currents, both ion-neutral and coulomb interaction effects result in virtual source size growth, added divergence and a lowering of the escape probability. Quite often, brightness loss can completely be attributed to trajectory displacement, but for the NAIS we find it is a combination of all three brightness constituting quantities.

Fig. 10 show the brightness versus beam current as selected by an aperture after acceleration to high voltage. These curves give more detailed information than Fig. 8. We see that the saturation of the 50% beam brightness when increasing the particle density also holds for the axial practical brightness (or say, 5% beam brightness). Note that we have transformed the particle density to the inverse ionic Knudsen number (as introduced in the previous chapter).

We may characterize the shape of the curves in Fig. 10 by the ratio between the 50% brightness and the 5% beam brightness. In an idealized configuration, where no particles are blocked by the membrane and no particle-particle interactions exist, we found that the ratio between the 50% brightness and the 5% beam brightness is 0.7 (see the chapter on Optics of ion emission from a volume). In this geometry, where we do include membrane blockage the membrane blockage, the ratio in the low current regime is also about 0.7 and only weakly changes to 0.6 in the particle interaction dominated regime. We conclude that it is a quite general result that using only a fraction of the beam gives a slightly higher brightness than

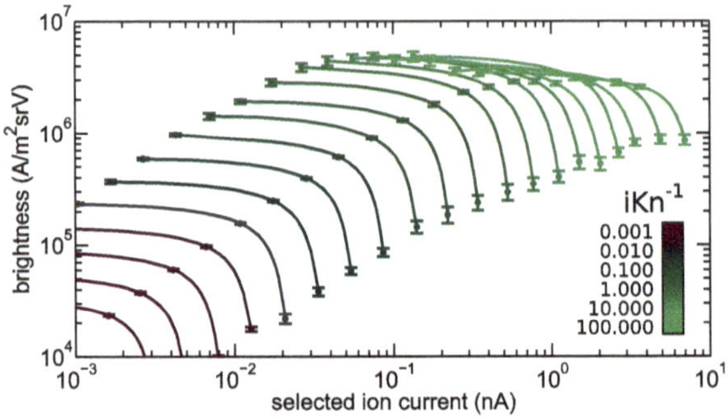

Figure 10 Brightness-current performance curves of a typical chip geometry for a wide range of particle densities as found by Monte-Carlo simulation. Ion-neutral scattering, ion-ion interaction and surface induced charge interaction are included. The error bars indicate the 95% confidence intervals of the statistical precision.

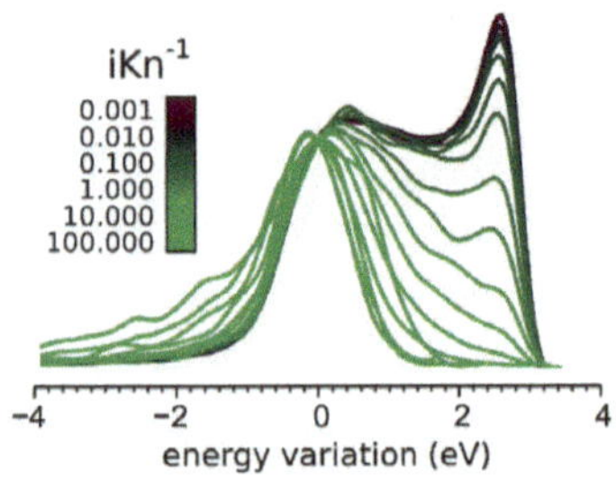

Figure 11 The energy intensity diagrams for various particle densities as indicated by the inverse ionic Knudsen number. The indicated energy is relative to the acceleration voltage. The distributions are arbitrarily normalized such that the shapes can be compared.

using 50%. Trying to use nearly 100% of the beam current gives a rapid drop of the average brightness though.

Besides brightness reduction, coulomb interactions may introduce energy spread via the Boersch effect. Fig. 11 shows the distribution of the energy for a wide range of particle densities. Keep in mind that the curves are normalized for distribution shape comparison; the higher iKn^{-1} beams contain more current. At low iKn^{-1}, the energy distribution is formed by the ion initial positions, the potential landscape and the escape probability. The width of the distribution is loosely speaking equal to the bias voltage. At high voltage, we loose some ions from the back and we transfer some of the acquired energy through neutral particle collisions. Both effects lower the right hand side of the distribution. The amount of ions made in front

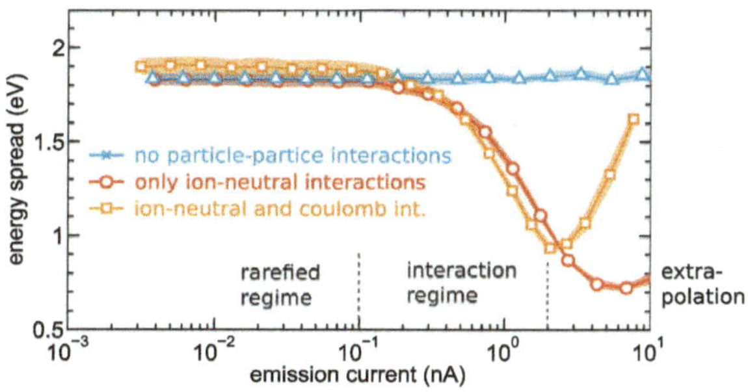

Figure 12 Monte-Carlo ray tracing result of 20k argon ions. The current in the beam is varied by using different particle densities in the chip. The shaded area indicates 95% confidence interval for the statistical precision of the simulation.

of the chip becomes relatively more important and a larger left-hand-side tail is formed.

The described effects are perhaps more clearly captured in Fig. 12. At the lower currents, the non-interacting regime, no particle density dependence is present. This regime corresponds to the darkest curves in Fig. 11. For increasing iKn^{-1}, we see a decrease as fewer ions from the back maintain their energy up to the exit. At the same time, when increasing the particle density, relatively more particles are generated in front of the chip and the contribution of the Boersch effects increases. The combination of the different effects lead to a minimum.

We have encountered a pleasant surprise; we seem to be able to decrease the energy spread by a factor of two by simply increasing the particle density. Increasing the particle density too much causes the Boersch effect to rapidly increase the energy spread.

3.3 Comparing chip geometries

We have studied a chip geometry that was considered nominal, but is it really the best choice? What happens if we can't produce these precise dimensions? In this section we explore the influence of geometrical variation of the chip.

The typical spot size of the electron beam in a NAIS optimized situation dictates typical aperture sizes of 100 to 500 nm. High fields give low coulomb interactions, low aberrations, and minimize gas scattering, so, preferably we maintain 10 kV/mm. In order to achieve a low energy spread and a high field strength, the spacing between the membranes should be

Table 2 Dimensions of the set of chips used for tray trace simulations. The first column is used as a reference to a particular geometry. The effective length is calculated using the neutral particle density distributions have been calculated (see the previous chapter for details).

chip	l nm	t nm	d nm	l_{eff} nm
S-1	100	20	100	155
S-2	100	100	100	226
M-1	300	100	150	428
M-2	300	100	300	498
L-1	1000	100	150	1111
L-2	1000	100	500	1409

kept small, typically smaller than 1 μm. The fabrication challenges dictate the minimal membrane thickness, for which 100 nm is certainly feasible.

We have chosen a set of six geometries that are considered sensible configurations to explore in more detail. The dimensions are given in Table 2. Chip S-2 is a fairly small chip and S-1 explores the potential benefits of very thin membranes. Chip M-1 is considered nominal and M-2 deviates by allowing more current to escape due to the bigger aperture. Chip L-1 and L-2 are exploring the boundaries of maximum chip size. In all simulations we set $E_1 = E_2 = 10$ kV/mm. The simulations are computationally costly so a more comprehensive full geometrical optimization is not easily achieved. Fortunately, this set of geometries is sufficient to get a good grasp of the relevant effects.

Fig. 13 shows the results of several ray tracing simulations of argon ions as a function of the inverse ionic Knudsen number. Because we are comparing different geometries, plotting as a function of particle density or emission current makes comparing rather unclear, while at equal ionic Knudsen number we are comparing the same amount of ion-neutral scattering. The beam properties are plotted together in one figure because the properties are linked and vary together. One can draw vertical lines and obtain a set of properties that is simultaneously achievable. Note that when $\text{iKn}^{-1} > 10$ one should be somewhat cautious in interpreting the results because the assumptions in the model become compromised, as already discussed in the previous section.

The easiest thing to look at first is the total current. Directly obvious is the continuous increase of the current with iKn^{-1}. One may expect that gas scattering, possibly in combination with coulomb repulsion, leads to a saturation or even a decrease. It turns out that the increase of produced current outweigh the additional loss mechanisms. We do see the increase getting more shallow, particularly for chip L-1, which has a small aperture to spacing ratio. An emission current of 1 nA seems feasible for any chip. 10 nA seems reachable too if the chip has a large enough aperture, but we are reaching the extrapolation regime of the model.

We can inspect the differences in emission current more easily by looking at the escape probability. The obvious notion is that chips with larger apertures allow more ions to escape. Secondly, we see that the aspect ratio of aperture-to-spacing can best be large for current maximization. The NAIS trajectories are not scalable because the initial conditions do not scale along when scaling the geometry and potentials. Therefore, it is predominantly the absolute aperture size and not the aspect ratio which determines the escape probability. Over a three orders of magnitude variation in iKn^{-1}, the escape probability is only a factor of two lowered at most. The reduction of escape probability due to gas scattering is not nearly as dominant as one may suspect.

Perhaps most important is the observation that $B > 1 \times 10^6$ $\mathrm{A\,m^{-2}\,sr^{-1}\,V^{-1}}$ seems to be achievable with any chip. All brightness curves show comparable behavior with respect to one another. The small differences between the chips in the lower current regime is a combination of differences in surface induced coulomb interactions and differences in the shape of the electric field. A noteworthy difference between chips is higher current for larger apertures at the cost of some brightness loss.

A little peculiar is the dipping behavior of chip L-2 and weakly that of chip M-2. Interestingly, when running the simulation without coulomb interactions, this dip remains. The straightforward explanation is that the decreasing mechanisms outweigh the increase in generated current. However, from a phase space point of view, this is not satisfying. When iKn^{-1} is increased, the phase space density can only increase. This is best illustrated by an example: if the inverse ionic Knudsen number is doubled, the front half of the chip now effectively generates the same current as the full chip at half the iKn^{-1}, and the amount of scattering in this front half is also the same. The phase space density produced by this front half is thus the same as the original situation, but in addition, the particles from the back half increase the density too, although less due to more scattering. A higher phase space density means a higher brightness. So the question is, how can the brightness ever decrease with increasing iKn^{-1}.

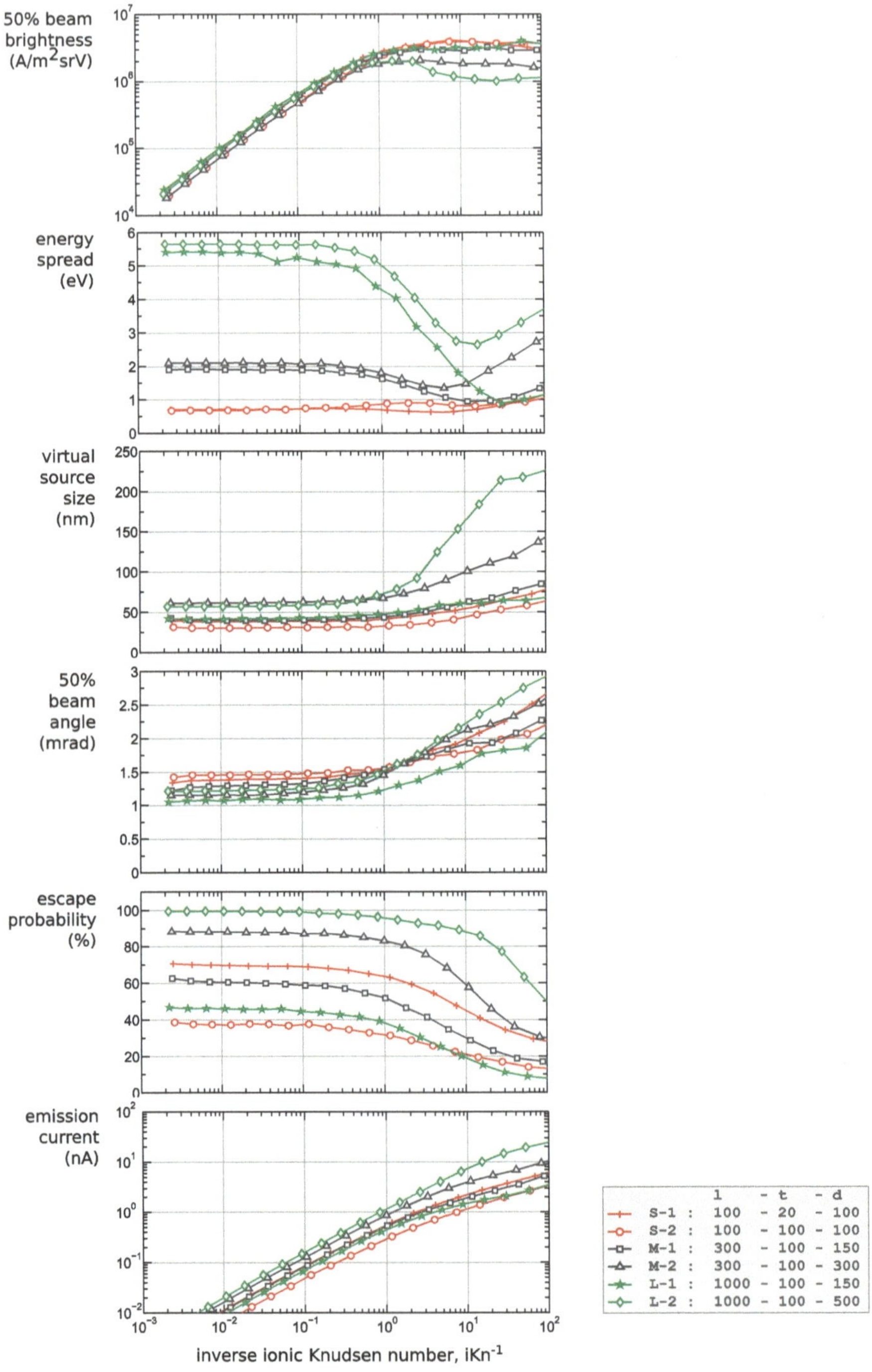
50% beam brightness (A/m2srV)
10^7
10^6
10^5
10^4
energy spread (eV)
6
5
4
3
2
1
0
virtual source size (nm)
250
200
150
100
50
0
50% beam angle (mrad)
3
2.5
2
1.5
1
0.5
0
escape probability (%)
100
80
60
40
20
0
emission current (nA)
10^2
10^1
10^0
10^-1
10^-2
10^-3
10^-2
10^-1
10^0
10^1
10^2
inverse ionic Knudsen number, iKn-1
l - t - d
S-1 : 100 - 20 - 100
S-2 : 100 - 100 - 100
M-1 : 300 - 100 - 150
M-2 : 300 - 100 - 300
L-1 : 1000 - 100 - 150
L-2 : 1000 - 100 - 500

Figure 13 Results of Monte-Carlo ray tracing of 20k argon ions for six different chips of characteristic dimensions. These simulations include the electric field in an outside the chip ($E_1 = E_2 = 10$ kV/mm), ion-neutral gas scattering, surface induced charge effects, and ion-ion coulomb interactions. All curves share the ionic Knudsen number on the x-axis, which is discussed in detail in the previous chapter; it represents roughly the number of ion-neutral scatter events for ions starting at the back of the chip. Varying the ionic Knudsen number is achieved by varying the neutral particle density. The brightness plotted is calculated by using 50% of the beam current as selected by an aperture down the column. The energy spread is measured as a fw50 of the 1D probability density function. The virtual source size is the 2D fw50 as formed by 50% of the total emitted current. As a measure for the beam divergence we use the angle with respect to the optical axis such that the inner cone contains 50% of the total beam current. The ion escape probability is an average for all starting positions. Any particle hitting the membrane is assumed to be neutralized and removed from the simulation.

Well, in fact, the differential brightness does increase. It is the practical brightness that is plotted here which decreases. The virtual source size grows due to scattering, and although the central phase space density does keep increasing, the overall average beam property, the practical brightness, does not. In normal beam lines we select the beam divergence angle, but use the full source size. However, we realize now it would be very convenient to do source area selection here as well, as by selecting the central part of the field, we select the high phase space density part and obtain a brighter beam.

Actually, we do have a real space capping aperture installed in a NAIS: the exit aperture itself. This explains why we don't see the dip in the smaller apertured configurations. It also explains why the dipping behavior is reversed when the large aperture starts to become limiting again. Note that the exit aperture is not only a real space limiting object, it also tends to cap off the larger angles. However, often we use only a small fraction of the central part of the beam. So, the exit aperture blocks current, but this isn't necessarily a bad thing; it mostly gets rid of unwanted current. It can be thought of as a phase space selecting window, with somewhat fuzzy boundaries. Due to this aperture capping, we never expect a virtual source size larger than the aperture size, except when trajectory displacement after the ions escaped the chip is important.

The energy spread shows a rather large difference between chips. The spread is mostly determined by the aperture spacing, as we keep the field constant. So, in general a larger energy spread is obtained for the larger spaced chips, although it can be reduced by working under reasonably high particle densities. The occurrence of a minimum was expected and dis-

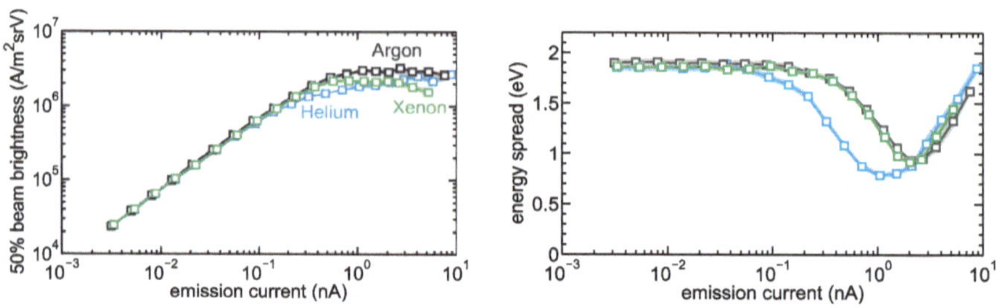

Figure 14 Monte-Carlo ray trace result for comparing the expected performance of helium, argon and xenon. The geometry used has spacing × thickness × aperture: 300 × 100 × 150 nm (chip M-1), and the fields $E_1 = E_2 = 10$ kV/mm. The statistical error (95% confidence interval) is indicated by the shaded areas.

cussed in the previous section. Note that the minimum does not occur for the smallest chips because the increased spread due to ions made in front of the chip is relatively more important.

Even though at high Knudsen numbers the simulation becomes a little unreliable, we are confident in the conclusion that brightness reaches a limit. Surely the inaccuracies lead to some shifting of the brightness curves but in whatever way the beam properties are formed, the coulomb interactions will limit the final beam. Looking at the specific shortcomings of the model, there is no reason to believe that the inaccuracies will cause a different conclusion. This is supported by the fact that fairly different chip geometries over a wide iKn^{-1} range lead to a comparable brightness.

3.4 Helium, argon, and xenon compared

One of the main assets of the NAIS concept is the ability to use different gases. It is not obvious that other gases give the same performance as argon, since the ionization cross section and the ion-neutral interaction are different. In the introductory chapter we concluded that in the rarefied limit, the brightness is proportional to the ionization cross section. This means a 10 times higher brightness for xenon compared to helium. When including the molecular diameter to account for ion-neutral interactions, the difference was reduced to a factor 3.5.

When investigating the higher particle density regime in a more comprehensive way, we find that the difference in attainable brightness between helium, argon, and, xenon is even less pronounced. Fig. 14 shows the result of a Monte-Carlo ray trace simulation including ion-neutral scattering, ion-ion repulsion, and surface induced charge interaction. The geometry was spacing × thickness×aperture: 300 × 100 × 150 nm (chip M-1).

We find the maximum attainable brightness to be close to 2.5×10^6 $\mathrm{A\,m^{-2}\,sr^{-1}\,V^{-1}}$ for all three species. It appears that the ion-neutral interaction and the coulomb interactions are balancing out the difference in ionization cross section. This precise balancing effect is not a general result, as different fields and a different geometry can cause larger discrepancies. Nevertheless, for most configurations we do not expect differences in brightness larger than a factor of two.

The energy spread does show different behavior for helium compared to argon and xenon. The dipping behavior of the energy spread occurs at a lower emission current. Since the ionization efficiency of helium is rather low, comparable gas scattering effects start becoming relevant at lower currents.

3.5 Increased electron beam current

So far we have assumed a 100 nA electron beam being focused in a 50 nm fw50 spot. This is based on a good electron column in combination with a Schottky electron gun. There may be tricks, however, to increase the current in the electron beam. We simulate one example to explore the benefits of engaging in the development of higher ebeam current. The same configuration as used for the simulations in Section 3.2 are used here, expect that the electron beam current is set to 300 nA instead of 100 nA.

The maximum acquired 50% beam brightness found is $B = 5.0 \pm 0.3 \times 10^6$ $\mathrm{A\,m^{-2}\,sr^{-1}\,V^{-1}}$, in combination with a total emission current of $I =$ 1.6 nA. The accompanying energy spread is $\Delta\epsilon = 1.6$ eV. The brightness has increased by a factor of about 1.7 compared to a 100 nA ebeam. One may expect the proportional factor 3, but because coulomb interactions play such an important role, the advantage is not nearly as great as in a non-interacting configuration.

Interestingly, the brightness is no longer attaining at a plateau. For currents $I > 1.6$ nA the brightness decreases with increasing current. For example, at $I = 10.3$ nA, the brightness is $B = 3.1 \pm 0.2 \times 10^6$ $\mathrm{A\,m^{-2}\,sr^{-1}\,V^{-1}}$. The energy spread slightly lowers to $\Delta\epsilon = 1.1$ eV at such a high current.

3.6 Comparing operation conditions

In the previous sections we have assumed up-to-spec electrostatic fields and an up-to-spec electron beam probe size. These conditions are realistic and can be considered nominal. For comparison we consider some different configurations in this section. We investigate weak fields inside and outside the chip, a fairly large geometry (spacing × thickness × aperture: 1300 × 200 × 400 nm), and a large ebeam probe size.

From earlier simulations we know that a decent choice for the inverse ionic Knudsen number is in the range 1 to 10 (see Fig. 13). We fix the ionization efficiency to $\eta_{tot} = 2.94\%$ yielding, for this particular chip geometry with effective length 1.593 um, an iKn^{-1} between 2.7 and 4.7. The required particle density maximum is 2.36×10^{24} m^{-3}. The results of a set of Monte-Carlo ray trace simulations using these settings is shown in Table 3. We take into account ebeam blockage by the top membrane, since a 200 nm ebeam fw50, and a 400 nm aperture diameter results in 6.3% of ebeam current loss.

Table 3 lists the results of a large set of different configurations simulated. The table can serve as a reference and we discuss only the most important conclusions.

If the ebeam is nominal, but we use relatively weak fields of $E_1 = E_2 =$ 1 kV/mm, we end up with a brightness of 2×10^5 $A\,m^{-2}\,sr^{-1}\,V^{-1}$, a tenfold decrease with respect to a 10 kV/mm situation. We can gain a lot by increasing only E_2 to 3 kV/mm, but preferably to 10 kV/mm. This indicates that brightness is mainly lost through trajectory displacement in the acceleration region. Other factors that reduce the brightness are the chip aberrations and in-chip coulomb aberrations are less important as increasing E_1 from 1 to 3 kV/mm as well hardly improves the brightness.

Rather surprisingly, $E_1 = 0$ kV/mm performs reasonably well. We note that the extraction current becomes rather low, such that coulomb interaction are less important. Furthermore, E_2 can penetrate the chip and offer some ion acceleration.

We should clearly not use $E_1 > E_2$. This leads to large chip aberrations, and it extracts a large current, that under a weak acceleration field, leads to coulomb interactions. On top of that, the energy spread becomes large.

We now consider the situation of a 200 nm ebeam fw50. If the fields are $E_1 = E_2 = 1$ kV/mm, we drop a factor two in brightness compared to 50 nm ebeam. This is much less of a deterioration than purely based on the ebeam current density drop, which would give a factor 16. Because the brightness is influenced by gas scattering and coulomb interactions, the influence of the ebeam size is less pronounced. When using stronger fields we see the same trends as for the 50 nm ebeam, but with a somewhat lower brightness overall.

Table 3 Monte Carlo ray trace results for chip geometry spacing × thickness × aperture: 1300 × 200 × 400 nm. The comparison is made between a nominal 50 nm ebeam fw50 and a 200 nm fw50 and a variety of in and out of chip fields are considered. Throughout these simulations the total ionization efficiency is set to 2.94 %, which is achieved by setting $n = 2.36 \times 10^{24}$ m^{-3}. An electron beam current of 100 nA at 1 keV landing energy was used.

d_e nm	E_1 V/μm	E_2 V/μm	iKn^{-1}	I nA	B_{50} A/m^2 sr V	$\Delta\epsilon$ eV	d_v nm
50	1	1	4.7	0.41 ± 0.01	$2.1 \pm 0.2 \times 10^5$	0.30 ± 0.02	139 ± 6
	1	3	4.7	0.52 ± 0.01	$5.4 \pm 0.3 \times 10^5$	0.41 ± 0.02	139 ± 3
	3	3	3.5	1.06 ± 0.02	$5.6 \pm 0.4 \times 10^5$	1.12 ± 0.02	135 ± 5
	1	10	4.7	0.72 ± 0.01	$7.1 \pm 0.5 \times 10^5$	0.61 ± 0.02	117 ± 3
	3	10	3.3	1.26 ± 0.02	$1.0 \pm 0.1 \times 10^6$	1.20 ± 0.02	180 ± 3
	10	10	2.7	2.17 ± 0.02	$1.6 \pm 0.0 \times 10^6$	4.40 ± 0.12	117 ± 2
	0	1	–	0.22 ± 0.01	$2.1 \pm 0.2 \times 10^5$	0.18 ± 0.01	113 ± 3
	0	10	–	0.40 ± 0.02	$6.0 \pm 0.8 \times 10^5$	0.80 ± 0.09	91 ± 3
	3	1	3.5	0.93 ± 0.02	$9.4 \pm 0.7 \times 10^4$	1.09 ± 0.03	235 ± 9
	10	1	2.7	1.94 ± 0.01	$2.8 \pm 0.1 \times 10^4$	4.44 ± 0.09	477 ± 9
200	1	1	4.7	0.39 ± 0.01	$1.2 \pm 0.1 \times 10^5$	0.36 ± 0.02	188 ± 8
	1	3	4.7	0.47 ± 0.02	$2.1 \pm 0.1 \times 10^5$	0.45 ± 0.02	204 ± 4
	3	3	3.5	0.83 ± 0.01	$3.0 \pm 0.1 \times 10^5$	1.28 ± 0.04	183 ± 5
	1	10	4.7	0.65 ± 0.01	$2.5 \pm 0.3 \times 10^5$	0.59 ± 0.03	131 ± 6
	3	10	3.5	1.02 ± 0.01	$3.5 \pm 0.2 \times 10^5$	1.26 ± 0.06	189 ± 3
	10	10	2.7	1.57 ± 0.02	$4.7 \pm 0.2 \times 10^5$	4.76 ± 0.14	187 ± 3
	0	1	–	0.21 ± 0.01	$7.9 \pm 0.9 \times 10^4$	0.21 ± 0.01	191 ± 6
	0	10	–	0.38 ± 0.01	$1.4 \pm 0.1 \times 10^5$	0.82 ± 0.08	123 ± 4
	3	1	3.5	0.75 ± 0.01	$8.8 \pm 0.7 \times 10^4$	1.25 ± 0.04	255 ± 7
	10	1	2.7	1.32 ± 0.01	$2.6 \pm 0.2 \times 10^4$	4.94 ± 0.16	470 ± 13

With this relatively large chip we can attain 1×10^6 A m^{-2} sr^{-1} V^{-1}, if we use $E_2 \approx 10$ kV/mm, $E_1 > 3$ kV/mm and a 50 nm ebeam probe. Obtaining simultaneously a sub-1 eV energy spread is rather tricky, and a smaller membrane spacing is needed to achieve this. Obtaining $B > 1 \times 10^5$ A m^{-2} sr^{-1} V^{-1} seems achievable for a wide range of operation conditions.

4. Ion-electron interactions

A rather special situation occurs for the NAIS, namely, we are dealing with a mixture of ions and electrons in the beam. At first glance one may ignore the influence of the electrons all together due to the large mass difference. Firstly, the velocity of the electrons is much higher than that of the ions, so the interaction time is very small. Secondly, the energy transfer can only be a fraction of the total energy due to momentum conservation. On the other hand, we should consider that every ion gets taken over by thousands of electrons during its acceleration and the collective effect may be of importance.

One way to investigate the effect is adding electrons to the Monte-Carlo simulation that was already established. It turns out that this is not so simple. The electrons are so fast in comparison with the ions, that we require extremely small time steps to properly discretize the interaction time. One ion encounters up to thirty thousand electrons and we need to simulate at least a few hundred ions to get sufficient statistics. Without any tricks to improve the current ray-tracing scheme, this isn't feasible. Therefore, we have chosen to use an analytical approach similar to the extended two-particle approximation, although this specific problem is much simpler.

4.1 Introduction to the method

An appropriate model is Rutherford scattering: a charged particle scatters from a stationary target with an interaction potential governed by the coulomb force. It was developed for alpha particles scattering of a nucleus, but the formalism works equally well for particles of opposite sign. In this model, the target particle is infinitesimally small and any (non-elastic) effects due to the details of the electronic structure of the atom are ignored. We will asses this assumption later on. The rest of this section is concerned with the longer ranged coulomb interaction of electron-ion scattering. Note that this assumption is implicitly made in practically any calculation of the coulomb interactions in a charged particle beam.

Rutherford scattering describes a single scatter event, but we are interested in the collective effect of several consecutive events. We will need to use statistical methods to determine the collective effect. We consider the statistics of a single on-axis ion to be representative for the full beam, a concept that is borrowed from the extended two-particle approximation (Jansen, 1988). The events can be assumed to be independent because we expect only small perturbations of the ion trajectory.

In the presented calculation we consider the simple case of an ion being accelerated in a uniform electric field $E\hat{\mathbf{z}}$, starting at the origin. The method described here is not necessarily limited to only this simple case, but it is the most important case at present.

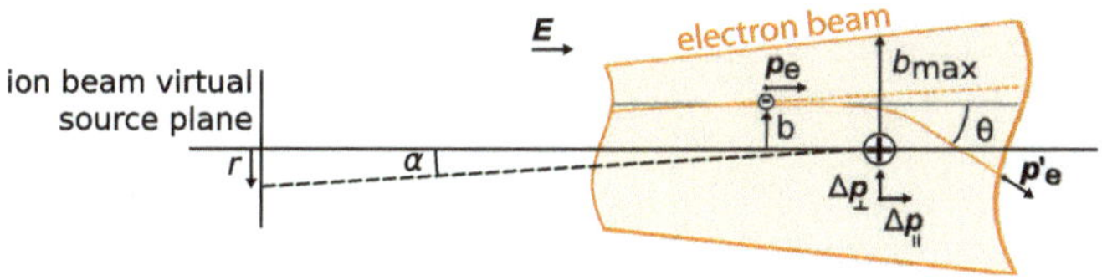

Figure 15 Electron-ion scattering. One particular electron trajectory is depicted; the shaded area indicates the range of possible electron trajectories.

We define the positions at which a scatter event occurs by z_i, with i the index of the scatter event. When assuming uniform spacing between the scatter events and neglecting the ion velocity relative to the electron velocity, the time between two scatter events is $\Delta t = e/I$. Uniform acceleration yields $z_i = \frac{1}{2}\frac{eE}{m}(\Delta t \cdot i)^2$, with m the ion mass and E the electric field strength. The last scatter event occurs when the electrons are just repelled back, so $z_N = \Phi_{e,0}/E$, where N is the total number of scatter events, and $e\Phi_{e,0}$ the initial electron energy. We can now write

$$z_i = \frac{e^3 E}{2mI_e^2} i^2, \qquad N = \sqrt{\frac{2m\Phi_{e,0}}{eE^2}} \frac{I_e}{e}. \tag{9}$$

We may replace i by $i - i_0$ where i_0 is a value between 0 and 1 to account for the discretization error of the locations of the scatter events. We found that the value i_0 does not influence the final result and was set to 0.5.

Some examples for the number of scatter events are instructive. Helium at 1 kV/mm acceleration voltage gives $N = 5.7 \times 10^3$, in contrast to $N = 32.6 \times 10^3$ for xenon. The number scales inversely proportional to the field so at 10 kV/mm we reduce the numbers to 0.6×10^3 and 3.3×10^3 respectively.

4.2 A single electron-ion scatter event

We will now look at the effect of a single scatter event. Fig. 15 accompanies the calculation presented hereafter. In the Rutherford model, the impact parameter b and the angle of the outgoing electron with respect to the z-axis θ are related by

$$\tan\frac{\theta}{2} = \left(\frac{m + m_e}{m}\right)^3 \frac{\kappa e}{2\Phi_e b} \approx \frac{\kappa e}{2\Phi_e b}. \tag{10}$$

If we assume an electron beam with uniform current density and radius b_{max}, then the probability density function of the impact parameter is $\rho(b) =$

$2b/b_{\max}^2$. We can change variables $b \to \theta$ using $|\rho(b)\mathrm{d}b| = |f(\theta)\mathrm{d}\theta|$, with the relation given by Eq. (10) to obtain the probability density function for θ:

$$f(\theta) = \left(\frac{\kappa e}{2\Phi_\mathrm{e} b_{\max}}\left(\frac{m+m_\mathrm{e}}{m}\right)^3\right)^2 \frac{\cos\frac{\theta}{2}}{\sin^3\frac{\theta}{2}} \approx \left(\frac{\kappa e}{2\Phi_\mathrm{e} b_{\max}}\right)^2 \frac{\cos\frac{\theta}{2}}{\sin^3\frac{\theta}{2}}. \tag{11}$$

One essential aspect of the calculation is finding an appropriate expression for $b_{\max}$. We will use the ebeam diameter and assume all the electron trajectories are parallel to the z-axis. This is an approximation, as for example the trajectory furthest away from the axis makes angle α_e with the optical axis. A more realistic maximum impact parameter might be $b_{\max} = r_{\max}\cos\alpha_\mathrm{e}$. This correction, however, is best avoided, as it leads to unrealistically small impact parameters for highly curved electron trajectories, something we expect when the electron beam is close to being repelled. Omitting this correction for weakly curved trajectories is acceptable as it will be a very small correction. Furthermore, the error is most pronounced far away from the axis, for which the contribution to the ion's momentum change is least. We note that at the electron beam focus, the trajectories are rather chaotic and the parallel beam assumption seems unsound. However, the cosine of a typical maximum skew angle of 40 mrad is 0.9992, so again, the error made is only small.

A second aspect of is the curvature of the electron trajectories due to the decelerating field. Consideration of this effect would require the formalism for Rutherford scattering to be modified. We argue that only in the vicinity of the ion, curvature could be relevant for the interacting forces, but fortunately the curvature is of much larger length scale such that in the vicinity of the ion the electron trajectories are straight.

We furthermore ignore the trajectory displacement of the electrons. This assumption was checked by considering the average θ deflection of a single scatter event and concluding that even at 10 to 100 scatter events, this is still much less than the typical beam divergence angle.

We do need to consider the evolution of $b_{\max}$ along the beam line. The ebeam follows parabolic curves in the decelerating electric field. A derivation for the trajectory equation is very similar to what is shown in the appendix of the chapter on Optics of ion emission from a volume. The result is

$$b_{\max}(z) = \frac{1}{2}d_\mathrm{e} + 2z\alpha_\mathrm{e}\sqrt{\frac{\Phi_{\mathrm{e},0}}{Ez}}\left(\sqrt{\frac{\Phi_{\mathrm{e},0}}{Ez}} \pm \sqrt{\frac{\Phi_{\mathrm{e},0}}{Ez} - 1}\right) \tag{12}$$

with the $\pm = -$ for the forward trajectory and $\pm = +$ for the backward trajectory. The energies of the electrons and the ions are simply given by

$$\Phi_{\mathrm{I}}(z) = \Phi_{\mathrm{I},0} + Ez, \qquad \Phi_{\mathrm{e}}(z) = \Phi_{\mathrm{e},0} - Ez. \tag{13}$$

It is not the outgoing angle of the electron but the properties of the ion that we are after, so we need to define the quantities of interest. If we ignore recoil, we can derive the momentum change of the ion caused by an electron using conservation of momentum. This gives $\Delta p_{\parallel} = \sqrt{2me\Phi_e}(1 - \cos\theta)$ parallel to the optical axis. We can derive the energy acquired by the ion in the axial direction by considering the energy transfer associated with the parallel momentum transfer. The result is

$$\epsilon_i = e\Phi_{\mathrm{e}}\frac{m_{\mathrm{e}}}{m}\left(1 - \cos\theta\sqrt{1 - \frac{m_{\mathrm{e}}}{m}\left(1 + \frac{m_{\mathrm{e}}}{m}\right)^{-2}\sin\frac{\theta}{2}}\right)^2 \approx e\Phi_{\mathrm{e}}\frac{m_{\mathrm{e}}}{m}(1 - \cos\theta)^2. \tag{14}$$

The more elaborate expression includes recoil, which is extremely small for the electron-ion combination.

We will also consider trajectory displacement, the most common way for coulomb interactions to cause loss of brightness. After a transverse momentum kick $\Delta p_{\perp} = \sqrt{2me\Phi_{\mathrm{e}}}\sin\theta$ in the transverse direction, the ion will acquire an angle α with the optical axis described by $\sqrt{2Me\Phi_{\mathrm{I}}}\tan\alpha = \Delta p_{\perp}$. In our uniform field acceleration situation, we know that the virtual source position is to first order located at $-2z$, so the displacement in the virtual source plane is $r_i = -2z\tan\alpha$. Putting this together gives

$$r_i = 2z\sqrt{\frac{m_{\mathrm{e}}}{m}\frac{\Phi_{\mathrm{e}}}{\Phi_{\mathrm{I}}}\left(1 - \frac{m_{\mathrm{e}}}{m}\left(1 + \frac{m_{\mathrm{e}}}{m}\right)^{-2}\sin\frac{\theta}{2}\right)}\sin\theta \approx 2z\sqrt{\frac{m_{\mathrm{e}}}{m}\frac{\Phi_{\mathrm{e}}}{\Phi_{\mathrm{I}}}}\sin\theta. \tag{15}$$

4.3 Combining multiple scatter events

Ultimately we are interested in the distribution of the random variable composed by the sum of many random variables. The total energy acquired by the ion due to electrons 1 to N is $\epsilon = \epsilon_1 + \epsilon_2 + \ldots + \epsilon_N$. The contributions are not identically distributed because the electron and ion energy is varying along z and the electron beam diameter is not constant, which prevents us from directly using the classical central limit theorem (CLT). Fortunately, we have a set of probability density functions that are non-zero on only a finite domain, which implies Lyapunov's condition. The result is that we can still employ a version of the CLT. We define

$$\epsilon = \sum_{i=1}^{N} \epsilon_i, \qquad \mu_\epsilon = \sum_{i=1}^{N} \mu_{\epsilon_i}, \qquad \sigma_\epsilon^2 = \sum_{i=1}^{N} \sigma_{\epsilon_i}^2 \tag{16}$$

and based on the Lyapunov-CLT we state that ϵ converges to a normal distribution with mean μ_ϵ and standard deviation σ_ϵ.

We find σ_{ϵ_i} from

$$\sigma_{\epsilon_i}^2 = \int f(\theta) \left(\epsilon_i(\theta) - \mu_{\epsilon_i}\right)^2 \mathrm{d}\theta \tag{17}$$

with f defined in Eq. (11). We prefer to find fw50 values and since we know the distribution is normal, we can use an appropriate constant to transform a standard deviation to a fw50. The final result for the collective energy spread of many scatter event is

$$\Delta\epsilon_{50} = 1.349 \sqrt{\sum_{i=1}^{N} \int_{\theta_{\min}}^{\theta_{\max}} f(\theta; z_i) \left(\epsilon_i(\theta; z_i) - \mu_{\epsilon_i}(z_i)\right)^2 \mathrm{d}\theta}. \tag{18}$$

A fwhm would lead to a prefactor of 2.355, while 1 gives the standard deviation. Note that for clarity we have indicated the dependence on z_i, while this was omitted earlier. $f(\theta; z_i)$ is the family of probability density functions that is parameterized by the z-coordinate z_i. $\theta_{\min}$ is determined by the maximum impact parameter, i.e. the electron beam radius, using Eq. (10). $\theta_{\max}$ is introduced to exclude scattering directly on the atom. This would require a more delicate approach and we want to exclude these effects.

We now develop the equation for the trajectory displacement. This is slightly more delicate than energy spread because we need to carefully deal with the 2-dimensional nature of the effect. We define the position of the ion in the virtual object plane as the vector $\mathbf{R}_i = (x_i, y_i) = (r_i(\theta)\cos\phi, r_i(\theta)\sin\phi)$. Note that due to rotational symmetry the ϕ angle has a uniform distribution. We define a vector $\mathbf{R} = \mathbf{R}_1 + \mathbf{R}_2 + \ldots + \mathbf{R}_N$ describing the collective effect. Since this is a bi-variate distribution we need to use a covariance matrix rather than a variance. We get

$$\mathbf{R} = \sum_{i=1}^{N} \mathbf{R_i}, \qquad C_R = \sum_{i=1}^{N} \begin{pmatrix} \langle x_i x_i \rangle & \langle x_i y_i \rangle \\ \langle x_i y_i \rangle & \langle y_i y_i \rangle \end{pmatrix} = \sum_{i=1}^{N} \begin{pmatrix} \sigma_{x_i}^2 & 0 \\ 0 & \sigma_{y_i}^2 \end{pmatrix}. \tag{19}$$

Note that $\mu_{\mathbf{R}} = 0$ and $\langle x_i y_i \rangle = 0$ due to rotational symmetry. x_i and y_i are not, however, independent. The CLT holds also in more dimensions in the sense that the distribution of $\mathbf{R}$ tends to the bi-variate normal distribution with covariance matrix C_R. By symmetry $\sigma_{x_i}^2 = \sigma_{y_i}^2$ and we can find an expression for this quantity by using the probability density $\rho(\theta, \phi) = (2\pi)^{-1} f(\theta)$ in combination with $x_i = r_i(\theta) \cos\phi$. We get

$$\sigma_{x_i}^2 = \frac{1}{2\pi} \iint f(\theta)\, r_i^2 \cos^2\phi \, d\theta d\phi = \frac{1}{2} \int f(\theta)\, r_i^2 \, d\theta. \tag{20}$$

Since the final distribution of the collective effect is a bi-variate normal distribution, the 2D fw50 diameter is equal to the diagonal elements of C_R up to a numerical factor. The final expression is

$$d_{50} = 1.806 \sqrt{\sum_{i=1}^{N} \int_{\theta_{\min}}^{\theta_{\max}} f(\theta; z_i)\, r_i^2(\theta; z_i) \, d\theta}. \tag{21}$$

The pre-factor is the same for a fwhm, while 0.769 gives the standard deviation.

4.4 Numerical evaluation

Eqs. (18) and (21) are computed numerically for the case of 1 keV electrons, focused into a 50 nm fw50 diameter. Upon evaluating the $\Delta\epsilon_{50}$ we found that the effect is essentially very small. In all cases we found $\Delta\epsilon_{50} < 1$ meV. This is much smaller than even the initial thermal energy and plots are omitted because the influence is irrelevant.

The trajectory displacement was not so clearly insignificant. Fig. 16 shows the numerical result for helium, argon and xenon on a typical range of field strengths. Notably, the effect can become comparable to the geometrical source size. We may however reduce the effect to well below 10 nm by using a 10 kV/mm acceleration field. This is in line with minimizing ion-ion type coulomb interactions.

Fig. 17 shows how the contribution of the trajectory displacement to the virtual source size is building up along the optical axis. In the first 100 nm after emission, no contribution can be made because the angular deviation is not transferred to a displacement, even though the ion energy is very low here. If we move further downstream, the effect becomes bigger and between 0.1 and 10 μm most of the contribution is formed. Further down the optical axis the ion velocity and the ebeam diameter are increased,

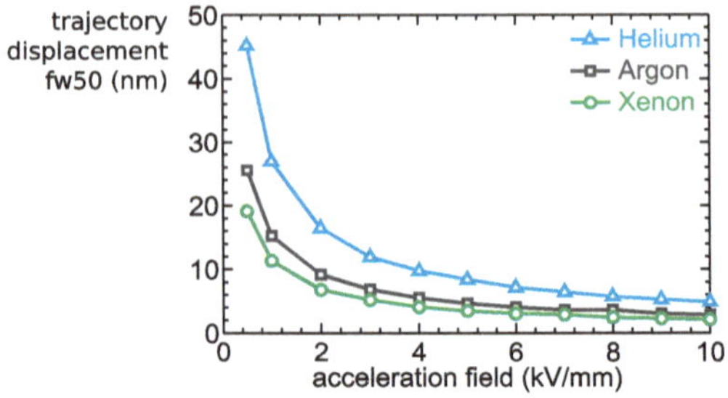

Figure 16 Trajectory displacement fw50 contribution to the virtual source as caused by many electrons scattering on the ions. The ebeam started with 100 nA of current and an initial energy of 1 keV focused down a 50 nm spot with half beam angle 40 mrad.

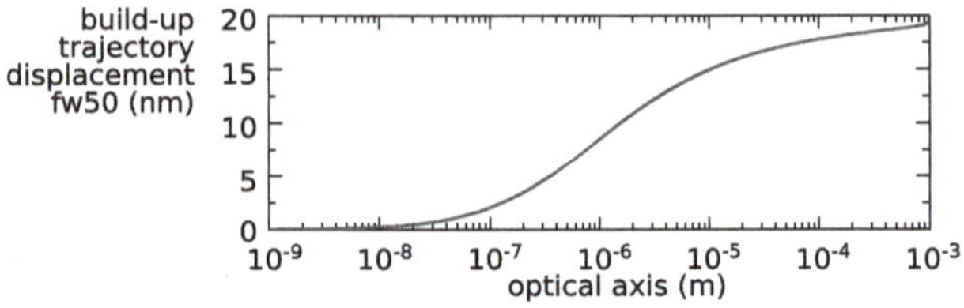

Figure 17 Accumulation of the trajectory displacement due to electrons scattering on argon ions, with $E = 1$ kV/mm.

leading to a weakening of the contribution. In the very final stage there is a tiny bump, which appears when the electrons are just repelled and have a long interaction time.

We found that lowering the initial ebeam energy increases the influence of the effect, even though this lowers the number of scatter events. This is attributed to the increased interaction time, which is reflected in Eq. (11).

We have experimented with values for $\theta_{\max}$, corresponding to minimum impact parameters between 1 and 100 pm and found no significant difference to the trajectory displacement calculation. The minimum impact parameter was fixed at 50 pm for further calculations. We conclude that the system is dominated by weak angle scattering and the assumption of no head-on collisions is substantiated. We can conclude that using the Rutherford model is thus acceptable. Mott cross sections or even more advanced models can describe electron-ion scattering in far more detail, but in our case such refinement won't be beneficial.

The trajectory displacement describes brightness loss by virtual source size growth, but neglects any added divergence in the beam. Usually this is a negligible contribution when compared to trajectory displacement. However, in an accelerating configuration, right in the beginning this may be important because the ion energy is very low. The possible influence was investigated by summing the transverse ion momentum. The resulting total momentum was much smaller than the thermal spread, hence this is of no concern.

We can state that the trajectory displacement to the ions caused by the electrons in the studied all-round NAIS configuration is not a dominating effect. In particular if the acceleration field is higher than 5 kV/mm the effect is smaller than 10 nm even for the most prone noble gas helium. Even though the effect is not dominating, the influence is only slightly smaller compared to a typical intrinsic virtual source size. We have learned that the effect can not in general be ignored a-priori and alternative configurations may suffer more problematically from the effect. The method described here can be used to study such situations. A decrease in electron energy, a smaller ebeam spot size, or an increase in ebeam current are all potential problematic changes. The contribution to the energy spread is negligible in all practical cases.

5. Conclusion

Simulations of ion emission were performed for realistic NAIS configurations including the electric fields, ion-neutral scattering, and coulomb interactions. Ultimately, the coulomb interactions pose a limit on the maximum attainable brightness. This limit can be extended by using strong electric fields of preferably 10 kV/mm. At such field strengths, a brightness between 1×10^6 and 5×10^6 $\mathrm{A\,m^{-2}\,sr^{-1}\,V^{-1}}$ seems achievable with an energy spread lower than 1 eV. For obtaining a sub-1 eV energy spread while maintaining the desired high electric fields it is required to use a membrane spacing of just a few hundred nanometer.

The ion-ion coulomb interactions were found to influence the beam in three ways: by trajectory displacement in the first microns after emission, by beam divergence growth inside the chip, and by increased membrane blockage.

When comparing chip geometries, the differences in performance were surprisingly small. This means that the tolerances on the chip fabrication can be relatively low. Some differences could be identified though. The larges spaced chips have larger energy spreads when using the same electric fields. These larger energy spreads are reduced when using a relatively high particle density. Chips with larger apertures yield higher currents, but typically at the cost of some brightness loss.

The beam performance was quite insensitive to the noble gas species. The differences in ionization efficiency, ion-neutral scattering and coulomb interaction seems to balance each other out reasonably well, at least in the simulated configurations.

When the operation conditions are sub-optimal, it is still possible to obtain $B \geq 1 \times 10^5\ \mathrm{A\,m^{-2}\,sr^{-1}\,V^{-1}}$ for a wide range of situations. However, $B \geq 1 \times 10^6\ \mathrm{A\,m^{-2}\,sr^{-1}\,V^{-1}}$ and sub-1 eV energy spread become mostly out of reach.

Surface induced charge forces were found to be important for NAIS emission, particularly for small aperture sizes of 100 nm or smaller and for weak in-chip fields. It seems best to use apertures of 150 nm or larger to avoid too much current to be lost by this effect. A lensing effect also became apparent, but this does not directly deteriorate the performance of the beam. The energy spread of the beam is hardly affected by the surface induced charge. A simple analytical approximation was found that gives decent accuracy and replaces a tedious field solving method.

The inevitable electron background passing through the ion beam gave rise to trajectory displacement of the ions. From the presented analytical calculations we can conclude that the effect is just small enough to be negligible, if the acceleration field is at least a few kV/mm. The contribution to the energy spread is negligible in all practical cases.

Appendix A. The extended two-particle approximation and the slice method

Here we briefly summarize the necessary equations for computing coulomb interactions using the slice method. Although the method is not explicitly used in this chapter, it is used in the introduction chapter.

The slice method can be an excellent tool; it is computationally quick, but also often insightful. For example, the slice method is used to investigate the coulomb interactions of electron emitters (Verduin, Cook, & Kruit, 2011; Cook, Verduin, Hagen, & Kruit, 2010). The appropriate literature (Kruit & Jansen, 2008; Jansen, 1988; Jiang, 1996) can be a bit too much in depth for some readers. The intention here is to just give the equations such that the interested reader can start calculating in no-time.

The extended two-particle approximation, initially developed by Jansen (1988), gives equations for the coulomb interaction effects for beam seg-

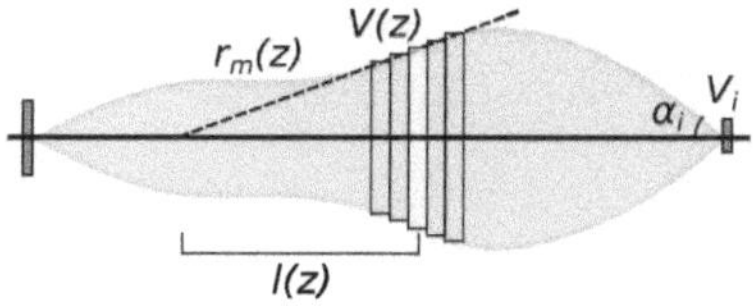

Figure 18 The slice method is based on an addition of contributions in thin beam segments for which the beam diameter and beam voltage can be considered constant.

ments with uniform properties. The slice method applies this method to beams that don't have such uniformity. By slicing the beam up into segments that can be considered uniform locally, the total effect can be found by summing over the contributions of the slices (Jiang, 1996) (see Fig. 18).

The contribution due to stochastic coulomb interactions can be calculated from the slice-integral

$$r_{\mathrm{td}} = \int M(z) l(z) \phi(z) dz. \tag{22}$$

$\phi(z)$ is the angular deflection per unit length. The multiplication with l transforms the angular deflection to lateral shift on the axis. M is the magnification of the virtual source position of that particular slice to the image or object position at which we want to calculate the contribution.

If the marginal ray trajectory $r_{\mathrm{m}}(z)$ is known, we can write

$$M(z) = \frac{r'_{\mathrm{m}}(z)}{\alpha_{\mathrm{i}}}\sqrt{\frac{V(z)}{V_{\mathrm{i}}}}, \qquad l(z) = \frac{r_{\mathrm{m}}(z)}{r'_{\mathrm{m}}(z)}. \tag{23}$$

The value for the angular deflection in a slice depends on the regime and is given by

$$\varphi(z) = 0.429 \frac{m^{3/2}}{\epsilon_0 q^{7/2}} \frac{I^3 r(z)}{V(z)^{5/2}} \qquad \text{Pencil} \tag{24}$$

$$\varphi(z) = 0.128 \frac{m^{1/3}}{\epsilon_0} \frac{I^{2/3}}{V(z)^{4/3} r(z)^{4/3}} \qquad \text{Holtzmark} \tag{25}$$

$$\varphi(z) = \left(\frac{T_1 D_{\mathrm{a}}^{18/7} D_{\mathrm{r}}^6 I^{18/7} r(z)^{6/7} V(z)^{-15/7}}{T_4 + T_2^{1/7} D_{\mathrm{r}}^6 D_{\mathrm{a}}^2 I^2 r(z)^2 V(z)^{-1}} \right)^{7/6} \qquad \text{Interpolation} \tag{26}$$

The constant in Jiang's interpolation formula are $T_1 = 4.618 \times 10^{-2}$, $T_2 = 2.041 \times 10^5$, $T_4 = 6.250 \times 10^{-2}$, $D_{\mathrm{r}} = (2\epsilon_0 \pi / q)^{1/3}$, $D_{\mathrm{a}} = \sqrt{m}/(\pi 2^{7/2} \epsilon_0 q^{1/2})$.

The contribution to the energy spread by the Boersch effect does not require any optical scaling and is given by

$$\Delta\epsilon = \int f_{\Delta\epsilon}(z)dz, \tag{27}$$

with the infinitesimal contribution in a slice given by

$$f_{\Delta\epsilon} = 0.642 * \frac{m}{\epsilon_0 q^2} \frac{I^2}{V(z)} \quad \text{Holtzmark} \tag{28}$$

$$f_{\Delta\epsilon} = 0.149 \frac{m^{1/3}}{\epsilon_0} \frac{I^{2/3}}{V(z)^{1/3} r(z)^{4/3}} \quad \text{Pencil} \tag{29}$$

$$f_{\Delta\epsilon} = \left(\frac{B_1 D_a^3 D_r^6 I^3 V(z)^{-3/2}}{1 + B_2^{1/4} D_r^6 D_a^2 I^2 r(z)^2 V(z)^{-1}} \right)^{2/3} \quad \text{Interpolation} \tag{30}$$

With $B_1 = 7.861 \times 10^{-2}$ and $B_2 = 6.828 \times 10^9$.

List of symbols

$\mathbf{E}_q$	surface induced self-field C^{-1}
$\Phi_q(\mathbf{r}_q)$	surface induced self-potential V
$\Phi(\mathbf{r})$	potential field V
$\mathbf{r}_q$	location of charge m
$\mathbf{r}_s$	location of surface element m
$\hat{\mathbf{a}}$	surface element unit vector
$\sigma(\mathbf{r}_s)$	surface charge density $\mathrm{C\,m^{-2}}$
κ	$1/4\pi\epsilon_0$ $\mathrm{N\,m^2\,C^{-2}}$
δ	Dirac delta function
e	elementary charge C
q	ion charge C
R	cylinder radius m
q_{im}	image charge C
r_{im}	image charge location m
Φ	ion acceleration voltage V
E	electric field $\mathrm{V\,m^{-1}}$
E_1	inside-chip electric field $\mathrm{V\,m^{-1}}$
E_2	outside-chip electric field $\mathrm{V\,m^{-1}}$
iKn	ionic Knudsen number
N	# simulated particles #
N	# electron-ion collisions #
I	ion beam current A
I_e	electron beam current A
d_v	virtual source size m
d_e	electron probe size m

I_e	electron beam current	A
Φ_e	electrostatic energy electrons	eV
$\Phi_{e,0}$	electr. landing energy electrons	eV
Φ_I	electrostatic energy ions	eV
$\Phi_{I,0}$	ion initial electrostatic energy	eV
m_e	electron mass	kg
m	ion mass	kg
b	impact parameter	m
b_{max}	electron beam envelope	m
φ	angle around optical axis	rad
θ	angle with optical axis	rad
$f(\theta)$	probability density function	rad^{-1}
ϵ	additional ion energy	eV
μ_ϵ	expectation value ion energy	eV
σ_ϵ	standard dev. ion energy	eV
$\mathbf{R}$	trajectory displacement	m
σ_x	stand. dev. traj. displacement	m
C_R	covariance matrix traj. disp.	m^2
θ_{min}	scatter angle cut-off	rad
θ_{max}	scatter angle cut-off	rad
z_i	position ith electron scatter	m
α	ion scatter angle	rad
α_e	electron divergence angle	rad
$\Delta p_\perp$	transverse ion mom. acquired	kg m/s
$\Delta p_\parallel$	parallel ion mom. acquired	kg m/s
$\Delta\epsilon_{50}$	fw50 electron-ion energy spread	eV
d_{50}	fw50 electron-ion traj. disp.	m

Acknowledgment

I would like to thank Pieter Kruit for his guidance and contributions. This work is part of the Industrial Partnership Program (IPP) of the Stichting voor Fundamenteel Onderzoek der Materie (FOM), which is financially supported by the Nederlandse Organisatie voor Wetenschappelijk Onderzoek (NWO). The IPP program is co-financed by FEI Company. This work is based on the dissertation titled *The Nano-Aperture Ion Source*, by Leon van Kouwen.

References

Birkhoff, R., Allen Jr., J., Anderson, V., Hubbell Jr., H., Becker, R., Ferrell, T., & Spears, D. (1980). The ultimate electron lens-deflection of low-energy electrons by the image potential in a pinhole. *Optik*, *55*(2), 157–164.

Bouwkamp, C. J., & De Bruijn, N. G. (1947). The electrostatic field of a point charge inside a cylinder, in connection with wave guide theory. *J. Appl. Phys.*, *18*(6), 562–577.

COMSOL (2013). *Comsol multiphysics 4.4 - electrostatics module.*

Cook, B., Verduin, T., Hagen, C. W., & Kruit, P. (2010). Brightness limitations of cold field emitters caused by coulomb interactions. *J. Vac. Sci. Technol. B*, *28*(6), C6C74–C6C79.

Griffiths, D. J. (1999). *Introduction to electrodynamics* (third edition). Prentice-Hall Inc.
Hagen, C. W., Fokkema, E., & Kruit, P. (2008). Brightness measurements of a gallium liquid metal ion source. *J. Vac. Sci. Technol. B*, *26*(6), 2091.
Hernandes, J., & Assis, A. (2005). Electric potential due to an infinite conducting cylinder with internal or external point charge. *J. Electrost.*, *63*(12), 1115–1131.
Jansen, G. (1988). *Coulomb interactions in particle beams* (Doctoral dissertation). Delft University of Technology. Retrieved from repository.tudelft.nl/.
Jiang, X. (1996). *Coulomb interactions in charged particle optical columns* (Doctoral dissertation). Delft University of Technology. Retrieved from repository.tudelft.nl/.
Kruit, P., & Jansen, G. (2008). Space charge and statistical coulomb effects. In J. Orloff (Ed.), *Handbook of charged particle optics* (second edition). CRC Press.
PulsarPhysics (2011). *General particle tracer, version 3.10*. Retrieved from www.pulsar.nl/gpt.
Radlicka, T., & Lencova, B. (2008). Coulomb interactions in GA LMIS. *Ultramicroscopy*, *108*(5), 445.
Steele, A. V., Knuffman, B., & McClelland, J. J. (2011). Inter-ion coulomb interactions in a magneto-optical trap ion source. *J. Appl. Phys.*, *109*(10), 104308. https://doi.org/10.1063/1.3585783.
ten Haaf, G., Wouters, S., van der Geer, S., Vredenbregt, E., & Mutsaers, P. (2014). Performance predictions of a focused ion beam from a laser cooled and compressed atomic beam. *J. Appl. Phys.*, *116*(24), 244301.
van der Geer, S. B., Reijnders, M. P., de Loos, M. J., Vredenbregt, E. J. D., Mutsaers, P. H. A., & Luiten, O. J. (2007). Simulated performance of an ultracold ion source. *J. Appl. Phys.*, *102*(11), 094312.
Verduin, T., Cook, B., & Kruit, P. (2011). Influence of gun design on coulomb interactions in a field emission gun. *J. Vac. Sci. Technol. B*, *29*(6).

CHAPTER TWELVE

Processes in the ionization volume of the nano-aperture ion source

Leon van Kouwen
Department of Imaging Physics, Delft University of Technology, Delft Netherlands
e-mail address: lvankouwen@gmail.com

Contents

1. Introduction

Our goal for the NAIS is making single-charged positive ions, but inevitably, many more processes occur in the ionization region. In this chapter we will identify possible other interactions and explore to what extent they affect the final beam composition. We use established theory to come to simple estimates for the final beam composition. We focus our attention to argon gas, but the method should be easily extendable to other gases.

1.1 What else happens in the ionization region?

The electrons interact in various ways with the neutral gas. We may remove more than one electron and acquire multiply charged ions, scatter elastically and transfer momentum, or bring the atom into an excited state. The last possibility can be sub-divided into excitation to a radiative level, excitation to a metastable level, and excitation to a Rydberg state, which is a highly

Advances in Imaging and Electron Physics, Volume 212
ISSN 1076-5670
https://doi.org/10.1016/bs.aiep.2019.09.005

excited state close to ionization. The reaction equations of said processes are

$X + e \rightarrow \tilde{X} + e$	momentum transfer	(1)
$X + e \rightarrow X^* + e$	(metastable) excitation	(2)
$X + e \rightarrow X^{**} + e$	Rydberg excitation	(3)
$X + e \rightarrow X^+ + 2e$	single ionization	(4)
$X + e \rightarrow X^{i+} + (1+i)e$	multiple ionization	(5)

We use X^+ for ions, $\tilde{X}$ for an energetic neutral, X^* for an excited atom (possibly metastable), and X^{**} for a Rydberg excitation.

The reaction products may engage in a plurality of interactions (Bogaerts & Gijbels, 1999; Lam, Zheng, Lo, Dem'yanov, & Napartovich, 2000), but only those processes that could be relevant in our configuration are considered here. The ions may scatter on the neutrals on the way out of the interaction volume. The most obvious reaction of an excited state atom is relaxation to the ground state by photon emission. We also consider the formation of molecular noble gas ions, a compound which has long been known (Tüxen, 1936) (see also for example Hornbeck and Molnar (1951) and Becker and Lampe (1965)). Two excited state atoms may form molecular ions via associative ionization. A second possibility is the encounter of a highly excited Rydberg atom and an unexcited atom, which is called Hornbeck-Molnar ionization. Molecular ion formation often needs to be considered in describing high density plasmas (Bogaerts & Gijbels, 1999; Lam et al., 2000). Furthermore, we consider the ionization of a neutral atom with high momentum due to an earlier elastic electron scatter event. We also consider the ionization of an already excited atom. The reaction equations are:

$X^* \rightarrow X + h\omega$	radiative decay	(6)
$\tilde{X} + e \rightarrow \tilde{X}^+ + 2e$	non-thermal ionization	(7)
$X^* + e \rightarrow X^+ + 2e$	step-wise ionization	(8)
$X^{**} + X \rightarrow X_2^+ + e$	Hornbeck-Molnar	(9)
$X^* + X^* \rightarrow X_2^+ + e$	associative ionization	(10)

Two elastic collisions that can be identified are

$X^+ + X \rightarrow X^+ + X$	momentum transfer collision	(11)
$X^+ + X \rightarrow X + X^+$	resonant charge exchange	(12)

Such elastic interactions are relevant for the NAIS, but do not alter the beam composition and are discussed in other chapters of this publication.

1.2 Influence on the performance

A good question is: do any of these reactions matter for the NAIS performance? At the end of the day, all we care about is the final ion beam. An excited atom that emits a photon and comes back to the ground state does not jeopardize the final beam quality for example. For the other processes this is not so clear though.

Elastic collisions of electrons can be problematic if the energy transfer is larger than the thermal energy. If this happens with great abundance, the gas is no longer thermally distributed and the brightness is affected. We first look at the possible energy transfer of an elastic collision event between an electron and an atom. The transferred energy from an electron to an atom at rest is $\epsilon_a = \epsilon_e 4m/M(1+m/M)^{-2}\sin\left(\frac{\theta}{2}\right)$, with m/M the electron to atom mass and ϵ_e the electron energy. For argon, $m/M = 1.37\times10^{-5}$, which leads to a maximum at $\theta = \pi$ of $\epsilon_a = 55$ meV, for an electron beam energy of 1 keV. On average the energy transfer is lower because of the probability of scattering at different angles. We conclude that the acquired energy is in the order of the thermal energy, which is just enough to affect the brightness.

Higher charged and molecular ions may lead to problems as well. In a fully electrostatic system, all trajectories are independent of the mass and the charge (except for the sign of the charge). This means that Ar^+, Ar^{2+}, and Ar_2^+ are focused into the same spot. The user, however, may not desire the heavier or the more energetic particles on the target. Secondly, the velocity differences can become a problem when scanning quickly. For example, if the scan electrodes are 100 mm above the sample and we consider a 10 keV argon beam, the difference in travel time for Ar^+, Ar^{2+}, and Ar_2^+ is in the 100 ns range. Since this is in the order of a possible pixel irradiation time, it can cause a lagging pixel blur. Another potential problem is magnetic field sensitivity. Magnetic deflection scales with $\sqrt{q/m}$, so if a magnetic field is present, we may end up with three different spots on the sample. Suppose an argon ion beam travels for 100 mm in a uniform 1 μT field at 10 keV beam energy, the deflection of the three species is Ar^{2+}: 77 nm, Ar^+: 55 nm, and Ar_2^+: 38 nm. That is more than enough to be unpleasant. The earth magnetic field is about 50 μT, so we would need excellent magnetic shielding or magnetic cancellation to prevent beam separation effects.

Due to the identified potential problems, a quantitative analysis of the produced species is required.

2. Calculation method

We will try to estimate the contributions of the various processes for argon gas. We begin by quantifying the relevant reactions by using total cross sections. The next step is considering the loss mechanisms of the reaction products.

2.1 Cross sections

Fig. 1 shows the cross section of argon for excitation, elastic momentum transfer, and ionization. The rate at which excited atoms can be made is proportional to the cross section. So, we can see directly that the rate of excited species generation is only about 10% of the total ionization rate for 1 keV electrons. Since ionization yields secondary electrons, which have a lower energy, they could engage in more efficient excitation. However, it was explained earlier that we mostly intend to operate the NAIS under the condition that most electrons do not engage in any ionization at all. That means that only about 1% of the primary current could become a slower electron. The cross section for excitation at lower energies, around the maximum, is about 10 times higher than at 1 keV, but since the secondary current is only about 1% of the primary current, the net influence of the secondary induced excitation is at most 10 times lower than the primary excitations. The double to single ionization ratio at 1 keV is 0.0470 and the single to triple ratio is 0.0083. Since these are direct processes there is not much we can do to tune this other than changing the electron beam impact energy.

We should be careful in considering what cross sections to use for what purpose. A total elastic cross section will give a huge overestimate of the momentum transfer because small angle scattering gives a big contribution to the inelastic cross section, while it hardly transfers momentum. The so-called momentum cross section accounts for this, but often includes non-elastic contributions. This is useful if you want to know what happens to the electron, but we are interested in the atom. What we need is the elastic momentum transfer cross section. Surprisingly, it is not so easy to find this particular cross section up to 1 keV. We have used a set composed by Hayashi (2003). This set was used and verified before by Nanbu and

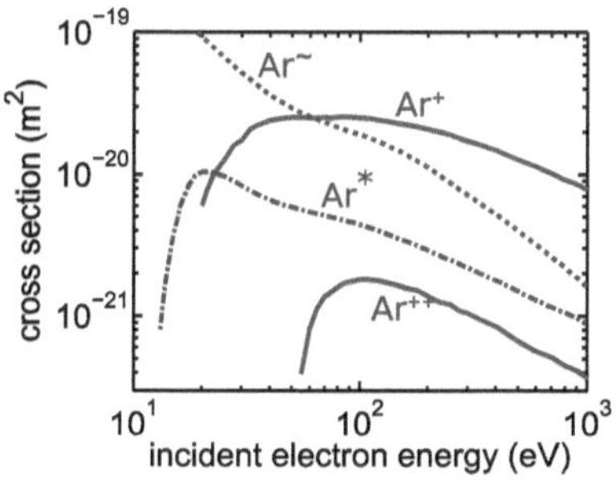

Figure 1 Cross sections for electron beam impact single ionization (Ar^+) (Rejoub, Lindsay, & Stebbings, 2002), double ionization (Ar^{2+}) (Rejoub et al., 2002), elastic momentum transfer ($Ar^{\sim}$) (Hayashi, 2003) and total excitation (Ar^*) (Yanguas-Gil, Cotrino, & Alves, 2005) of electrons scattering on argon atoms.

Kageyama (1996) and supposedly discussed by Kosaki and Hayashi (1992) (in Japanese).

Fig. 1 shows the total cross section for excitation, however, for the Hornbeck-Molnar process (Hornbeck & Molnar, 1951; Bultel & Vervisch, 2002) we require the atoms to be in a highly excited state, i.e. a Rydberg state. The energy required to form a molecular argon ion from two ground state atoms is 14.71 eV (Bogaerts & Gijbels, 1999) and the ionization potential of argon ions 15.76 eV (Yanguas-Gil et al., 2005). In principle, any state with energy above 14.71 eV could participate in the Hornbeck-Molnar process and Bogaerts and Gijbels (1999) argue that this is indeed the case. We neglect the influence of kinetic (thermal) energy as it is relatively low at room temperature. The relevant excited states for the Hornbeck-Molnar are therefore all excited states above $3d'[3/2]_1$, starting from the $4d[1/2]_1$ state. The cross sections for this level range are not directly available, but the individual level cross sections for the first 25 excited states are (Yanguas-Gil et al., 2005). We subtract the sum of all states up to $4d[1/2]_1$ from the total to get the applicable cross section. See Fig. 2 for the result. We note that somewhat different cross sections for the total (de Heer, Jansen, & van der Kaay, 1979) and the individual levels (Hayashi, 2003) can be found elsewhere. Yanguas-Gil et al. (2005) have considered a variety of sources and tested the cross sections for consistency with swarm parameter measurements and we use the set suggested by them.

2.2 Loss mechanisms

From the cross sections alone we can not estimate the rate of reaction products. The reaction constituents may diffuse away or relax to the ground state

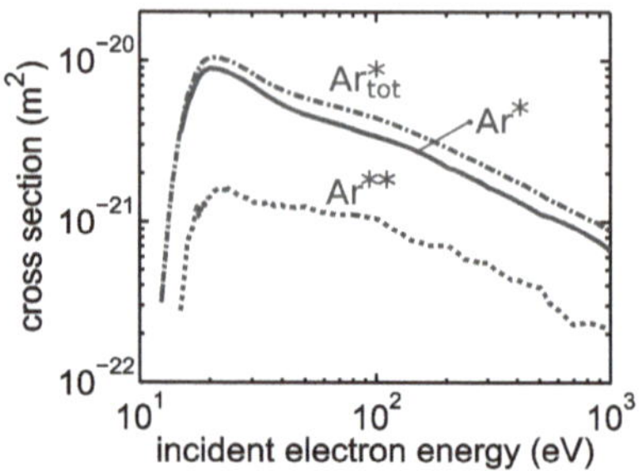

Figure 2 Total excitation cross sections of electron impact on argon gas and a summation of all individual level cross section below the $4d[1/2]_1$ state. Data from Yanguas-Gil et al. (2005). The difference between these two cross sections represents the cross section for excitation to highly excited state that may participate in the Hornbeck-Molnar process.

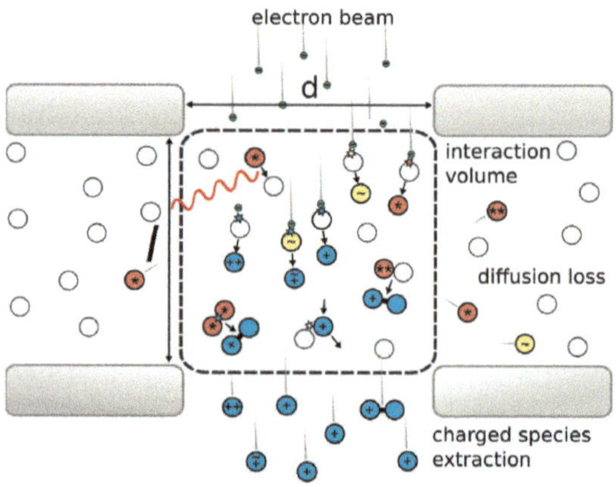

Figure 3 A sketch of the interaction volume showing the processes under consideration. We model the system as a cylinder of height *l* which is equal to the membrane spacing, and diameter *d* being equal to the aperture size. All means of ion formation are assumed to happen within this volume and all ionic species are assumed to escape the double membrane structure. Diffusion may occur in all directions.

before engaging in a reaction. We therefore need to calculate the concentration of reaction constituents. We will simplify the chip by assuming all reactions happen inside a cylinder which we will call the interaction volume. The height is equal to the membrane spacing and the diameter equals the aperture size. Fig. 3 illustrates this model. We assume that all charged species are extracted and become a part of the final beam. Excited species may engage in secondary reactions to form charged species, or may be lost due to diffusion or radiation. The rate equation for the total number of a particular constituent is

$$\dot{N}^x = \dot{N}^x_{\text{production}} - \dot{N}^x_{\text{reaction}} - \dot{N}^x_{\text{diffusion}} - \dot{N}^x_{\text{decay}} \tag{13}$$

N^x is the total number of particles in the interaction volume of constituent x.

Since we will deal with particle numbers and time scales, it is good to get acquainted with some typical values for the NAIS first. The electron current is around 100 nA, so one electron every 1.6 ps, or about 600 billion electrons per second passing through the ionization volume. The typical flight time through a chip of, say 250 nm length, is 10 fs, so a snapshot of the ionization region will typically not contain any electrons. We operate at a typical ion current of 1 to 2 nA, which amounts to about one ion every 100 ps. We tune the extraction field such that the residential time is smaller than 100 ps, as we preferably have only one ion at a time in the interaction region to prevent ion-ion interaction. A field of 0.5 kV/mm across a few hundred nanometers of spacing leads to these time scales (Tondare, 2006). The number of molecules in the interaction volume is $N = nAl$. Since different chip sizes can be acceptable for the NAIS, the range of this value is rather broad, typically between 1×10^3 and 1×10^5.

Diffusion is one mechanism for loosing excited state species before engaging in a reaction. For the RMS diffusion length in N_{dim} dimensions, it is well known that $\langle x^2 \rangle - \langle x \rangle^2 = 2N_{\text{dim}}Dt$. For the diffusion coefficient we can use $D = \frac{1}{3}\lambda \langle v_t \rangle$, with $\langle v_t \rangle$ the mean thermal velocity of the particles. If after some time τ the RMS distance due to diffusion is about equal to the typical size of the interaction volume s, we have lost a significant amount of atoms by diffusion. This can be interpreted as the diffusion time constant. So, we set $\sqrt{\langle x^2 \rangle - \langle x \rangle^2} = s$ and get for the diffusion time constant

$$\tau_{\text{diff}} = \frac{3s^2}{2N_{\text{dim}}\lambda \langle v_t \rangle} \tag{14}$$

We can interpret this diffusion time as the average residential time of an excited atom in the interaction volume. A typical chip geometry may give for the interaction volume $l = 250$ nm and $d = 100$ nm. The particle density is such that $\lambda = 125$ nm. If we consider 1 D diffusion along the cylinder ($s = l$) and 2 D diffusion radially outward ($s = d$), we obtain for τ_{diff} in the length direction 2.15 ns and 172 ps in the radial direction. We may inversely add them to obtain a single diffusion time of 159 ps for argon.

An estimate for the rate of atoms lost by diffusion is simply $N^x(\tau_{\text{diff}})^{-1}$. Note that we could of course solve Fick's law instead, which seems natural and more accurate. However this leads to complications on how to model the geometry. Spherical symmetry is easily solved but isn't quite the right geometry for our cylindrically symmetric system. Solving in 1D and 2D separately leads to the problem that the membranes do not truly confine, but probably adsorb the excited states, so radially it can't be interpreted truly 2D. As solving Fick's law leads to comparable concessions, we simply use the diffusion time constant. Only a full 3D simulation, carefully taking the geometry into account would offer good accuracy, but for the purposes of simple estimation as presented here, this is avoided.

Relaxation to the ground state by photon emission is a second mechanism for loss of excited state atoms. For the lowest excited states of argon the lifetime is 2.2 ns (Sahin, Tapan, Ozmutlu, & Veenhof, 2010), but higher levels have much longer lifetimes up to 300 ns (Aguilera, Blanco, Campos, & Ortiz, 1992; Das & Karmakar, 2005). Note that two metastable levels of argon ($4s[3/2]_2$, $4s[3/2]_1$) have lifetimes of 55.9 s and 22.9 s (Filipovic, Marinkovic, Pejcev, & Vuskovic, 2000). We can conclude that this mechanism takes place on a time scale much larger than diffusion and optical relaxation only plays a minor role. This means we do not need to distinguish between metastable states and radiative states as essentially all states outlive diffusion. Only for relatively large chips at high pressures do we get diffusion times in the order of optical relaxation times, but only for the lowest excited levels.

3. Estimates of the beam composition

If we consider no reactions but only diffusion and radiation, we can estimate an upper bound for the number of excited state atoms. We use Eq. (13) in steady-state to get

$$\dot{N}^* = \dot{N}_e n \sigma_* l - \frac{N^*}{\tau_{\text{diff}}} - \frac{N^*}{\tau_{\text{rad}}} = 0 \quad \rightarrow \quad N^* = \frac{\dot{N}_e n \sigma_* l}{\tau_{\text{diff}}^{-1} + \tau_{\text{rad}}^{-1}} \tag{15}$$

For our estimate we take $l = 250$ nm, $d = 100$ nm and $\lambda = 125$ nm, giving $\tau_{\text{diff}} = 159$ ps. The ebeam is 1 keV and $I_e = 100$ nA. For the radiative lifetime we use 20 ns, which is a little arbitrary, but this is also hardly of influence.

We find for the number of excited state atoms in the interaction volume $N^* = 0.39$, and $N^*/N = n^*/n \approx 1 \times 10^{-5}$. This relative concentration of excited argon atoms to ground-state atoms is so low, that associative ionization is negligible. Mind that the rate of this process is proportional to N^{*2}. In other words, two excited atoms will never find each other before being diffused away. Step-wise ionization is also highly unlikely, as it is proportional to the relative concentration of excited to ground state atoms. The concentration of non-thermal atoms is $\tilde{N} = 0.72$; very low as well, so non-thermal ionization gives a negligible contribution too. This particular example illustrates a more general result we have obtained from considering a variety of chip configurations. Namely, that step-wise ionization, non-thermal ionization and associative ionization are negligible processes for any reasonable NAIS configuration (contributions are smaller than 0.001% of the total current).

The Hornbeck-Molnar process cannot be disregarded on this basis. For the reaction rate we use $k_{\mathrm{HM}} = 2.0 \times 10^{-15}\ \mathrm{m^3\,s^{-1}}$, as suggested by Bogaerts and Gijbels (1999), without distinction in excited level. It is important to note though, that his reaction rate should be used with a concentration of excited states above 14.71 eV, and not with the total concentration of all the excited states. It is interesting to derive a cross section and a time scale from this reaction rate. $\sigma_{\mathrm{HM}} = k_{\mathrm{HM}} / \langle v_t \rangle = 3.7 \times 10^{-18}\ \mathrm{m^2}$ and $\tau_{\mathrm{HM}} = (kn)^{-1} = 43$ ps if $n = 1 \times 10^{25}\ \mathrm{m^{-3}}$. The cross section is large, which may be attributed to the fact that Rydberg atoms have a large physical size. From the time scale we conclude that the reaction may very well compete with diffusion.

The rate at which the Hornbeck-Molnar process produces molecular argon ions is $\dot{N}^{Ar_2^+} = k_{\mathrm{HM}} n N^{**}$. Substitution in Eq. (13) gives for the concentration of highly excited atoms

$$N^{**} = \frac{\dot{N}_e n \sigma_{**} l}{\tau_{\mathrm{diff}}^{-1} + \tau_{\mathrm{rad}}^{-1} + k_{\mathrm{HM}} n}. \tag{16}$$

The radiative lifetime of the highly excited states needs to be taken, for which we use 200 ns. Effectively, radiation is negligible. The dissociation energy of molecular argon ions is 1.23 eV (Smirnov, 2014). Since this energy is more than we typically have available in the interaction volume, thermally or by electric acceleration, we assume that all molecular ionic argon that is formed becomes a part of the final beam. The molecular argon current is given by

$$I[Ar_2^+] = \frac{I_e k_{\mathrm{HM}} n^2 \sigma_{**} l}{\tau_{\mathrm{diff}}^{-1} + \tau_{\mathrm{rad}}^{-1} + k_{\mathrm{HM}} n}. \tag{17}$$

Table 1 Calculation of different ionic species contributions for a few combinations of particle density and chip geometry. The main component of the current is in all cases singly charged argon. For this calculation the electron beam was 100 nA at 1 keV.

l nm	d nm	λ nm	n m^{-3}	τ_{diff} ps	$I[Ar^{1+}]$ nA (%)	$I[Ar^{2+}]$ nA (%)	$I[Ar^{3+}]$ nA (%)	$I[Ar_2^+]$ nA (%)
250	100	125	1.8×10^{25}	159	3.5 (87.6)	0.33 (8.2)	0.087 (2.2)	0.082 (2.0)
250	100	250	9.0×10^{24}	80	1.8 (88.1)	0.17 (8.3)	0.044 (2.2)	0.028 (1.4)
1000	200	500	4.5×10^{24}	169	3.5 (88.1)	0.33 (8.3)	0.087 (2.2)	0.058 (1.4)
1000	500	100	2.3×10^{25}	4776	17.6 (87.3)	1.66 (8.2)	0.437 (2.2)	0.478 (2.4)

We find that this current is in general not negligible and ionic molecular argon is expected to be formed.

The current from ions formed directly by impact ionization is given by

$$I[Ar^{i+}] = i \cdot I_e n \sigma_{i+} l. \tag{18}$$

On the way out, a double-charged ion may attain one electron from a neutral atom, resulting in two single-charged ions. The cross section for this process is $\sigma = 0.11 \times 10^{-20}$ m^2 at a center of mass energy of 0.4 eV (Okuno, 1986) (at 1 eV it lowers to 0.051×10^{-20} m^2). Since this cross section is smaller than the typical electron impact ionization cross section, we can directly conclude that this process is negligible because most electrons do not engage in ionization at all. This is a tricky argument so a numerical example is in place. The fraction of generated double-charged argon that engages in such a reaction is roughly $n\sigma l$, which gives for a typical configuration with, say $n = 2.5 \times 10^{25}$ m^{-3} and $l = 200$ nm , a fraction of 0.6%. This was a worst case scenario estimate so quite confidently we can neglect redistribution of double-charged ions. We did not find a cross section for triple ionized argon at such a low energy, but we expect the effect to be limited as well.

Table 1 summarizes the results for a few typical NAIS configurations. The relevant contributions to the final beam that have been identified are direct single, double and triple ionization, and Hornbeck-Molnar molecular ion formation. Other contributions are not listed as they are insignificant.

4. Conclusion

Besides creating single charged ions via electron impact ionization, double and triple charged impact ionization and the formation of molecular argon via the Hornbeck-Molnar process are identified as relevant processes in the NAIS. Other processes involving excited or energized atoms are negligible because these atoms diffuse away before engaging in a charge-producing interaction.

Calculations for a few typical situations indicate that most of the beam is comprised of singly charged argon, the species the NAIS is designed for. Besides single charged argon, the beam current consist for about 8.2% out of double charged argon, 2.2% out of triple charged argon, and for roughly 2% out of molecular argon. There is no obvious way, if at all, to prevent the production of such species.

We should comment on the fact that we have simplified the geometry severely, notably by using a top-hat particle density distribution. One result of this is that the calculated currents at a particular particle density and gas region length are not very accurate. This should not change the relative contributions though. The simple method of describing diffusion will give an inaccuracy in the molecular argon contribution, and this number should therefore be considered only as a rough estimate. We have also neglected the blockage of ions due to the membranes. Hence, the calculated currents are generated currents. Again, the relative amounts should not be affected too much. Neglecting any process besides electron impact ionization and Hornbeck-Molnar ionization is believed to be a sound assumption despite the simplifications made. The estimates of other contributions are so convincingly low, that an improvement in the accuracy can not lead to a different conclusion.

List of symbols

$\tilde{X}$	energetic atom
X^*	excited atom
X^{**}	highly excited atom
X^+	single-charged ion
X^{i+}	multiple-charged ion
X_2^+	diatomic ion
i	ion charge number #
D	diffusion coefficient $\mathrm{m^2\,s^{-1}}$
I_e	electron beam current A
n	neutral particle density $\mathrm{m^{-3}}$
l	gas length m
d	aperture diameter m
s	typical length m
N_{dim}	# diffusive dimensions #
$\langle v_t \rangle$	mean thermal velocity $\mathrm{m\,s^{-1}}$
τ_{diff}	diffusion time constant s

τ_{rad}	radiative life time	s
λ	mean free path	m
$\dot{N}$	particle production/loss	s^{-1}
N	# of particles	#
N^*	# of excited particles	#
N^{**}	# of highly excited part.	#
n^*	excited particle density	#
$\dot{N}_e$	electron rate	s^{-1}
σ_{**}	highly excited cross section	m
σ_{HM}	Hornbeck-Molnar cross section	m
k_{HM}	Hornbeck-Molnar reaction rate	$m^3\,s{-}1$
τ_{HM}	Hornbeck-Molnar time constant	s
$I[Ar_2^+]$	diatomic argon current	A
$I[Ar^{i+}]$	argon ion current	A

Acknowledgment

I would like to thank Pieter Kruit for his guidance and contributions. This work is part of the Industrial Partnership Program (IPP) of the Stichting voor Fundamenteel Onderzoek der Materie (FOM), which is financially supported by the Nederlandse Organisatie voor Wetenschappelijk Onderzoek (NWO). The IPP program is co-financed by FEI Company. This work is based on the dissertation titled *The Nano-Aperture Ion Source*, by Leon van Kouwen.

References

Aguilera, J. A., Blanco, F., Campos, J., & Ortiz, M. (1992). Lifetimes of some 6p levels of neutral argon. *Phys. Rev. A*, *45*(5), 2753–2756.

Becker, P., & Lampe, F. (1965). Mass-spectrometric study of the bimolecular formation of diatomic argon ion. *J. Chem. Phys.*, *42*(11), 3857–3863.

Bogaerts, A., & Gijbels, R. (1999). Role of Ar2+ and Ar+2 ions in a direct current argon glow discharge: A numerical description. *J. Appl. Phys.*, *86*(8), 4124–4133.

Bultel, A., & Vervisch, P. (2002). The Hornbeck-Molnar process in argon. *J. Phys. B, At. Mol. Opt. Phys.*, *35*(1), 111–124.

Das, M. B., & Karmakar, S. (2005). Radiative lifetimes of some excited states of neutral argon. *Phys. Scr.*, *71*(3), 266.

de Heer, F. J., Jansen, R. H. J., & van der Kaay, W. (1979). Total cross sections for electron scattering by Ne, Ar, Kr and Xe. *J. Phys. B, At. Mol. Opt. Phys.*, *12*(6), 979.

Filipovic, D. M., Marinkovic, B. P., Pejcev, V., & Vuskovic, L. (2000). Electron-impact excitation of argon: II. The lowest resonance 4s[3/2]1 and metastable 4s[3/2]2 and 4s'[A1/2]0 states. *J. Phys. B, At. Mol. Opt. Phys.*, *33*(11), 2081.

Hayashi, M. (2003). *Bibliography of electron and photon cross sections with atoms and molecules published in the 20th century – Argon* (Tech. Rep.). National Institute for Fusion Science (Report No. NIFS-DATA-72).

Hornbeck, J. A., & Molnar, J. P. (1951). Mass spectrometric studies of molecular ions in the noble gases. *Phys. Rev.*, *84*(4), 621.

Kosaki, K., & Hayashi, M. (1992). Denkigakkai Zenkokutaikai Yokoshu. *Prepr. Natl. Meet. Inst. Electr. Eng. Jpn.* (in Japanese).

Lam, S. K., Zheng, C.-E., Lo, D., Dem'yanov, A., & Napartovich, A. P. (2000). Kinetics of Ar*2 in high-pressure pure argon. *J. Phys. D, Appl. Phys.*, *33*(3), 242–251.
Nanbu, K., & Kageyama, J. (1996). Detailed structure of dc glow discharges – Effects of pressure, applied voltage, and y-coefficient. *Vacuum*, *47*(6–8), 1031–1033.
Okuno, K. (1986). Charge transfer of Ar2+ and Kr2+ in their own gases studied by the beam guide technique. *J. Phys. Soc. Jpn.*, *55*(5), 1504–1515.
Rejoub, R., Lindsay, B. G., & Stebbings, R. F. (2002). Determination of the absolute partial and total cross sections for electron-impact ionization of the rare gases. *Phys. Rev. A*, *65*(4), 042713.
Sahin, O., Tapan, I., Ozmutlu, E. N., & Veenhof, R. (2010). Penning transfer in argon-based gas mixtures. *J. Instrum.*, *5*(5), P05002.
Smirnov, B. M. (2014). *Theory of gas discharge plasma*. Springer International Publishing.
Tondare, V. (2006). *Towards a high brightness, monochromatic electron impact gas ion source* (Doctoral dissertation). Delft University of Technology. Retrieved from repository.tudelft.nl.
Tüxen, O. (1936). Massenspektrographische Untersuchungen negativer Ionen in Gasentladungen bei höheren Drucken. *Z. Phys.*, *103*(7), 463–484.
Yanguas-Gil, A., Cotrino, J., & Alves, L. L. (2005). An update of argon inelastic cross sections for plasma discharges. *J. Phys. D, Appl. Phys.*, *38*(10), 1588–1598.

Index

T

U

V

Z

FSC
www.fsc.org
MIX
Papier | Fördert
gute Waldnutzung
FSC® C083411

Kolibri 360 GmbH
Ferdinand-Jühlke-Straße 7
99095 Erfurt, Deutschland
produktsicherheit@kolibri360.de